A First Course in Linear Algebra
Third Edition

by David Easdown

Cover: String Theory Triptych, by Clare Easdown (2007)

Pearson Australia
Unit 4, Level 3
14 Aquatic Drive
Frenchs Forest NSW 2086
Ph: 02 9454 2200
www.pearson.com.au

Managing Editor: Jill Gillies
Senior Consultant: Danielle Woods
Production Controller: Barbara Honor

Printed and bound in Australia by The SOS Print + Media Group

1 2 3 4 5 15 14 13 12 11

ISBN: 978 1442 548251

An imprint of Pearson Australia

In memory of Larry Blakers (1917–1995)

Contents

Preface to the Third Edition

I have been touched by the overwhelmingly positive reaction from students to the second edition, and it is a pleasure to fine-tune the material further to make it even more useful in this third edition. The arithmetic of complex numbers is always present, like a benevolent but silent mountain, ready to support any foray into understanding the subtle and mysterious behaviour of matrices with real number entries. It was always my intention to write an appendix on complex numbers, but ran out of time till now, and am pleased that finally it is complete and able to be included in this new edition.

The layout, general philosophy of the book, and the ways in which it is intended to be used, are explained in the prefaces to the second and first editions, which are reproduced in the following pages, so I will use this space to comment about the new features of this edition.

The appendix on complex numbers aims to develop the reader's fluency to the point of understanding Stephen Smale's beautiful proof of the Fundamental Theorem of Algebra. There are many proofs of this theorem in the literature, but the one given here is my favourite and combines conceptual clarity with an algorithm. On the face of it the proof is existential, being a proof by contradiction, but the main idea easily converts to a formula for improving the approximation of a first guess towards finding the root of a polynomial. This exploits the derivative of a polynomial and becomes the formula associated with the usual Newton's Method from ordinary calculus. The idea of travelling down a tangent line resonates with the Linear Algebra Principle, so it appeals to me as a fitting way to complete a first course in linear algebra.

The Smale proof is full of subtleties and relies on a continuity argument and the Intermediate Value Theorem. These particular ingredients are also used in the proof, in Chapter 4, of the invariance of the Right-Hand Rule for the cross product of geometric vectors. These aspects may not be fully appreciated until the reader has been exposed to more advanced calculus or an introduction to mathematical analysis. Nevertheless it is productive to look ahead and start to see connections and the ways different branches of mathematics intertwine and depend on each other. Another subtle point of the Smale proof is the implicit fact that the number of roots of a polynomial is bounded by the degree. There are several ways of seeing this, and I have included a short proof (in the Hints and Solutions section) of an exercise involving the elegant Vandermonde determinant.

The exercises have been amplified in places. In particular, there are some more challenging exercises on determinants, leading to mathematical jewels such as the Cayley-Hamilton Theorem, Cramer's Rule and a proof that no permutation of a finite set can be both odd and even.

Again, I would particularly like to thank the University of Sydney and its students, for their collective responsiveness, enthusiasm and positive energy in the classroom. For this edition, I would especially like to thank my co-lecturers, Bob Crossman, Bill Gibson, James East and Alex Molev, for their timely help and suggestions, and colleagues including Maria Athanassenas, Koo-Guan Choo, Ruth Corran, Andrew Crisp, Daniel Daners, Holger Dullin, Alex Fun, Terry Gagen, Victoria Gould, Mark Lawson, Sonia Morr, Nitsa Movshovitz-Hadar, James Parkinson, Brad Roberts, Neil Saunders, Lev Shneerson, John Van Rooyen and Neville Weber for greatly appreciated continuing support, encouragement and careful reading that led to improvements in the text. Finally, as ever, I would like to thank the publishing team at Pearson Education Australia, including Jill Gillies, Danielle Woods and Michael Porteous, and my daughter Clare for providing yet another variation on a beautiful cover design.

David Easdown
September 2010

Preface to the Second Edition

The second edition is intended to be a self-contained introductory course in linear algebra of about one semester's length, suitable especially for students entering university for the first time, or for mature-age students returning after a period of absence and looking for a gentle lead-in. Students who complete this course will have a solid basis for undertaking more advanced topics in linear algebra with applications to physical and biological sciences and engineering.

The Introduction and twelve chapters are each intended to be read or browsed in a sitting and correspond to about one week of a thirteen-week lecture course. Each chapter has some introductory material followed by bite-size sections, the last of which is usually more difficult and suitable for an advanced student. Following on from each chapter is

- a concise list of important ideas and useful facts, which serves to put that part of the course in a nutshell
- graded exercises, intended to give the reader an opportunity for a thorough workout, the level of difficulty of which is indicated by one or more stars.

Students can work at their own pace on the exercises, ranging from routine to very difficult. The book also has

- hints and solutions to most exercises near the back of the book, which can be used as a quick check or as a stimulus to get started on a particularly hard problem
- an appendix on the theorem of Pythagoras
- an appendix on mathematical implication, which is like a mini-course on propositional logic, explaining the truth values for implication, and as a ready reference for many of the variations in mathematical nomenclature
- a comprehensive index to assist in the fast location of particular facts or definitions.

The book is designed to be handy and especially easy to pluck off the shelf, to carry around and use at the desk or in the classroom. Lecturers should find it useful in the classroom or lecture theatre, obviating the need for copious note-taking, so that students can concentrate on understanding. The author used the first edition when teaching linear algebra in first semester of 2007 at the University of Sydney, and referred directly to all but the advanced subsections of each chapter. This left some sections as extra reading for more experienced or interested students. The second edition is accompanied by a DVD of "learning nuggets", with live exposition by the author of some key aspects or details of the course. Each nugget lasts not more than a few minutes.

Again, I would particularly like to thank the University of Sydney and its students, for their comments, questions, responsiveness and making teaching in the classroom such a pleasure. For this edition, I would especially like to thank my co-lecturers, Bill Gibson and Adrian Nelson, for their timely help and suggestions, and Koo-Guan Choo for his remarkable enthusiasm and careful proofreading. I would also like to acknowledge with deep gratitude the wonderful and continuing support and encouragement of Nalini Joshi, Head of the School of Mathematics and Statistics, Rosanne Taylor, Faculty of Veterinary Science, and Christine Asmar, Institute of Teaching and Learning, each of whom acted as mentors and generously gave of their time to observe my lecturing and provide invaluable comments. Finally, as ever, I would like to thank the publishing team at Pearson Education Australia, including Jill Gillies, Sonia Wilson and Cameron Craig, the film crew and production team from the Faculty of Business and Economics, including Stephen Watson and Paul Henry, and of course my daughter Clare for providing another beautiful cover design.

David Easdown
August 2007

Preface to the First Edition

This book is the culmination of many years teaching mathematics, and linear algebra in particular, at several universities in Australia. It is quite different in style from most standard mathematics texts, and makes no attempt to be encyclopedic or comprehensive in its range of material or topics. The marvellous and altruistic Wikipedia and other online resources have created a revolution in the way we access information, rendering obsolete the more traditional forms of encyclopedias and instruction manuals. This book therefore focuses on just a couple of important themes, and aims for a certain depth of understanding through simple examples, mainly drawn from geometry of the plane. Paul Halmos enunciated the principle that all good theories have accessible examples and it is the role of the teacher always to bring them to the forefront of the developing mind of the student. Following his advice I have tried to choose examples which are easy to follow, but pithy and strike to the heart of the most important concepts in linear algebra. Classical results about the geometry of triangles seem to me to be an endless resource, not only for illustrating the utility of concepts such as linear independence and determinants, but almost every aspect of proof exploration and construction. Linear transformations of the plane, such as reflections, rotations and shears are a launching point for the main ideas and variations behind eigenvectors, eigenvalues and the Jordan canonical form.

The book is intended primarily to be a bridge for students entering university for the first time, and a resource that will guide them over some difficult thresholds in learning mathematics. Linear algebra is hard because it is so abstract. It needs to be abstract in order to capture such a wide diversity of phenomena. The purpose of this book is to provide a solid intuitive background and basis for the new student, who can then tackle the more advanced and abstract concepts with confidence and robustness.

The style of the book is largely conversational and informal. It was written very quickly, so I apologise if there are any errors or if I take too many liberties, particularly in my tendency to jump to conclusions of a metamathematical nature. The Plateau Principle, the Conjugation Principle and the Principle of Linear Algebra are my own silly inventions, but I personally find them useful and students are polite in not raising objections in the classroom. As for the last one, I had trouble settling on the right wording. I would have liked something snappy like "lines rule ok", but decided the colloquialism was too dated, so chose instead the very mundane "arrange matters to move linearly". If people have suggestions or objections, I would be pleased to hear them, or indeed any other comments or criticisms of my book. For example, I planned to write appendices on mathematical implication and complex numbers, but ran out of time for this edition.

I would particularly like to thank the University of Sydney and its remarkable students, who never cease to amaze me, for contributing towards such a stimulating, friendly and rewarding teaching environment since I arrived in 1990. I would also like to thank my colleagues, for their comments, advice, encouragement, ideas, exemplary examples and dedication to the craft of excellent teaching, in particular Simon Borg-Olivier, Sandra Britton, Steve Carnie, Koo-Guan Choo, Ruth Corran, Clio Cresswell, Daniel Daners, Brian Davey, Tom Dickson, Chris Durrant, Jane Reid Easdown, Terry Gagen, Bill Gibson, Stephen Glasby, Georg Gottwald, Tom Hall, Jenny Henderson, Bob Howlett, John Howie, Deborah Hughes-Hallett, Nalini Joshi, David Kohel, John Konrads, Lai Fong Low, John Luxton, Bianca Machliss, Andrew Mathas, Douglas Munn, Charlie Macaskill, Anne-Marie Mirza, Adrian Nelson, Mike Newman, Leon Poladian, Cheryl Praeger, Gordon Preston, Eugene Seneta, Jamie Simpson, Don Taylor, Neville Weber and Karl Wehrhahn. Finally I would like to thank the wonderful and enthusiastic publishing team at Pearson Education Australia, including Natalie Crouch, Sonia Wilson and Cameron Craig, and of course my daughter Clare for providing such a beautiful cover design.

David Easdown
January 2007

0 Introduction

- **Introducing linear algebra** Time: 5.07

0 Introduction

This book is a first course in *linear algebra*, one of the cornerstones of mathematics. Another cornerstone is *calculus*. Both calculus and linear algebra are studied by almost all mathematics students at university, though usually labelled as distinct subjects. Gradually students discover that linear algebra and calculus are inseparable (but not identical) twins that intertwine to form the backbone of almost all applications of mathematics to physical and biological sciences and engineering.

So what do we mean by *linear algebra*? The adjective *linear* has something to do with lines, and we will explain this shortly. The noun *algebra* is derived from the Arabic expression *al-jabr*, which means literally *the reuniting of broken bones* or, less literally, *the science of restoration* or *balancing*. The expression was used in this more abstract sense in the famous treatise

Hisab Al-Jabr wal Mugabalah (Book of Calculations, Restoration and Reduction),

written by Abu Abdallah Muhammad ibn Musa Al-Khawarizmi, who lived approximately 780 to 850 AD in and around Baghdad, acknowledged as one of the greatest mathematicians of all time and the founder of modern algebra. What exactly are we "restoring" or "balancing"? Pupils even at primary school will be familiar with sequences of implications like the following:

$$\begin{array}{rcl} & 2x+5 & = \ 19 \\ \Longrightarrow & 2x & = \ 2x+5-5 \ = \ 19-5 \ = \ 14 \\ \Longrightarrow & x & = \ \frac{1}{2}(2x) \ = \ \frac{1}{2}(14) \ = \ 7 \end{array}$$

Here, all of the quantities are numbers (integers in fact) and x is an unknown or "hidden" quantity described by the first equation. The above sequence of implications reveals or "restores" to us the value of x using

algebraic manipulation of equations.

Though this illustration is rudimentary, the reader should appreciate that profound ideas are involved:

- known and unknown quantities and operations are expressed symbolically
- equals signs replace "balancing scales"
- there exists a systematic and foolproof procedure for unravelling the first equation using arithmetic operations.

To us, in retrospect, finding x is no more difficult than exposing a clothed foot, by first removing the shoe and then the sock. But being able to exploit abstract connections between algebraic manipulation and familiar physical actions like dressing or undressing is one of the most powerful contributions of mathematics! Algebra draws out the essence of familiar experience and exploits it to create a language with which we can explore the unfamiliar, with useful and often surprising consequences. We will return to this theme many times in this book.

Now, linear algebra has something to do with *lines*, and straight lines are the simplest curves in geometry. (Indeed, straight lines do not "curve" at all!) The invention of calculus stems from the idea of approximating a complicated curve by a tangent line at a given point:

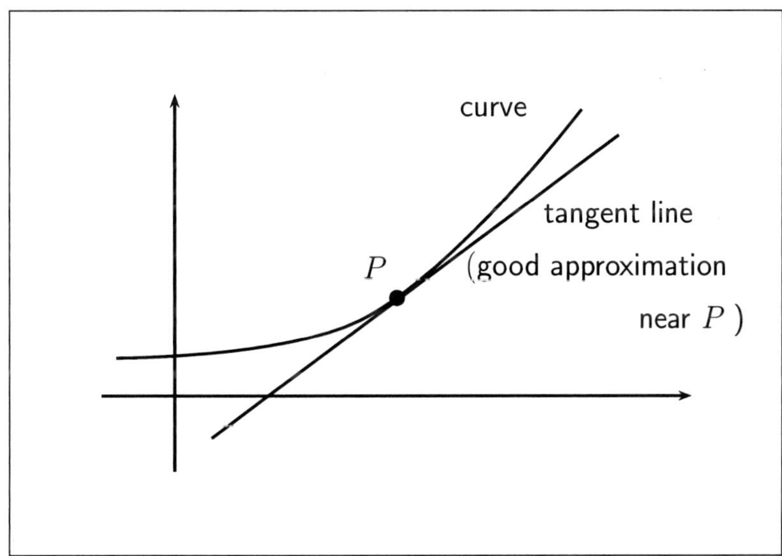

The slope of the tangent line is called the *derivative* of the curve at that point. It is one of the miracles of mathematics that, by thinking in the right way about infinity and limiting processes, we can "calculate" and manipulate derivatives, leading to the subject *differential calculus*. Here is an example of a line in the xy-plane having positive slope m and y-intercept $c > 0$:

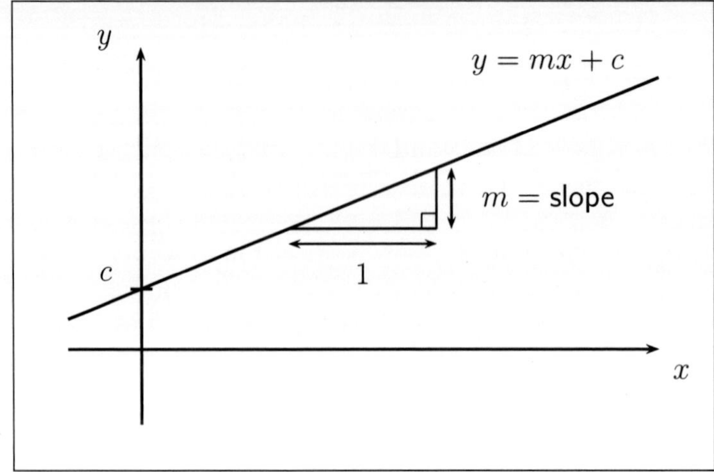

Of course, this is not the most general form of a line in the plane, because we could have drawn it sloping downwards ($m < 0$) or even horizontally ($m = 0$), or passing through the y-axis at or beneath the origin ($c \leq 0$). The equation

$$y = mx + c$$

where m and c are any real numbers would capture all of these possible diagrams. However, this equation treats x as an *independent* variable and y as a *dependent* variable, and no choice of constants m and c will capture a vertical line, parallel to the y-axis:

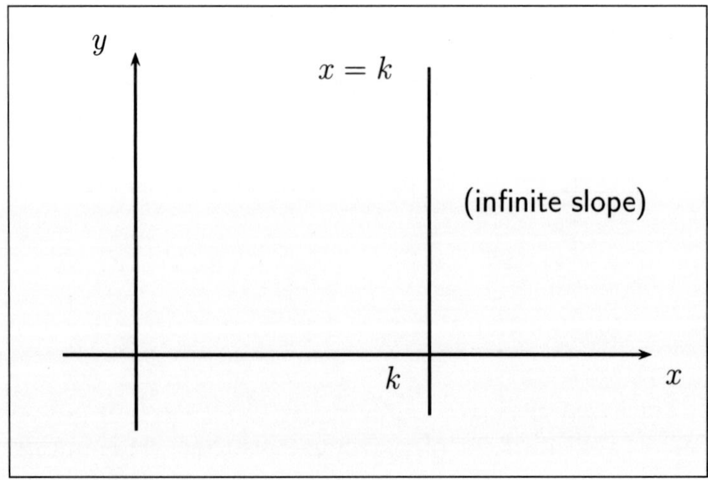

If we ignore any distinction between dependent and independent variables, but give x and y the same "status", we can easily combine all possibilities in a single class of equations, beautiful in its universality and implicit symmetry in the variables x and y:

$$\boxed{ax + by = c}$$

where now a, b and c are arbitrary constants. We call a and b the *coefficients* of x and y respectively. Here, if $b = 0$ and $c = k/a$ then we recover the vertical line $x = k$, whilst if $b = 1$ and $a = -m$ then we recover the line $y = mx + c$. As a bonus, there are two extra degenerate cases: "empty lines", when $a = b = 0$ and $c \neq 0$; and the whole xy-plane, when $a = b = c = 0$. The expression

$$\boxed{ax + by}$$

is called a *linear combination* of x and y, and involves just two multiplications followed by a single addition. The central theme of calculus is to use tangent lines to reduce sophisticated mathematics to simple arithmetic (up to approximation).

But why stop at two variables? If we introduce a third independent variable z we can form, by analogy, the *linear combination*

$$\boxed{ax + by + cz}$$

where a, b and c are arbitrary constants, and the equation of a *plane* in space

$$\boxed{ax + by + cz = d}$$

where d is yet another constant. Here, we are thinking of x, y and z as labelling three axes in space. Any two axes are perpendicular to each other:

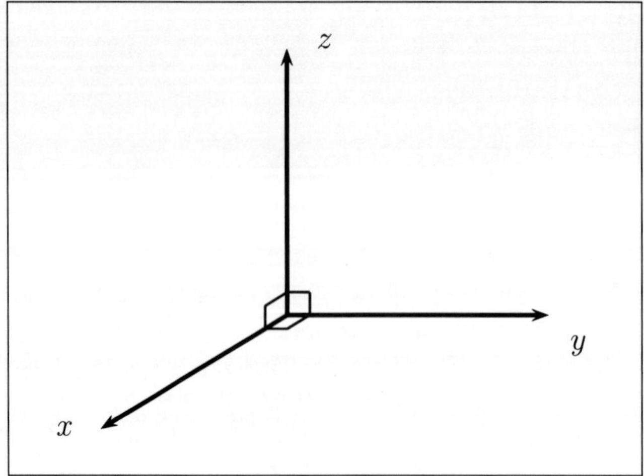

The labelling of the axes in the previous diagram follows a right-handed convention (explained later in the book). The equation describes triples (x, y, z) of numbers that satisfy the constraint $ax + by + cz = d$. Each triple represents a point in space. There are infinitely many such points, which collectively yield a plane. For example, the special case $z = 0$ (when $a = b = d = 0$ and $c = 1$) produces a copy of the usual xy-plane sitting in space consisting of points $(x, y, 0)$ with no restriction whatsoever on x and y. Planes in space play the same role in the calculus of two variables as lines do in the calculus of one variable. Well-behaved surfaces or "landscapes" are approximated arbitrarily well by tangent planes. It is of central importance to all students of mathematics to become thoroughly conversant with the algebra of planes (and, later, hyperplanes).

We think of space as *three*-dimensional, because of the three independent directions in which we may move. We regard planes as *two*-dimensional. In terms of the equation $ax + by + cz = d$, this makes sense, because we can arbitrarily choose two of the values of x, y and z, and then, by algebraic manipulation, determine the third (provided the coefficient is nonzero). There are *two degrees of freedom*. For example, consider the plane described by the following equation:

$$\boxed{x + 2y + 3z \ = \ 4}$$

If we choose, say,

$$x = 1\,,\ y = -1\,,$$

then the value $z = \frac{5}{3}$ is forced by the equation:

$$\begin{aligned} 1+2(-1)+3z &= 4 \\ \implies \quad 3z &= 4-1+2 = 5 \\ \implies \quad z &= \frac{5}{3} \end{aligned}$$

Similarly, we could arbitrarily assign values to y and z and then the value of x is forced, or assign values to x and z and then y is forced.

The notion that a line in the xy-plane is *one*-dimensional also makes sense in terms of the following equation:

$$ax + by = c$$

If one of x, y is chosen arbitrarily then the other is determined by the equation (provided the relevant coefficient is nonzero). A line has *one degree of freedom.*

In applying mathematics to physics, biology, economics, or any other field of human endeavour, there is nothing special about one, two or three variables, dimensions or degrees of freedom. We would like our mathematics to be able to deal with any number of variables or quantities. So let n be a positive integer, which we can think of as being as large as we like. Consider the n variables x_1, x_2, x_3, $\ldots$, x_n. (For $n = 4$, most people find it convenient to persist in a slightly perverse alphabetic convention and call the variables x, y, z, w instead.) Now consider the *linear combination*

$$a_1x_1 + a_2x_2 + a_3x_3 + \cdots + a_nx_n$$

where a_1, a_2, a_3, $\ldots$, a_n are constants, called *coefficients.* Evaluating a linear combination involves just simple arithmetic: n multiplications followed by addition. A *linear equation in n variables* has the form

$$a_1x_1 + a_2x_2 + a_3x_3 + \cdots + a_nx_n = b$$

where b is a further constant. This equation now has $n-1$ degrees of freedom, in the sense that if we arbitrarily choose values for $n-1$ variables, then the value of the remaining variable is completely determined by algebraic manipulation of the equation (assuming the coefficient of this variable is nonzero). Think of the *n-tuple*

$$(x_1\,,\; x_2\,,\; x_3\,,\; \ldots\,,\; x_n)$$

as a "point" in n-space. (The terminology n-tuple comes from generalising the idea of duple, triple, quadruple, quintuple, sextuple and so on.) The set of points satisfying the previous equation form a configuration called a *hyperplane*, which has dimension $n-1$, sitting inside n-space (which is itself n-dimensional). This generalises the earlier examples of two-dimensional planes sitting inside 3-space, and one-dimensional lines sitting inside 2-space. Of course, we cannot visualise what is going on for more than three dimensions, but

geometrical intuition in three dimensions is central to the way mathematicians think in any number of dimensions.

If the reader can be lured into using n variables (in place of one, two or three variables), then it is but a small step to consider not just one equation, but a *system* of, say, m linear equations:

$$\begin{array}{ccccccccccc}
a_{11}x_1 & + & a_{12}x_2 & + & a_{13}x_3 & + & \cdots & + & a_{1n}x_n & = & b_1 \\
a_{21}x_1 & + & a_{22}x_2 & + & a_{23}x_3 & + & \cdots & + & a_{2n}x_n & = & b_2 \\
\vdots & & \vdots & & \vdots & & & & \vdots & & \vdots \\
a_{m1}x_1 & + & a_{m2}x_2 & + & a_{m3}x_3 & + & \cdots & + & a_{mn}x_n & = & b_m
\end{array}$$

Each equation can be thought of as a "linear constraint" on the n variables. A *solution* of the system will be an n-tuple $(x_1\,,\; x_2\,,\; \ldots\,,\; x_n)$ of values for which all of the constraints are simultaneously satisfied. It is far from clear how solutions behave, or even if they exist for a given system. We can form a *matrix* of coefficients

$$M = \begin{bmatrix} a_{11} & a_{12} & a_{13} & \cdots & a_{1n} \\ a_{21} & a_{22} & a_{23} & \cdots & a_{2n} \\ \vdots & \vdots & \vdots & \vdots & \vdots \\ a_{m1} & a_{m2} & a_{m3} & \cdots & a_{mn} \end{bmatrix}$$

and *column vectors* of variables and constants respectively

$$\mathbf{x} = \begin{bmatrix} x_1 \\ x_2 \\ x_3 \\ \vdots \\ x_n \end{bmatrix} \qquad \mathbf{b} = \begin{bmatrix} b_1 \\ b_2 \\ \vdots \\ b_m \end{bmatrix}$$

(drawn here as if m is smaller than n, but of course it could be the other way around, or they could be equal). A matrix is just a rectangular array of numbers (or other objects depending on the context) with a certain number of rows and columns. A column vector is the special case when the matrix has just one column. All information in the original system of m linear equations can be summarised by the single *matrix equation*:

$$\boxed{M\mathbf{x} = \mathbf{b}}$$

One needs to understand the definition of *matrix multiplication* to see this. It is an extraordinary phenomenon that we are able to manipulate matrix equations in a manner similar to ordinary primary school arithmetic. There are certain provisos and a novice gradually learns to recognise particular features of ordinary arithmetic which apply. Careful study will be amply rewarded: the arithmetic of matrices is weird, wonderful and powerful. *En route* to developing fluency, the student will be introduced to such delights as the theory of determinants, eigenvectors and eigenvalues.

Now it may come as something of a surprise that we can gain insight about a system of equations, for which there may be a seemingly overwhelming myriad of variables and coefficients, just by manipulating a simple matrix equation. But when one drives a car, or uses a computer, does one pause to think about the details of what lies under the bonnet, or the circuitry, or the programming language behind the software? At times of course it is necessary to delve inside, but mathematicians develop an ability to change focus and, when it suits, to operate at a higher level which takes advantage of intelligent packaging of information. This is the

> **Plateau Principle:** Look for and be prepared to use a variety of plateaus as starting points for a mathematical investigation.

(A mathematician invokes this principle every time he or she uses a theorem without thinking about why the theorem is true.) There are certain conceptual leaps in the development of mathematics. The invention of zero is certainly one of them and is documented in the treatise by Al-Khawarizmi referred to near the beginning of this Introduction. The idea of a function as an object is another. Treating an array of numbers as a kind of "number" itself is one of the most powerful ideas in mathematics. (Indeed, the last two examples of functions and matrices are closely connected: matrices encode certain types of functions called *linear transformations*.)

The early chapters are concerned with *geometric vectors* (directed line segments) in the plane and in space. We look for and develop connections with algebraic operations. Our geometric intuition will be extremely valuable when we pass to higher dimensions. There is a multitude of ways to combine and manipulate geometric vectors. We proceed gently, but it is important to practise, as each step is important and leads to a seminal idea in higher mathematics. The notions of orthogonality and projection, in particular, touch and permeate almost every area of pure and applied mathematics, numerical analysis and statistics.

Each chapter is intended to be read in one sitting or study session, equivalent to one to two lectures, followed by exercises. The unstarred exercises are designed to give the reader a thorough workout, and are suitable for students of all backgrounds. Exercises which are starred are intended to be more difficult, and suitable for advanced students. To illustrate, we finish this Introduction with some exercises, the first of which perhaps is the easiest, and the last perhaps the most difficult in the book. Who is advanced, or not advanced, of course is relative to experience and, to make this point, two of the starred exercises on the next page are identical to unstarred exercises in a later chapter! Good luck and, above all, have fun doing mathematics!

Exercise 0.1 Manipulate an algebraic expression to explain the following phenomenon:

> I am a mind reader. Think of a number from 1 to 21. Double it and add 4. Halve your answer. Take away the number you first started with. You are now thinking of the number 2.

Exercise 0.2 Consider the line $2x + 3y = 6$. Find the y- and x-intercepts, and draw the line in the xy-plane. Find the slope of the line regarding the x-axis as horizontal (as usual). Find the slope instead regarding the y-axis as horizontal.

Exercise 0.3 Find the equation of the line obtained by reflecting the line $2x + 3y = 6$ in the line $y = x$. Describe the relationship between the slopes of the original and the reflected lines.

Exercise 0.4 Find the equation of the line obtained by rotating the line $2x + 3y = 6$ ninety degrees anticlockwise about the origin. Describe the relationship between the slopes of the original and the rotated lines.

Exercise 0.5 Find the point of intersection of the lines $2x + 3y = 6$ and $3x + 2y = 6$.

Exercise 0.6* Find the equation of the line obtained by reflecting the line $ax + by = c$ in the line $y = x + k$.

Exercise 0.7* Find the equation of the line obtained by rotating the line $ax + by = c$ ninety degrees anticlockwise about the point (x_0, y_0).

Exercise 0.8* Prove that the lines $ax + by = k$ and $cx + dy = \ell$ intersect in a single point if and only if $ad - bc \neq 0$.

Exercise 0.9* Look at the corner of the room. The walls are two planes which meet in a line. Follow the line upwards towards the ceiling. Where it meets the ceiling is the point of intersection of three planes. Find the point (x, y, z) of intersection of the following three planes:

$$\begin{array}{rcrcrcr} 2x & + & 3y & + & 4z & = & -4 \\ 5x & + & 5y & + & 6z & = & -3 \\ 3x & + & y & + & 2z & = & -1 \end{array}$$

Exercise 0.10[**] Manipulate an algebraic expression to explain the following phenomenon:

> I am a mind reader. Think of an integer from 1 to 21. Double it. Take the square root and add a half. Throw away everything to the right of the decimal point, and the decimal point. You are thinking of an integer X. Now consider the following sequence, and move from left to right the same number of steps as the integer you first thought of:
>
> $$1, 2, 2, 3, 3, 3, 4, 4, 4, 4, 5, 5, 5, 5, 5, 6, 6, 6, 6, 6, 6$$
>
> (For example, if you move 10 steps you will reach 4 in the sequence, and after 11 steps you will reach 5.) Call the number you have reached Y. Take Y away from X. You are now thinking of the most important number in mathematics (and the label of the chapter you are on the point of leaving).

1 Geometric Vectors

WATCH IT NOW on DVD VIDEO

- **Geometric vectors**
  Time: 4.49
- **Vector addition** (Section 1.1)
  Time: 7.31
- **Scalar multiplication and subtraction**
  (Sections 1.2, 1.3)
  Time: 6.17
- **Vectors in geometry and a proof template**
  (Section 1.5)
  Time: 9.19

1 Geometric Vectors

In this chapter we will describe a new arithmetic, which extends the usual arithmetic of real numbers. Of course, it is not really new: what we will describe is now regarded as part of classical mathematics. But it may be new for the reader. Or perhaps the reader has seen these ideas before, in the course of doing physics or other related topics in science which rely heavily on mathematics, but has not seen this material treated systematically, or from an algebraic point of view.

By the way, the "usual arithmetic of real numbers" is not at all a trivial thing, and we could write a first course about that alone! It is a good example of the Plateau Principle that we will assume familiarity with the real numbers, how to add and multiply them, and many of their wonderful properties, such as each having an infinite decimal expansion, and collectively being represented geometrically as an infinite, continuous "real line".

An arithmetic has two essential features: firstly there are *objects*, the prime focus of our attention; and secondly there are *operations*, which combine objects to create or build other objects. Most operations, like the familiar addition and multiplication of real numbers, are *binary*, that is, they operate on *two* objects at a time to create a third object. But one can also have *unary* operations which take just *one* object and transform it into another. For example, negation (forming the negative) is a unary operation on real numbers; inversion (forming the reciprocal) is a unary operation on nonzero real numbers.

In the arithmetic of this chapter, there are two kinds of objects. The first and simplest kind of objects comprise the familiar real numbers. These are called *scalars*, for a geometric reason related to "scaling" which we will come to shortly.

> The collection of all real numbers is denoted by $\mathbb{R}$.
>
> Real numbers are also referred to as *scalars*.

We can have other types of scalars, such as *complex numbers*, and they will be mentioned in advanced topics in later chapters. Basic definitions and important properties of complex numbers are explained in Appendix 3.

The real number system is a model *par excellence* for dealing with a vast range of quantities encountered in real life, such as mass, time, distance, temperature, speed and bank account balances. All such quantities have a magnitude (size), but, at best, a primitive sense of direction – they can only move up or down. In applications we frequently need an appropriate model for manipulating quantities such as force, displacement, velocity and acceleration, which not only have a magnitude, but can point in an infinity of directions in the plane or in space.

These considerations lead to an expanded, more sophisticated arithmetic, with a second "higher order" class of objects. These are **directed line segments** or **arrows**, which one can think of as "floating" or "suspended" in the plane or in space:

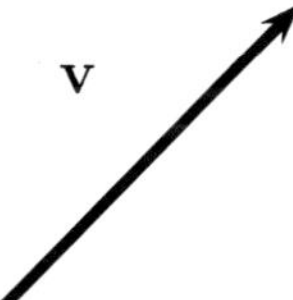

We call $\mathbf{v}$ above a **geometric vector** (or just **vector** for short), and characterise it by two properties:

- its *magnitude* = length, a nonnegative scalar, denoted by $|\mathbf{v}|$
- its *direction*.

We deliberately wrote **vector**, **directed line segments**, **arrows** in **bold type**, to reinforce the convention adopted in printed mathematics of symbolising these objects with **boldface** letters (such as $\mathbf{u}$, $\mathbf{v}$, $\mathbf{w}$).

How do we recognise when two vectors represent the same object?

> Vectors are *equal* if they have the same magnitude and direction, regardless of position in the plane or in space.

In the following diagram all of the vectors shown are equal, and we write

$$\mathbf{v} = \mathbf{w}.$$

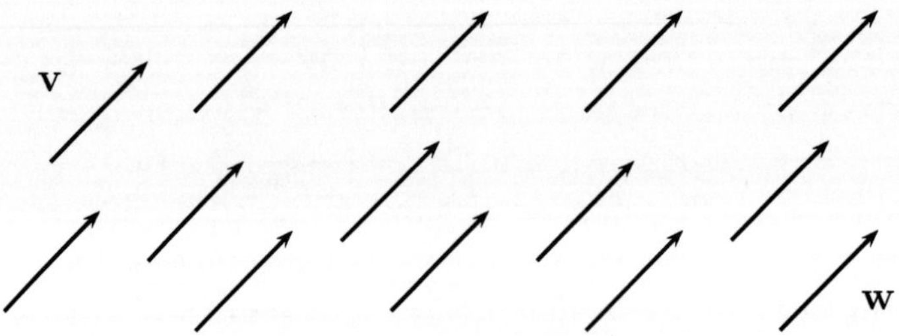

Thus $\mathbf{v} = \mathbf{w}$ means that $\mathbf{v}$ can be moved *parallel to itself* to coincide exactly with $\mathbf{w}$.

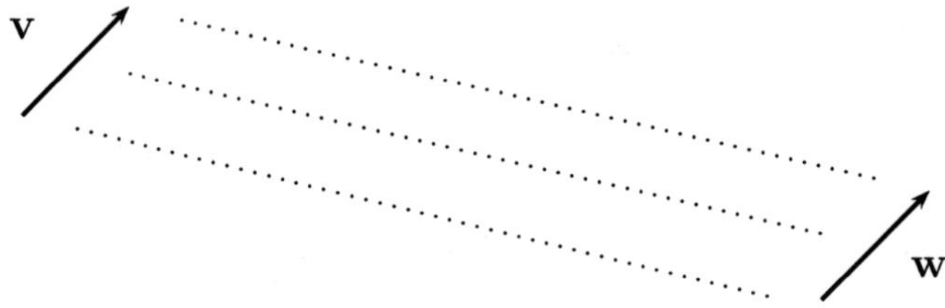

There is one important, and extremely useful, exception to the boldface convention for vectors. If P and Q are points in the plane or in space then

$$\overrightarrow{PQ}$$

denotes the directed line segment (vector) that points from P to Q, in that order:

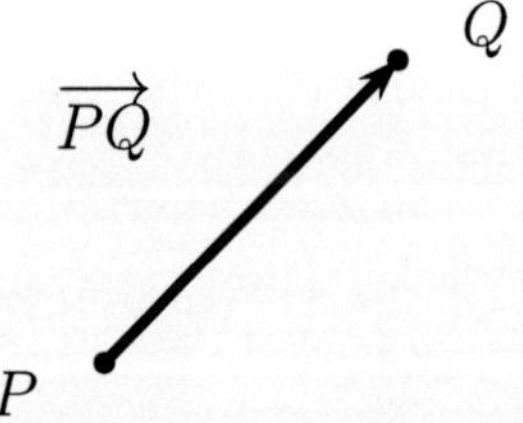

When one gets used to this notation, it works like a charm, especially in proofs in geometry. But it is easy to slip into error:

Warning: One of the most common errors students make is to write or use $\overrightarrow{QP}$ when $\overrightarrow{PQ}$ is intended.

In fact, $\overrightarrow{QP}$ and $\overrightarrow{PQ}$ have the same magnitude, but *point in opposite directions.*

Now, so far it has been very easy. But we should pausc bricfly to point out that if the definitions seem obvious, it is because the reader is standing on a plateau. We do not want to undermine this plateau – indeed, it is very solid. But it is important to be aware of hidden assumptions. What does it mean to move a vector parallel to itself? How do we tell if two vectors point in the same direction? For the purposes of this book, we will assume that we know the answers, namely, just look and see! However, the notion of *being parallel* is sophisticated. As soon as one starts to think about it, one stumbles on interesting mathematics. We think of two lines in the plane as being parallel if they "stay the same positive distance apart forever". Certainly, then, parallel lines never meet, by definition, since if they met then the distance between them would become zero. But let's look and see! Here is a straight road that goes forever on our infinite plane:

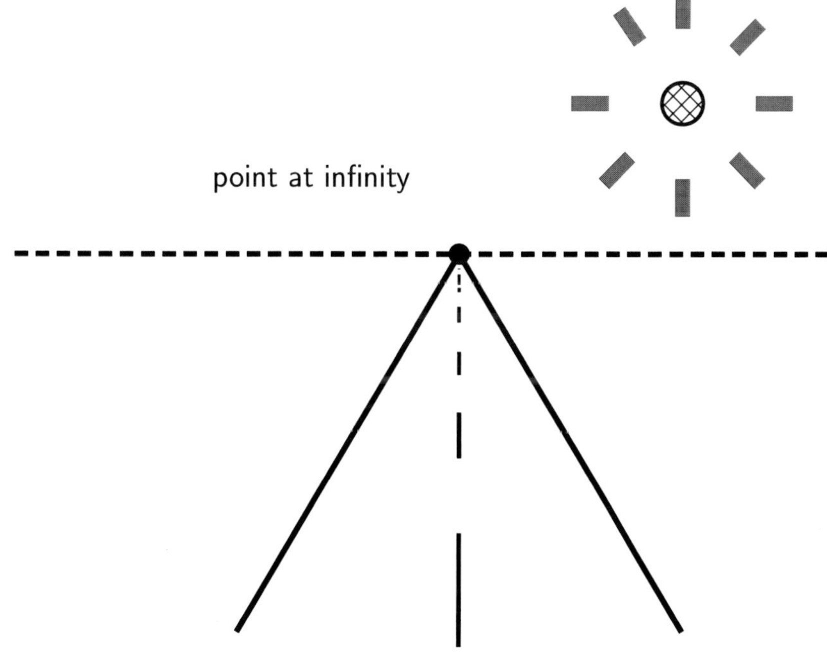

The opposite sides of the road are parallel, remaining always a positive road-width apart. However, when we look, we see them converging, meeting at some point far away in the distance. This is *not* an illusion, nor an artefact of our lack of precision in representing reality. In the mathematical model, with a truly infinite plane (where the earth is flat and the sun is an ornament), and truly infinite parallel lines, whose images are projected onto a "flat screen", the lines *really do meet* (Exercise 6.15)! The horizon may be regarded as a copy of the real line, and *in the projection* the sides of the road meet at a *limit point* on that real line. Limit points exist because, in our model, the real number system is *complete*. Completeness is a technical term with which advanced students of calculus come to grips, and we shall not pursue this any further here. We are touching on a subject called *projective geometry*, in which parallel lines meet at "points at infinity". It provides a wonderful model, for example, for modern applications of mathematics to computer graphics and virtual reality, where three-dimensional space needs to be represented by two-dimensional images. *Projection* is a central idea throughout all of mathematics, and we will carefully develop some of its key features in the third chapter.

1.1 Addition of geometric vectors

We now introduce the first of several binary operations, gradually increasing in sophistication. The simplest is called *addition of geometric vectors.*

Vectors float in the plane or in space, and are happy to move, at our convenience, parallel to themselves without changing. If $\mathbf{u}$ and $\mathbf{v}$ are vectors then we can

move them "tip to tail"

so that the point at the tip of $\mathbf{u}$ coincides with the point at the end of the tail of $\mathbf{v}$. The *vector sum* $\mathbf{u}+\mathbf{v}$ then is the directed line segment, which is the "net effect" of moving from the end of the tail of $\mathbf{u}$ to the tip of $\mathbf{v}$:

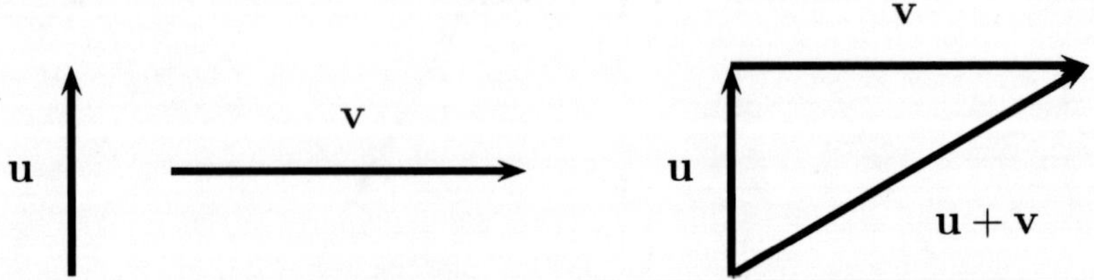

Clearly, we can repeat this procedure, adding together a cascade of vectors to find the net effect of lining them up tip to tail. Here is an illustration with three vectors:

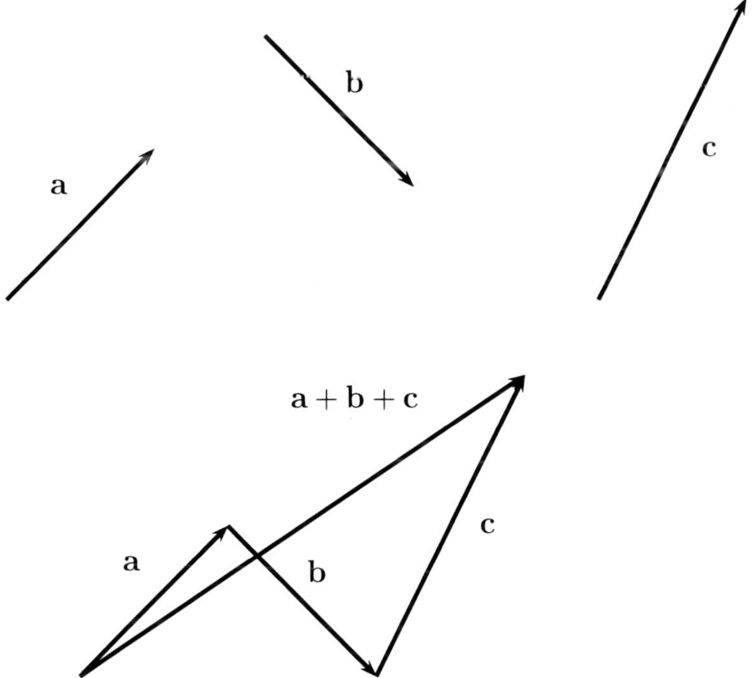

The order in which we line up the vectors tip to tail does not matter, because we have a

> **Parallelogram Law of Addition:** The vector sum $\mathbf{v}+\mathbf{w}$ is represented by the diagonal of the parallelogram formed using sides $\mathbf{v}$ and $\mathbf{w}$.

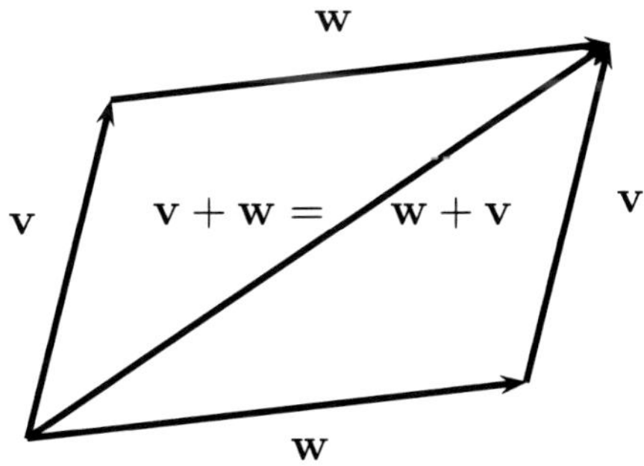

Clearly, from the diagram, $\mathbf{v} + \mathbf{w} = \mathbf{w} + \mathbf{v}$, so that the order of addition does not matter. This property is called the *Commutative Law*, and it is important to add this terminology to one's mathematical vocabulary.

Commutative Law of Addition: for any vectors $\mathbf{v}$ and $\mathbf{w}$,

$$\mathbf{v} + \mathbf{w} \;=\; \mathbf{w} + \mathbf{v}$$

Another very important property is the *Associative Law* which says that *bracketing* does not matter, the verification of which is left as an exercise. (We were using it implicitly before when we added three vectors together, and it is usual to delete brackets altogether, except for emphasis, or if there is a possibility of ambiguity.)

Associative Law of Addition: for any vectors $\mathbf{u}$, $\mathbf{v}$ and $\mathbf{w}$,

$$(\mathbf{u} + \mathbf{v}) + \mathbf{w} \;=\; \mathbf{u} + (\mathbf{v} + \mathbf{w})$$

We finish this section by observing that addition of vectors is captured succinctly using the point-to-point notation:

If P, Q and R are points then

$$\overrightarrow{PQ} + \overrightarrow{QR} \;=\; \overrightarrow{PR}\,.$$

1.2 Multiplication by a scalar

The next binary operation again takes two objects to produce a third. However, this time we start with a scalar and a vector to produce a vector. In the positive case, the idea is to "scale up" or "scale down" the length of the vector, without changing direction. This is why scalars are called scalars!

Let λ be a scalar (real number), and $\mathbf{v}$ a vector. We form the *scalar multiple* $\lambda\mathbf{v}$, another vector, using rules which depend on whether λ is positive, negative or zero.

> **Positive Case:** $\lambda > 0$. Then $\lambda\mathbf{v}$ points in the same direction as $\mathbf{v}$ but its length is the result of multiplying the length of $\mathbf{v}$ by λ.

Thus, if $\lambda > 1$ then we "stretch" $\mathbf{v}$ to form $\lambda\mathbf{v}$. If $0 < \lambda < 1$ then we "shrink" $\mathbf{v}$ to form $\lambda\mathbf{v}$. If $\lambda = 1$ then there is no change:

$$1\mathbf{v} \;=\; \mathbf{v}$$

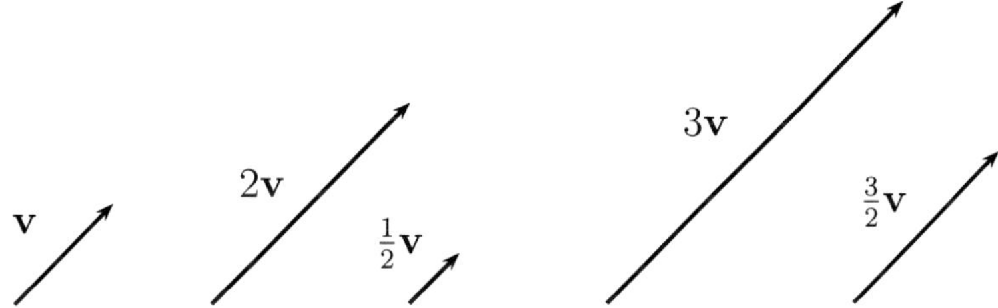

At this point it is worth reminding the reader that if λ is a real number, then the *absolute value* or *magnitude* of λ is just the distance from λ to 0 on the real number line:

$$|\lambda| \;=\; \begin{cases} \lambda & \text{if } \lambda \geq 0 \\ -\lambda & \text{if } \lambda < 0 \end{cases}$$

For example,

$$|-2| \;=\; -(-2) \;=\; 2\,.$$

We now describe multiplication of a vector by a negative scalar:

> **Negative Case:** $\lambda < 0$. Then $\lambda\mathbf{v}$ points in the opposite direction as $\mathbf{v}$, but its length is the result of multiplying the length of $\mathbf{v}$ by $|\lambda|$.

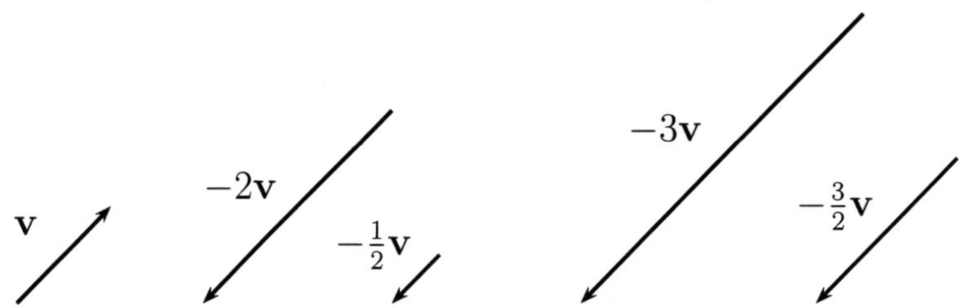

Important Special Case:

$$(-1)\mathbf{v} = -\mathbf{v}$$

is called the **negative** of $\mathbf{v}$ obtained by reversing $\mathbf{v}$.

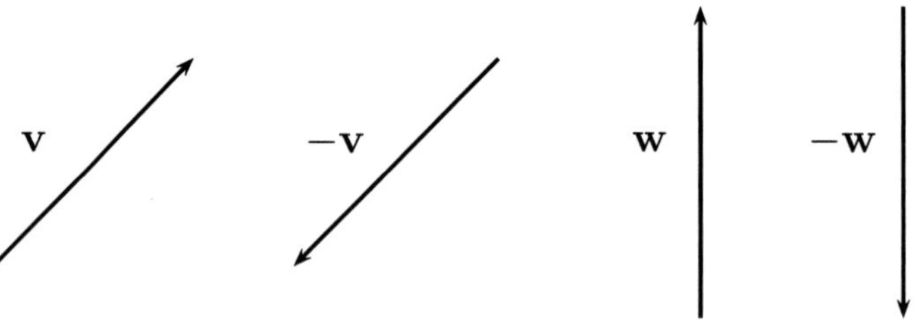

What happens if we add a vector to its negative, tip to tail?

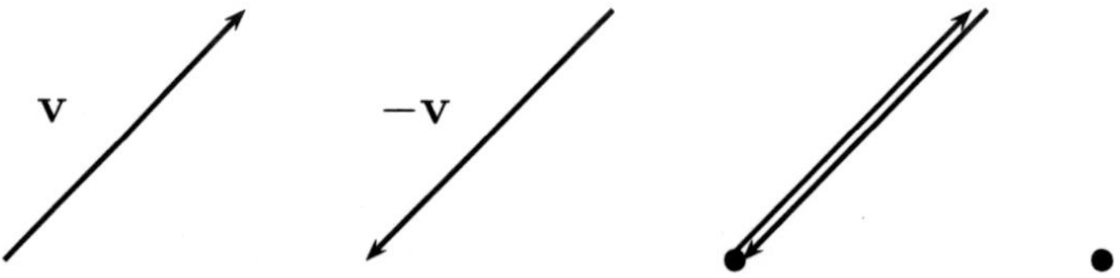

The configuration evaporates to a point, which we can imagine floating harmlessly in space. We obtain the *zero vector*, denoted by $\mathbf{0}$. This is formally distinct from the scalar 0, which is the *length* of the zero vector. (Remember, a geometric vector has *two* properties: length and direction.)

The zero vector $\mathbf{0}$ has zero length and points in every direction.

The zero vector enables simplification of vector expressions. For example, if we are working with points P and Q in the plane or space

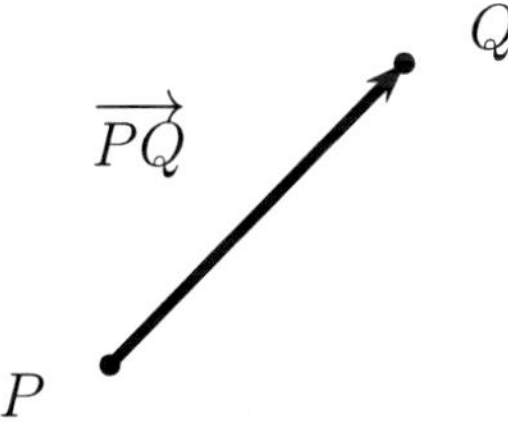

then $\overrightarrow{PQ} + \overrightarrow{QP} = \overrightarrow{PP} = \mathbf{0}$ and $\overrightarrow{QP} + \overrightarrow{PQ} = \overrightarrow{QQ} = \mathbf{0}$, so that the following holds:

$$-\overrightarrow{PQ} = \overrightarrow{QP} \quad \text{and} \quad -\overrightarrow{QP} = \overrightarrow{PQ}$$

The negative of any vector $\mathbf{v}$ has the following important properties:

$$-(-\mathbf{v}) = \mathbf{v} \quad \text{and} \quad \mathbf{v} + (-\mathbf{v}) = (-\mathbf{v}) + \mathbf{v} = \mathbf{0}$$

We can finally complete the definition of multiplication by a scalar:

Zero Case: Multiplying $\mathbf{v}$ by the zero scalar gives the zero vector:

$$0\mathbf{v} = \mathbf{0}$$

1.3 Subtraction of vectors

Our third binary operation, *subtraction of vectors*, is a neat combination of constructions from the previous two sections:

For any vectors $\mathbf{v}$ and $\mathbf{w}$, define their *difference* $\mathbf{v} - \mathbf{w}$ by adding the negative of $\mathbf{w}$ to $\mathbf{v}$:

$$\mathbf{v} - \mathbf{w} = \mathbf{v} + (-\mathbf{w})$$

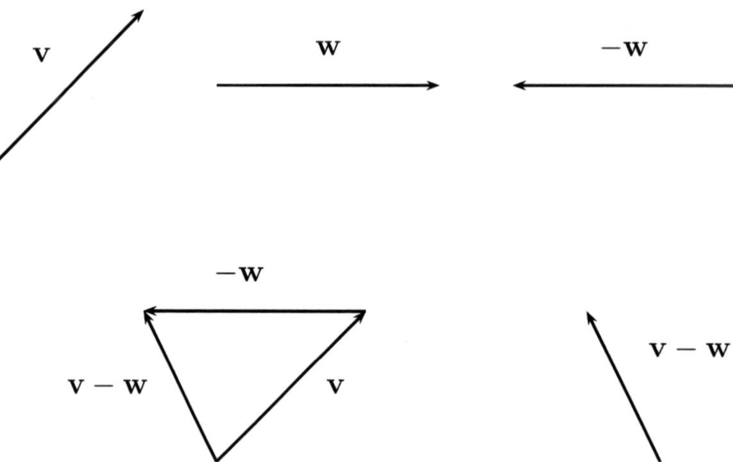

It is easy to make a mistake, especially to produce a vector pointing in the wrong direction! A simple check:

If in doubt, remember that $\mathbf{v} - \mathbf{w}$ must be the vector which added to $\mathbf{w}$ gives back $\mathbf{v}$.

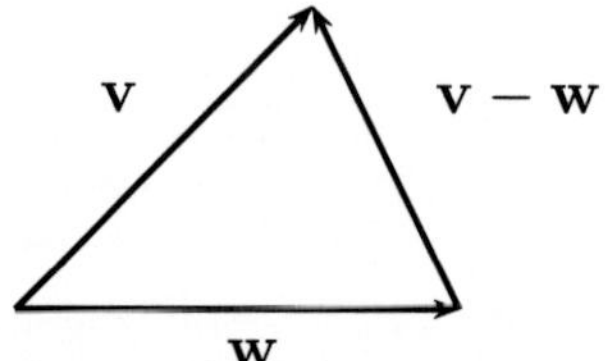

Of course, we expect that the claim in the previous box can be verified algebraically, using the definition and properties of vectors:

$$\begin{aligned}
\mathbf{w} + (\mathbf{v} - \mathbf{w}) &= \mathbf{w} + (\mathbf{v} + (-\mathbf{w})) \\
&= \mathbf{w} + ((-\mathbf{w}) + \mathbf{v}) \\
&= (\mathbf{w} + (-\mathbf{w})) + \mathbf{v} \\
&= \mathbf{0} + \mathbf{v} \\
&= \mathbf{v}
\end{aligned}$$

For example, the first equality is by definition of subtraction of vectors. Which of these equalities uses the associative law? the commutative law?

1.4 List of useful properties

We pause here to collect together a number of properties of vector addition and scalar multiplication, many of which we have already discussed. You should have no difficulty drawing diagrams to convince yourself that the others are true. Here, $\mathbf{u}$, $\mathbf{v}$, $\mathbf{w}$ are vectors and λ, μ are scalars:

$$\mathbf{v} + \mathbf{w} = \mathbf{w} + \mathbf{v}\,, \qquad (\mathbf{u} + \mathbf{v}) + \mathbf{w} = \mathbf{u} + (\mathbf{v} + \mathbf{w})$$

$$\mathbf{v} + \mathbf{0} = \mathbf{0} + \mathbf{v} = \mathbf{v}\,, \qquad -(-\mathbf{v}) = \mathbf{v}$$

$$\mathbf{v} + (-\mathbf{v}) = \mathbf{v} - \mathbf{v} = \mathbf{0}\,, \qquad \lambda(\mu\mathbf{v}) = (\lambda\mu)\mathbf{v}$$

$$\lambda(\mathbf{v} + \mathbf{w}) = \lambda\mathbf{v} + \lambda\mathbf{w}\,, \qquad (\lambda + \mu)\mathbf{v} = \lambda\mathbf{v} + \mu\mathbf{v}$$

$$1\mathbf{v} = \mathbf{v}\,, \qquad (-\lambda)\mathbf{v} = -(\lambda\mathbf{v})\,, \qquad (-1)\mathbf{v} = -\mathbf{v}$$

1.5 The geometry of parallelograms

We finish this chapter with an illustration of the elegance of vector arithmetic in attacking problems in geometry. In plane geometry, a *quadrilateral* $PQRS$ is a figure with four sides, joining points P to Q to R to S and back to P.

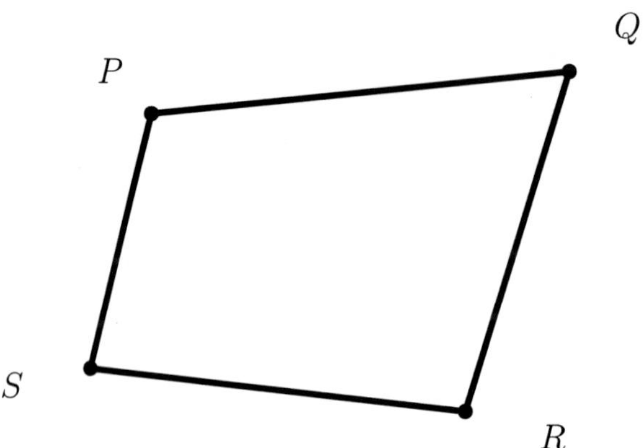

We denote a side joining points P and Q by PQ or QP without any implied direction. Sides are *adjacent* if they are joined by a point, and *opposite* otherwise. For example, in the previous figure PQ and QR are adjacent, and PQ and RS are opposite. If opposite sides are parallel and have the same length then the quadrilateral is called a *parallelogram*, as in the following figure:

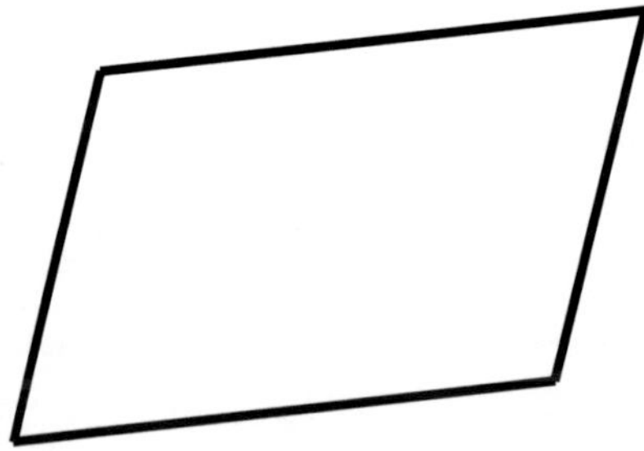

In fact, to check that a quadrilateral is a parallelogram it is enough to check the condition for just one pair of opposite sides:

> A quadrilateral $PQRS$ is a parallelogram if sides PQ and RS are parallel and have the same length.

We can explain why this is true using vector arithmetic. We are supposing, for our quadrilateral $PQRS$ above, that the opposite sides PQ and RS are parallel and have the same length. Our task is to verify that the remaining opposite sides QR and SP are also parallel and have the same length. We can formulate this in the language of vectors:

> Our assumption is that $\overrightarrow{PQ} = \overrightarrow{SR}$ and we need to show $\overrightarrow{QR} = \overrightarrow{PS}$.

The trick is to "expand" $\overrightarrow{QR}$ to incorporate all of the vertices of the quadrilateral, exploit the assumption $\overrightarrow{PQ} = \overrightarrow{SR}$, and then "contract" or "distil" everything down to $\overrightarrow{PS}$:

$$\begin{aligned}
\overrightarrow{QR} &= \overrightarrow{QP} + \overrightarrow{PS} + \overrightarrow{SR} \\
&= -\overrightarrow{PQ} + \overrightarrow{PS} + \overrightarrow{PQ} \\
&= \overrightarrow{PQ} - \overrightarrow{PQ} + \overrightarrow{PS} \\
&= \mathbf{0} + \overrightarrow{PS} \\
&= \overrightarrow{PS}
\end{aligned}$$

It is sound investment to examine this proof carefully. Finding proofs in mathematics can be like looking for "needles in haystacks". Gradually, with time and practice, the reader will develop a repertoire of natural techniques and approaches. In the previous proof, of course one should be certain of each step in the vector arithmetic. On a "broader brush-stroke" level, however, it is important to notice a

> **Proof Template:** expand – apply something – contract.

In a sense we can make precise, expansion and contraction are "mutually inverse" operations. The previous proof is an instance of the *Conjugation Principle*, explained fully in later chapters on matrix inversion and diagonalisation. Interested readers might look ahead, and even search the Internet for a branch of algebra called *group theory*, in which the Conjugation Principle becomes an artform!

We finish with another example. Consider again our quadrilateral $PQRS$ and add *diagonals* PR and QS:

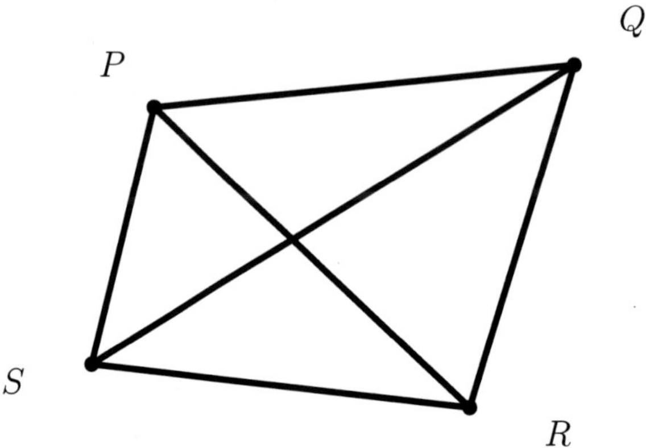

Suppose that the diagonals PR and QS bisect each other, that is, cut each other into perfect halves. Call their point of intersection T as in the following diagram:

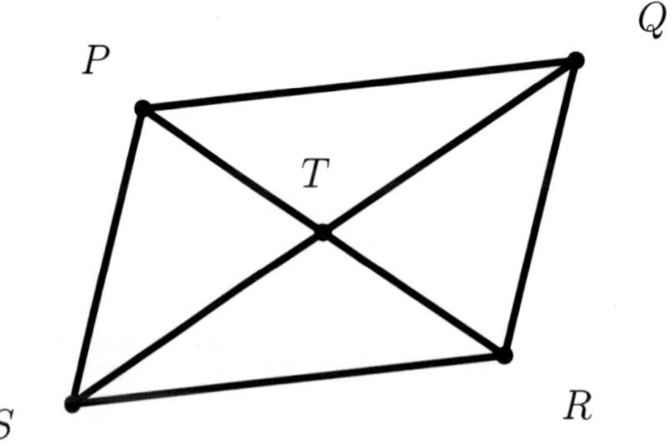

Claim: The quadrilateral $PQRS$ is a parallelogram.

Of course, the previous figure looks like a parallelogram, but how can we be sure? As we showed earlier it is enough to check that the opposite sides PQ and RS have the same length and are parallel. In vector notation, it is enough to prove $\overrightarrow{PQ} = \overrightarrow{SR}$. Let's try out our proof template: expand – apply something – contract. Indeed, it works if we incorporate the new vertex T and observe that

$$\overrightarrow{PT} = \overrightarrow{TR} \quad \text{and} \quad \overrightarrow{TQ} = \overrightarrow{ST}$$

since T bisects each diagonal:

$$\begin{aligned} \overrightarrow{PQ} &= \overrightarrow{PT} + \overrightarrow{TQ} \\ &= \overrightarrow{TR} + \overrightarrow{ST} \\ &= \overrightarrow{ST} + \overrightarrow{TR} \\ &= \overrightarrow{SR} \end{aligned}$$

We have proved the "if" half of the following proposition in plane geometry:

Proposition: A quadrilateral is a parallelogram if and only if the diagonals bisect each other.

The converse, that is, the "only if" part of this proposition, and a generalisation are left as starred exercises. Vector proofs in geometry will be taken further, in the next chapter, after we have introduced the notion of *linear independence* of a pair of vectors.

Chapter 1: Important Ideas and Useful Facts

1.1 A *geometric vector* $\mathbf{v}$ is a directed line segment in space, described by its length $|\mathbf{v}|$ and direction. Two vectors are *equal* if they have the same magnitude and direction, regardless of their position in space.

1.2 In this context, a *scalar* λ is a real number. The *scalar multiple* $\lambda\mathbf{v}$ has length $|\lambda||\mathbf{v}|$ and the same direction as $\mathbf{v}$ if λ is positive, and opposite direction if λ is negative.

1.3 If P and Q are points in space then $\overrightarrow{PQ}$ denotes the vector pointing from P to Q. The *position vector* of the point P is the vector $\overrightarrow{OP}$ where O denotes the origin in space.

1.4 A *parallelogram* is a quadrilateral such that two opposite sides are parallel and have the same length (which implies that the other two opposite sides are also parallel and have the same length).

1.5 Parallelogram Law of Vector Addition: The *vector sum* $\mathbf{v}+\mathbf{w}$ is represented by the diagonal of the parallelogram formed using sides labelled by $\mathbf{v}$ and $\mathbf{w}$.

1.6 Commutative Law of Addition: $\mathbf{v}+\mathbf{w} = \mathbf{w}+\mathbf{v}$

1.7 Associative Law of Addition: $\mathbf{u}+(\mathbf{v}+\mathbf{w}) = (\mathbf{u}+\mathbf{v})+\mathbf{w}$

1.8 The zero vector $\mathbf{0}$ has zero length and points in every direction. We have $\mathbf{0}+\mathbf{v}=\mathbf{v}$ and $0\mathbf{v}=\mathbf{0}$ for every vector $\mathbf{v}$.

1.9 The *negative* of $\mathbf{v}$ is $-\mathbf{v} = (-1)\mathbf{v}$ with the same length as $\mathbf{v}$ but pointing in the opposite direction. If P and Q are points in space then $\overrightarrow{QP} = -\overrightarrow{PQ}$.

1.10 The *vector difference* $\mathbf{v}-\mathbf{w}$ equals $\mathbf{v}+(-\mathbf{w})$ and has the property that $\mathbf{w}+(\mathbf{v}-\mathbf{w})=\mathbf{v}$.

1.11 If $\mathbf{v}$ and $\mathbf{w}$ are vectors and λ and μ are scalars then the following hold:

$$\lambda(\mu\mathbf{v}) = (\lambda\mu)\mathbf{v}\,, \quad \lambda(\mathbf{v}+\mathbf{w}) = \lambda\mathbf{v}+\lambda\mathbf{w}\,, \quad (\lambda+\mu)\mathbf{v} = \lambda\mathbf{v}+\mu\mathbf{v}$$

$$-(-\mathbf{v}) = \mathbf{v}\,, \quad \mathbf{v}-\mathbf{v} = \mathbf{0}\,, \quad 1\mathbf{v} = \mathbf{v}\,, \quad (-\lambda)\mathbf{v} = -(\lambda\mathbf{v})$$

1.12 A quadrilateral is a parallelogram if and only if the diagonals bisect each other.

Exercise 1.1 Solve for $\mathbf{x}$ in terms of $\mathbf{u}$, $\mathbf{v}$ and $\mathbf{w}$:

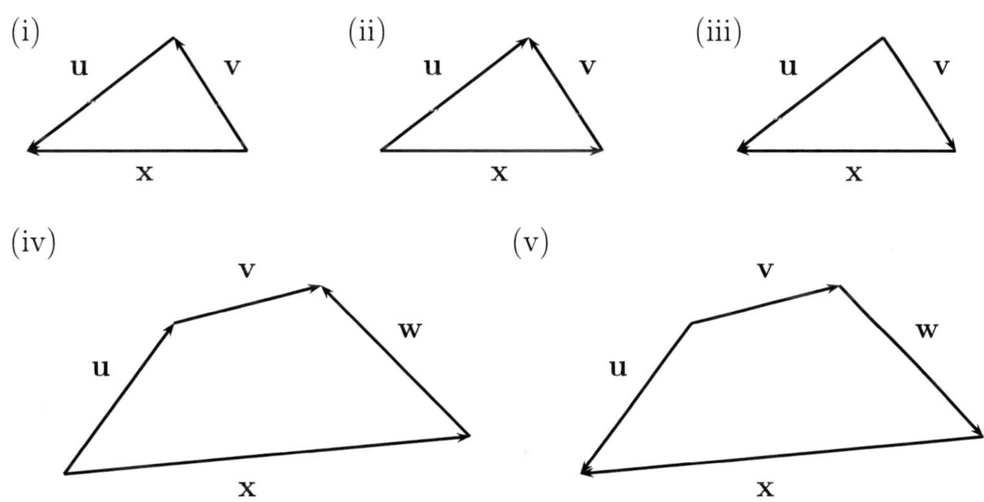

Exercise 1.2 Solve for $\mathbf{x}$ in terms of $\mathbf{a}$, $\mathbf{b}$ and $\mathbf{c}$:

(i) $2\mathbf{x}+\mathbf{b} = 2\mathbf{a}-\mathbf{b}$ (ii) $2\mathbf{x}+\mathbf{b}-\frac{1}{6}\mathbf{c} = \frac{1}{2}(\mathbf{x}-2\mathbf{b})+\frac{1}{3}(\mathbf{x}+3\mathbf{c})$

Exercise 1.3 Express $\frac{1}{2}\mathbf{v}$ in terms of $\mathbf{a}$, $\mathbf{b}$ and $\mathbf{c}$ and simplify when $\mathbf{v} = -\mathbf{x} - 2\mathbf{y} + 3\mathbf{z}$ and the following hold:

$$\mathbf{x} = 4\mathbf{a} - \mathbf{c}\,, \qquad \mathbf{y} = -\mathbf{a} + 3\mathbf{c}\,, \qquad \mathbf{z} = 4\mathbf{b} - 3\mathbf{c}$$

Exercise 1.4 Draw a diagram to illustrate the associative law of vector addition. (Don't worry about the fact that your diagram is not general.)

Exercise 1.5 Use the $\overrightarrow{PQ}$ notation to provide a diagram-free proof of the associative law of vector addition. (This is now a general proof.)

Exercise 1.6 Draw a diagram to illustrate the rule $\lambda(\mathbf{v}+\mathbf{w}) = \lambda\mathbf{v}+\lambda\mathbf{w}$ where $\mathbf{v}$ and $\mathbf{w}$ are vectors and λ is a positive scalar. What geometric property makes the rule work? What geometric property of diagrams makes the rule work if λ is now a negative scalar?

Exercise 1.7 Give a geometric interpretation of the *triangle inequality*:

$$|\mathbf{a}+\mathbf{b}| \leq |\mathbf{a}|+|\mathbf{b}|$$

When does equality occur?

Exercise 1.8 Draw a triangle OPQ and let R be the midpoint of the side PQ. Verify that $\overrightarrow{OR} = \frac{1}{2}(\overrightarrow{OP} + \overrightarrow{OQ})$.

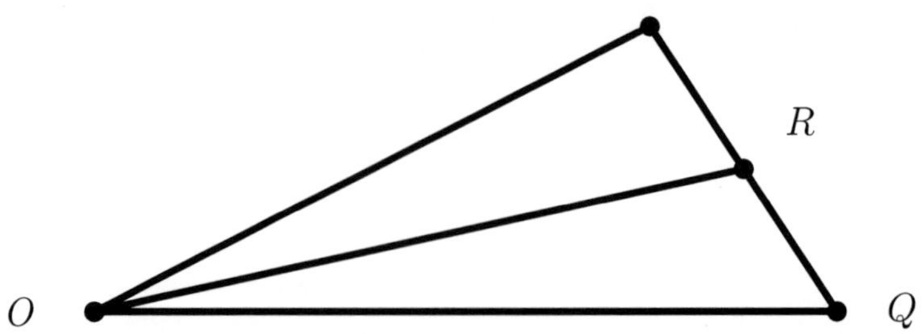

Exercise 1.9* The *median* of a triangle is the line joining a vertex to the midpoint of the opposite side. Prove that, for any triangle, the medians can be shifted parallel to themselves to form another triangle.

Exercise 1.10* Prove that the midpoints of the sides of a quadrilateral form a parallelogram.

Exercise 1.11* Prove that the diagonals of a parallelogram bisect each other.

Exercise 1.12** Draw a parallelogram $PQRS$. Let T divide the side SR in the ratio $r : s$ and let U be the point of intersection of the diagonal PR with the line QT. Find the ratio in which U divides the diagonal. (As a check, your answer should be such that when $r = 0$ the ratio becomes $1 : 1$, recovering the result of the previous exercise.)

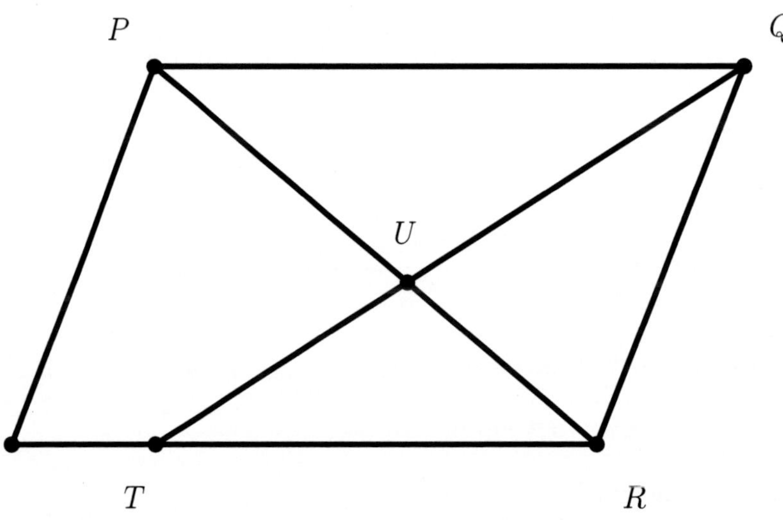

2 Position Vectors and Components

WATCH IT NOW on DVD VIDEO

- **Unit vectors and hat notation** (Section 2.1)
  Time: 7.59
- **Parallel vectors** (Section 2.2)
  Time: 5.23
- **Position vectors and components** (Section 2.3)
  Time: 14.07
- **Finding components from coordinates** (Section 2.3)
  Time: 3.50
- **Length of a vector** (Section 2.4)
  Time: 5.26

2 Position Vectors and Components

In this chapter we introduce position vectors and components. These lead to coordinate-wise operations that correspond to the geometric operations of the previous chapter. Along the way we will discuss unit vectors, the very useful "hat" notation, and the notion of linear independence for a pair of vectors. (Linear dependence and independence generalise to any number of vectors, but the concepts are difficult to master. It is important to start gently and not be in a hurry.)

2.1 Magnitude, unit vectors and hat notation

Recall that a geometric vector $\mathbf{v}$ has a length, also called its *magnitude*, denoted by $|\mathbf{v}|$. Some authors also use v to denote the length of $\mathbf{v}$, which is useful to reduce the clutter of complicated expressions. However, in this book, we will stick to the more cumbersome $|\mathbf{v}|$, because of the advantage that it is incapable of being misinterpreted.

Now a scalar (real number) λ also has a magnitude or absolute value $|\lambda|$. There is a nice relationship between magnitudes of scalars and vectors, which follows immediately from the definition of multiplication of a vector by a scalar (in Section 1.2):

Simple fact: For any geometric vector $\mathbf{v}$ and any scalar λ,

$$|\lambda\mathbf{v}| \quad = \quad |\lambda||\mathbf{v}|\,.$$

Call a vector $\mathbf{u}$ a *unit vector* if $|\mathbf{u}| = 1$. Clearly, there are infinitely many unit vectors in the plane or in space:

Unit vectors have length one and there is always a unit vector pointing in any given direction.

Consider any nonzero vector $\mathbf{v}$. Then $|\mathbf{v}| \neq 0$ so that the reciprocal $1/|\mathbf{v}|$ exists as a real number and can be used for purposes of scalar multiplication. In particular, we can form a new vector, called *the hat of* $\mathbf{v}$:

$$\widehat{\mathbf{v}} = \frac{1}{|\mathbf{v}|}\mathbf{v} = \frac{\mathbf{v}}{|\mathbf{v}|}$$

Notice that we are using a convenient and commonsense convention:

Division by nonzero scalars: write $\dfrac{\mathbf{v}}{\lambda}$ for $\dfrac{1}{\lambda}\mathbf{v}$

whenever λ is a nonzero scalar. Division by nonzero scalars has all of the usual algebraic properties corresponding to scalar multiplication.

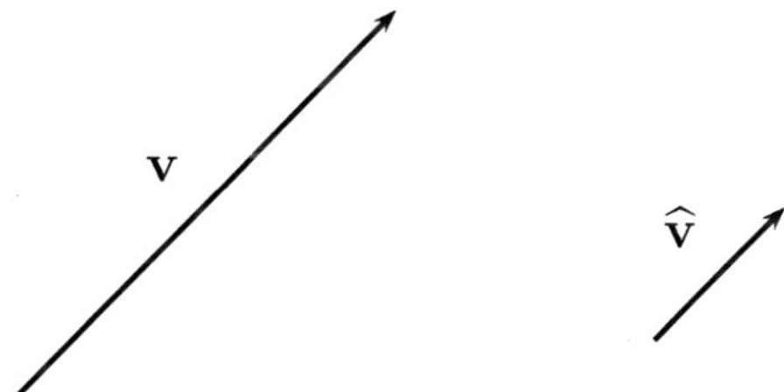

Since $\widehat{\mathbf{v}}$ is a positive scalar multiple of $\mathbf{v}$, both vectors point in the same direction. Observe (by the simple fact of the previous page) that

$$|\widehat{\mathbf{v}}| = \left|\frac{1}{|\mathbf{v}|}\mathbf{v}\right| = \left|\frac{1}{|\mathbf{v}|}\right||\mathbf{v}| = \frac{1}{|\mathbf{v}|}|\mathbf{v}| = 1.$$

This verifies that

$\widehat{\mathbf{v}}$ is the unit vector pointing in the direction of $\mathbf{v}$.

Note further that

$$\boxed{\mathbf{v} = |\mathbf{v}|\,\widehat{\mathbf{v}}}$$

since, by definition and properties of scalar multiplication,

$$|\mathbf{v}|\,\widehat{\mathbf{v}} = |\mathbf{v}|\left(\frac{1}{|\mathbf{v}|}\mathbf{v}\right) = \left(|\mathbf{v}|\frac{1}{|\mathbf{v}|}\right)\mathbf{v} = 1\mathbf{v} = \mathbf{v}\,.$$

Examples of explicit calculations of $\widehat{\mathbf{v}}$ are given in Section 2.4.

2.2 Parallel vectors

Two vectors are said to be *parallel* if they point in the same or in opposite directions. Since the zero vector points in every direction, we have that

> the zero vector $\mathbf{0}$ is parallel to every vector.

For nonzero vectors we can reformulate this in terms of scalar multiplication:

> **Simple fact:** Nonzero vectors $\mathbf{v}$ and $\mathbf{w}$ are parallel if and only if
>
> $\mathbf{v} = \lambda\mathbf{w}$ for some nonzero scalar λ.

It is instructive, and useful practice in understanding and manipulating the hat notation, to think carefully through why this is true. On the one hand, suppose $\mathbf{v} = \lambda\mathbf{w}$ for some nonzero scalar λ. If $\lambda > 0$ then by definition of scalar multiplication $\mathbf{v}$ is obtained by stretching or shrinking $\mathbf{w}$ without changing direction. If $\lambda < 0$ then the same is true except that the direction of $\mathbf{w}$ has been reversed. Thus, $\mathbf{v}$ points in the same or the opposite direction of $\mathbf{w}$, so by definition they are parallel.

Suppose, on the other hand, that $\mathbf{v}$ and $\mathbf{w}$ are parallel. If they point in the same direction, then their unit vectors are identical:

$$\widehat{\mathbf{v}} = \widehat{\mathbf{w}}$$

so that the following holds:

$$\mathbf{v} = |\mathbf{v}|\,\widehat{\mathbf{v}} = |\mathbf{v}|\,\widehat{\mathbf{w}} = \frac{|\mathbf{v}|}{|\mathbf{w}|}\,\mathbf{w}$$

If they point in opposite directions, then their unit vectors are negatives of each other:

$$\widehat{\mathbf{v}} = -\widehat{\mathbf{w}}$$

so that the following holds:

$$\mathbf{v} = |\mathbf{v}|\,\widehat{\mathbf{v}} = |\mathbf{v}|\,(-\widehat{\mathbf{w}}) = -\frac{|\mathbf{v}|}{|\mathbf{w}|}\,\mathbf{w}$$

Thus, $\mathbf{v} = \lambda\mathbf{w}$, taking $\lambda = |\mathbf{v}|/|\mathbf{w}|$ in the first case and $\lambda = -|\mathbf{v}|/|\mathbf{w}|$ in the second case. The proof is now complete.

But, for nonzero λ,

$$\mathbf{v} = \lambda\mathbf{w} \iff \mathbf{w} = \frac{1}{\lambda}\mathbf{v}$$

so that we can express the parallel criterion alternatively in words:

> Nonzero vectors $\mathbf{v}$ and $\mathbf{w}$ are parallel if and only if they are scalar multiples of each other.

2.3 Position vectors and components

The unit vectors pointing in the directions of the three positive axes in space are used so frequently that all mathematicians agree to give them names:

> The unit vectors $\mathbf{i}$, $\mathbf{j}$ and $\mathbf{k}$ point in the directions of the positive x-, y- and z-axes respectively.

The unit vectors $\mathbf{i}$ and $\mathbf{j}$ have double lives, firstly in the xy-plane:

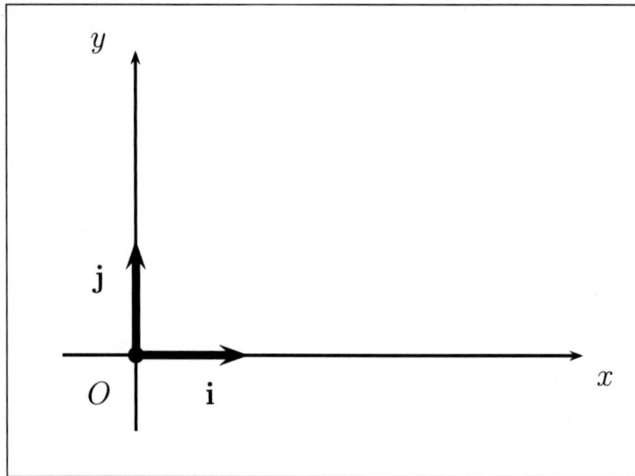

and secondly in space:

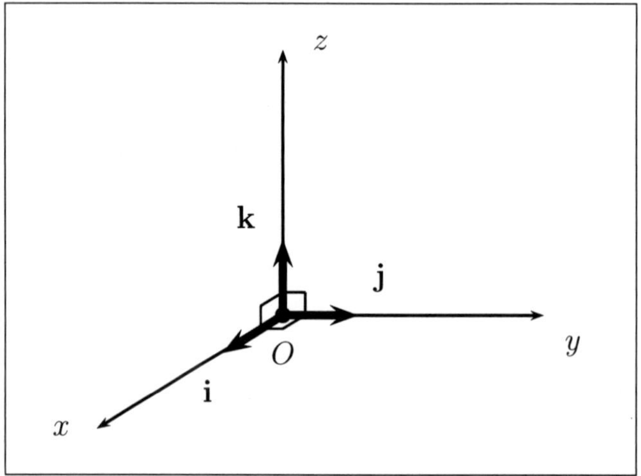

We have drawn them above emanating from the origin O, but of course they are free to float anywhere in the plane or in space as our purpose suits. Every geometric vector "decomposes" in terms of $\mathbf{i}$, $\mathbf{j}$ and $\mathbf{k}$. To see this we first introduce *position vectors*:

> The *position vector of a point* P in the plane or in space is the vector
>
> $$\mathbf{v} = \overrightarrow{OP}$$
>
> which is the directed line segment emanating from the origin and terminating at P.

For example, in the plane, $\mathbf{i}$ and $\mathbf{j}$ are the position vectors of the points $(1,0)$ and $(0,1)$ respectively. In space, $\mathbf{i}$, $\mathbf{j}$ and $\mathbf{k}$ are the position vectors of the points $(1,0,0)$, $(0,1,0)$ and $(0,0,1)$ respectively.

In the plane:

> If $\mathbf{v}$ is the position vector of the point $P(a,b)$ then
>
> $$\mathbf{v} = a\,\mathbf{i} + b\,\mathbf{j},$$
>
> called the *Cartesian form* or *decomposition* of $\mathbf{v}$.

This decomposition follows by the definitions of vector addition and scalar multiplication from the previous chapter, as one can see from the following diagram. This is not a general diagram, for we have placed $P = P(a,b)$ in the first quadrant. But the reader can easily see that the Cartesian form holds regardless of where P lies.

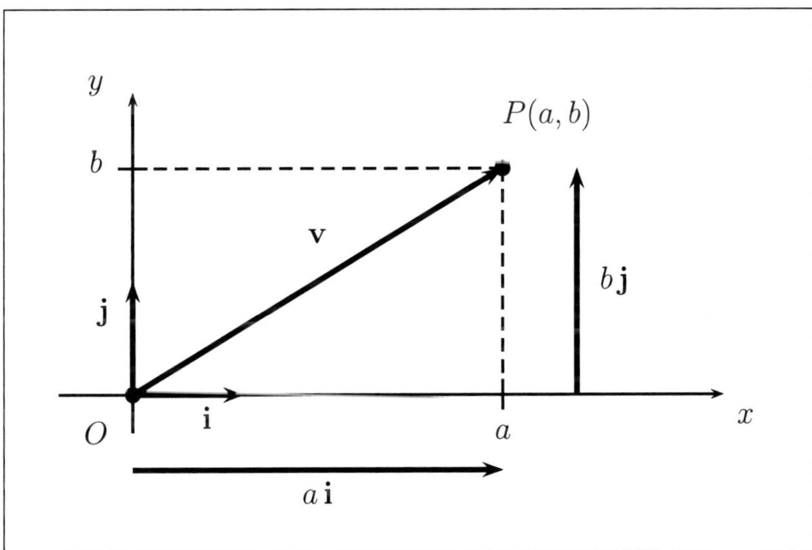

The adjective *Cartesian* has been universally accepted in honour of the great seventeenth century French mathematician and philosopher René Descartes (1596–1650), who invented the coordinate system that we now almost take for granted, but at the time was a revolutionary conceptual advance. (His surname is Latinised to *Cartesius*.)

In space:

> If $\mathbf{v}$ is the position vector of the point $P(a, b, c)$ then
>
> $$\mathbf{v} = a\,\mathbf{i} + b\,\mathbf{j} + c\,\mathbf{k},$$
>
> again called the *Cartesian form* or *decomposition* of $\mathbf{v}$.

Again one can see that this is true using diagrams. Below we have placed $P = P(a, b, c)$ in the first octant (where a, b, c are positive), but the Cartesian form holds regardless of where P lies.

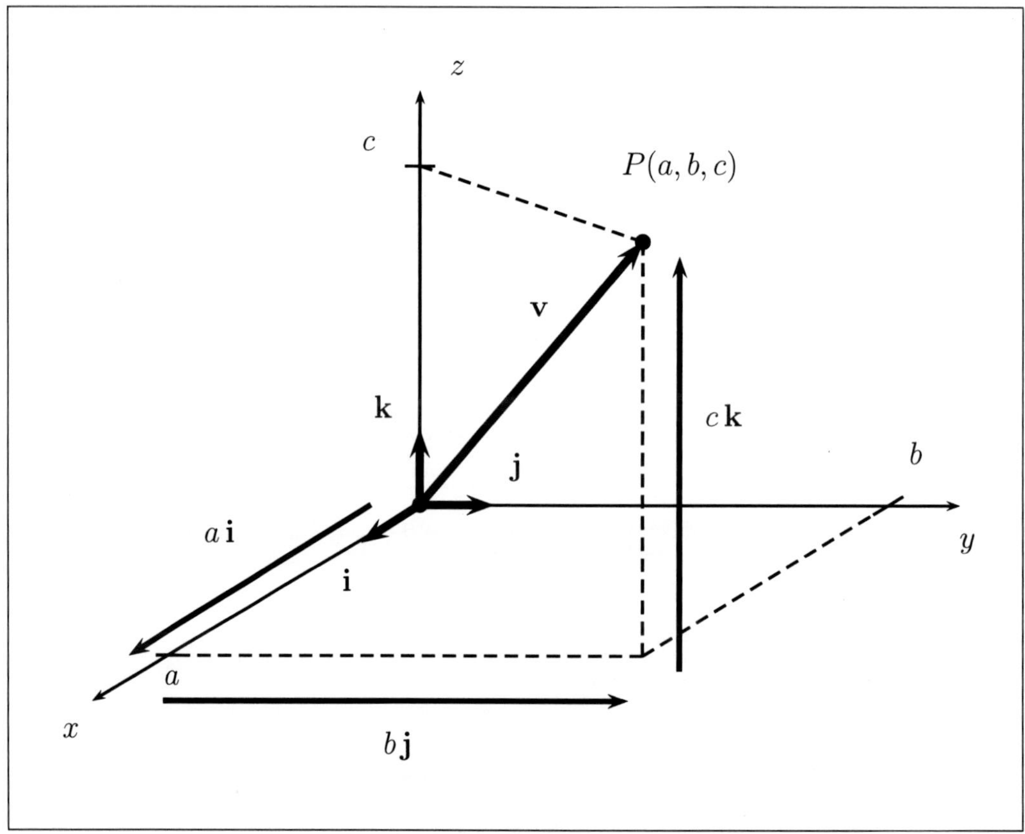

> We call (a, b, c) the *components* or *coordinates* of $\mathbf{v} = a\,\mathbf{i} + b\,\mathbf{j} + c\,\mathbf{k}$.

Operations with vectors are easy using components!

Let's see what happens with vector addition. If

$$\mathbf{v_1} = a_1\mathbf{i} + b_1\mathbf{j} + c_1\mathbf{k}\,, \quad \mathbf{v_2} = a_2\mathbf{i} + b_2\mathbf{j} + c_2\mathbf{k}$$

then, using properties from Section 1.4, the following holds:

$$\begin{aligned} \mathbf{v_1} + \mathbf{v_2} &= a_1\mathbf{i} + b_1\mathbf{j} + c_1\mathbf{k} + a_2\mathbf{i} + b_2\mathbf{j} + c_2\mathbf{k} \\ &= a_1\mathbf{i} + a_2\mathbf{i} + b_1\mathbf{j} + b_2\mathbf{j} + c_1\mathbf{k} + c_2\mathbf{k} \\ &= (a_1 + a_2)\mathbf{i} + (b_1 + b_2)\mathbf{j} + (c_1 + c_2)\mathbf{k} \end{aligned}$$

This demonstrates the following rule:

To add vectors, simply add respective components.

For example, if $\mathbf{v} = 2\mathbf{i} + 3\mathbf{j} - 5\mathbf{k}$ and $\mathbf{w} = 3\mathbf{i} - 2\mathbf{j} - \mathbf{k}$ then

$$\mathbf{v} + \mathbf{w} = (2+3)\mathbf{i} + (3 + (-2))\mathbf{j} + (-5 + (-1))\mathbf{k} = 5\mathbf{i} + \mathbf{j} - 6\mathbf{k}\,.$$

The middle step is included for emphasis. In practice one leaps straight to the final answer. What about the other operations? Following a pattern, one would expect, in this example, for subtraction,

$$\mathbf{v} - \mathbf{w} = -\mathbf{i} + 5\mathbf{j} - 4\mathbf{k}\,,$$

for negation,

$$-\mathbf{v} = -2\mathbf{i} - 3\mathbf{j} + 5\mathbf{k}\,,$$

and for scalar multiplication (by 7, say),

$$7\mathbf{v} = 14\mathbf{i} + 21\mathbf{j} - 35\mathbf{k}\,.$$

All of these are correct, because of the following rules:

To subtract vectors, simply subtract respective components.

To negate a vector, simply negate components.

To multiply a vector by a scalar, multiply the components by the scalar.

The formal justification of each of these is left as an exercise, just mimicking the derivation we gave for vector addition.

The ease with which we perform vector operations this way presupposes that we know what the components are. So, how do we find them? The answer is to locate the coordinates of the points at the tip and at the end of the tail of a given vector, and then perform a subtraction *in the correct order.* (Don't worry, everyone gets the order wrong at least once in their lives!)

Let $P(a_1, b_1, c_1)$ and $Q(a_2, b_2, c_2)$ be points in space. Then

$$\mathbf{v} = \overrightarrow{PQ}$$

has components

$$(a_2 - a_1, b_2 - b_1, c_2 - c_1)\,,$$

that is,

$$\mathbf{v} = (a_2 - a_1)\mathbf{i} + (b_2 - b_1)\mathbf{j} + (c_2 - c_1)\mathbf{k}\,.$$

For example, if

$$P = (1, 2, 3)\,, \quad Q = (2, -1, 4)\,, \quad R = (7, 0, -2)$$

then

$$\overrightarrow{PQ} = \mathbf{i} - 3\mathbf{j} + \mathbf{k}\,, \quad \overrightarrow{PR} = 6\mathbf{i} - 2\mathbf{j} - 5\mathbf{k}\,, \quad \overrightarrow{QR} = 5\mathbf{i} + \mathbf{j} - 6\mathbf{k}\,.$$

2.4 Length of a vector

It is important to be able to recover the length of a vector from its components, and we record the relevant formulae here.

> If $\mathbf{v} = a\mathbf{i} + b\mathbf{j}$ is a vector in the plane then
>
> $$|\mathbf{v}| = \sqrt{a^2 + b^2}\,.$$

This follows from the Theorem of Pythagoras (explained in Appendix 1).

> If $\mathbf{v} = a\mathbf{i} + b\mathbf{j} + c\mathbf{k}$ is a vector in space then
>
> $$|\mathbf{v}| = \sqrt{a^2 + b^2 + c^2}\,.$$

This follows from a further application of the Theorem of Pythagoras and is a good exercise in drawing and interpreting diagrams in three dimensions. To illustrate these formulae, if

$$\mathbf{u} = 4\mathbf{i} - 3\mathbf{j}\,, \quad \mathbf{v} = -\mathbf{i} + 3\mathbf{j} + 4\mathbf{k}\,, \quad \mathbf{w} = 3\mathbf{i} - \mathbf{k}$$

then

$$|\mathbf{u}| = \sqrt{4^2 + (-3)^2} = \sqrt{16+9} = \sqrt{25} = 5\,,$$
$$|\mathbf{v}| = \sqrt{(-1)^2 + 3^2 + 4^2} = \sqrt{1+9+16} = \sqrt{26}\,,$$
$$|\mathbf{w}| = \sqrt{9+1} = \sqrt{10}\,,$$

so that

$$\widehat{\mathbf{u}} = \frac{4}{5}\mathbf{i} - \frac{3}{5}\mathbf{j}\,, \quad \widehat{\mathbf{v}} = -\frac{1}{\sqrt{26}}(\mathbf{i} - 3\mathbf{j} - 4\mathbf{k})\,, \quad \widehat{\mathbf{w}} = \frac{1}{\sqrt{10}}(3\mathbf{i} - \mathbf{k})\,.$$

Using points, if $P = (1, 1, -3)$ and $Q = (4, -5, 3)$ then $\overrightarrow{PQ} = 3\mathbf{i} - 6\mathbf{j} + 6\mathbf{k}$, the distance from P to Q is

$$|\overrightarrow{PQ}| = \sqrt{9+36+36} = 9\,,$$

and a unit vector pointing from P towards Q is

$$\frac{1}{9}(3\mathbf{i} - 6\mathbf{j} + 6\mathbf{k}) = \frac{1}{3}(\mathbf{i} - 2\mathbf{j} + 2\mathbf{k})\,.$$

2.5 Linear independence for two vectors

We finish this chapter with an introduction to the concept of linear independence, just for two vectors, and a nontrivial application to geometry.

The definition of *linear independence* (below) may seem vague and abstract, but it is central to more advanced linear algebra (the next course after you have finished this book). It leads to notions of *basis* and *dimension* that arise in a plethora of situations and applications that at first sight appear to have little to do with geometric vectors. One of the great surprises in mathematics is how such a diversity comes within the ambit of a geometrical way of thinking. Nevertheless, there is a certain threshold of abstraction that must be overcome before the ideas flow easily, and the definition of linear independence is a common stumbling block for students.

Recall, from the Introduction, the idea of a *linear combination*

$$\boxed{ax + by}$$

of variables x and y, related to the general form of the equation of a line in the plane. Replace the variables by vectors $\mathbf{x}$ and $\mathbf{y}$ in the plane (by the simple expedient of converting to boldface) and call the resulting linear combination $\mathbf{v}$:

$$\boxed{\mathbf{v} \;=\; a\mathbf{x} + b\mathbf{y}}$$

Now, changing our point of view even further, let's fix $\mathbf{x}$ and $\mathbf{y}$ and allow the scalars a and b to vary over all real numbers. What can be reached by $\mathbf{v}$? Now, if $\mathbf{x}$ and $\mathbf{y}$ were chosen to be *parallel*, then one would be a scalar multiple of each other (Section 2.2), say,

$$\mathbf{y} \;=\; \lambda\mathbf{x}\,,$$

and then $\mathbf{v}$ would become

$$\mathbf{v} \;=\; a\mathbf{x} + b\mathbf{y} \;=\; a\mathbf{x} + b\lambda\mathbf{x} \;=\; (a + b\lambda)\mathbf{x} \;=\; \mu\mathbf{x}\,,$$

where $\mu = (a + b\lambda)$, which is just another scalar multiple of $\mathbf{x}$. No matter how hard we tried, we could only build something of the form $\mu\mathbf{x}$, which always points in the direction or opposite direction of $\mathbf{x}$. Regarded as position vectors, rooted at the origin, the tips of the vectors $\mathbf{v}$ would "span" a line passing through the origin, and no more.

Suppose now that $\mathbf{x}$ and $\mathbf{y}$ are *not* parallel. If we anchor them at the origin, like the unit vectors $\mathbf{i}$ and $\mathbf{j}$ of Section 2.3, then because we have access to two nonparallel directions, the linear combinations $\mathbf{v} = a\mathbf{x} + b\mathbf{y}$ above produce position vectors whose tips "span" the entire plane!

This is easy to see from a diagram, and we even get analogues of the usual quadrants, depending on whether a and b are positive, negative or mixtures of both, and whether the angle (between $0°$ and $180°$) from $\mathbf{x}$ to $\mathbf{y}$ is measured anticlockwise or clockwise. In the following diagram the angle from $\mathbf{x}$ to $\mathbf{y}$ is measured clockwise and examples of linear combinations are exhibited in each "quadrant":

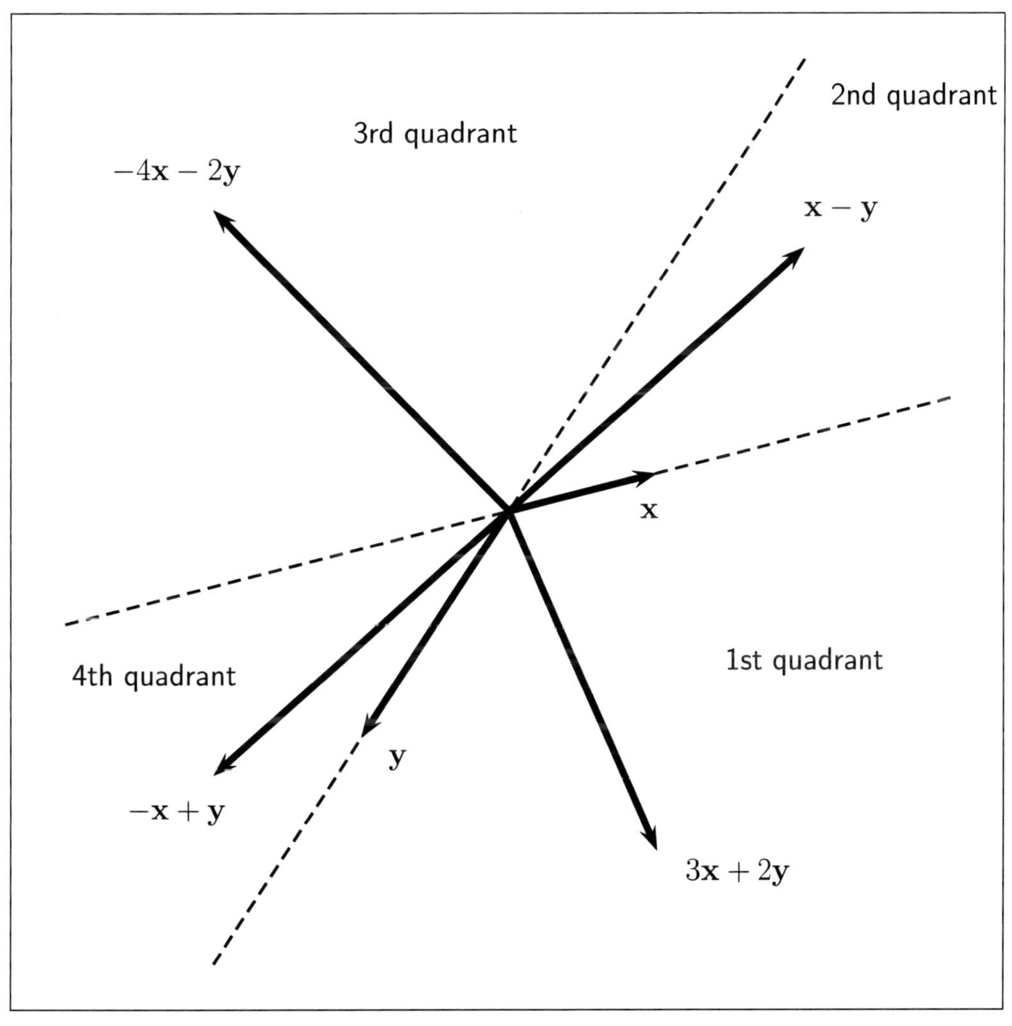

From the point of view of manufacturing new vectors from old ones, the property of *not being parallel* is desirable! There is a surprisingly useful criterion, captured in the following definition:

Two vectors $\mathbf{x}$ and $\mathbf{y}$ are said to be *linearly independent* if the equation

$$a\mathbf{x} + b\mathbf{y} \;=\; \mathbf{0}$$

implies that the scalars a and b must both be zero.

Thus, the only way one can manufacture the zero vector $\mathbf{0}$ from linearly independent vectors $\mathbf{x}$ and $\mathbf{y}$ is by forming the *trivial* linear combination:

$$0\mathbf{x} + 0\mathbf{y} \;=\; \mathbf{0}$$

If the condition of the definition fails then $a\mathbf{x} + b\mathbf{y} = \mathbf{0}$ for some scalars a and b such that either $a \neq 0$ or $b \neq 0$, whence, by vector arithmetic,

$$\mathbf{x} = -\frac{b}{a}\mathbf{y} \qquad \text{or} \qquad \mathbf{y} = -\frac{a}{b}\mathbf{x}\,.$$

In both cases, $\mathbf{x}$ and $\mathbf{y}$ point in the same or opposite directions, so are parallel!

On the other hand, if $\mathbf{x}$ and $\mathbf{y}$ are parallel, so that, without loss of generality,

$$\mathbf{x} \;=\; \lambda\mathbf{y}$$

for some nonzero scalar λ, then

$$(1)\mathbf{x} + (-\lambda)\mathbf{y} \;=\; \mathbf{0}\,,$$

so that we can express $\mathbf{0}$ as a nontrivial linear combination of $\mathbf{x}$ and $\mathbf{y}$. This proves the following:

Two vectors are linearly independent if and only if they are not parallel.

What is the use of this? Well, in geometry of the plane, nonparallel vectors are ubiquitous. Whenever one draws a (nondegenerate) triangle, two sides yield vectors which are not parallel. In a (nondegenerate) parallelogram, adjacent sides are not parallel. We illustrate how the criterion can be exploited to prove the following theorem:

> **Theorem:** The medians of a triangle intersect in a point.

Exploration leading to a proof: Consider a triangle PQR with medians PA, QB and RC, where A, B and C are the midpoints of the sides opposite P, Q and R respectively.

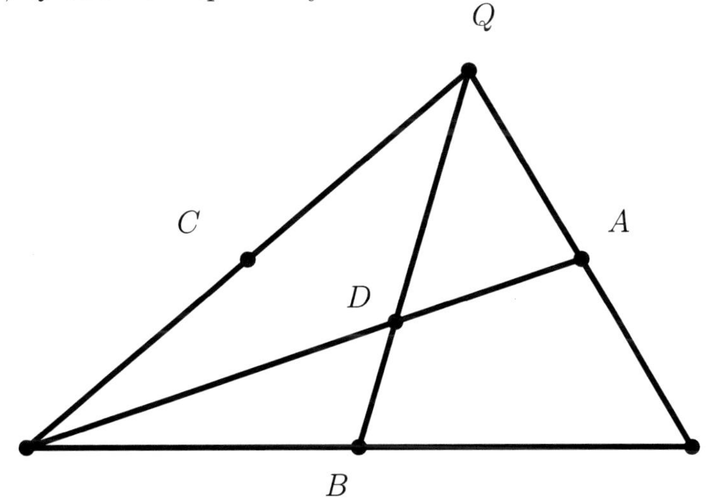

Our task is to show that PA, QB and RC intersect in a single point in the interior of the triangle. Let D be the point of intersection of the medians PA and QB (ignoring RC for the time being). The line segments PD and QD are some fractions of PA and QB respectively, so that in the language of vectors

$$\overrightarrow{PD} = \alpha \overrightarrow{PA} \qquad \text{and} \qquad \overrightarrow{QD} = \beta \overrightarrow{QB}$$

for some scalars α and β. (At the moment we do not know the values of α and β, though one could make a guess based on the diagram.)

We should be able to exploit the fact that the vectors $\overrightarrow{PQ}$ and $\overrightarrow{QR}$ are not parallel, so are linearly independent. Our strategy then is to try to express the zero vector as a linear combination of these two vectors. Firstly, we try an expansion of the first equation above:

$$\overrightarrow{PQ} + \overrightarrow{QD} = \overrightarrow{PD} = \alpha \overrightarrow{PA} = \alpha \left(\overrightarrow{PQ} + \overrightarrow{QA}\right) = \alpha \left(\overrightarrow{PQ} + \frac{1}{2}\overrightarrow{QR}\right),$$

aiming always to collect together copies of $\overrightarrow{PQ}$ and $\overrightarrow{QR}$, whence

$$\overrightarrow{QD} = \alpha \left(\overrightarrow{PQ} + \frac{1}{2}\overrightarrow{QR}\right) - \overrightarrow{PQ} = (\alpha - 1)\,\overrightarrow{PQ} + \frac{\alpha}{2}\,\overrightarrow{QR}.$$

But, by Exercise 1.8,

$$\overrightarrow{QD} = \beta\,\overrightarrow{QB} = \frac{\beta}{2}\left(\overrightarrow{QP} + \overrightarrow{QR}\right) = \frac{\beta}{2}\left(\overrightarrow{QR} - \overrightarrow{PQ}\right)$$

so that

$$\frac{\beta}{2}\left(\overrightarrow{QR} - \overrightarrow{PQ}\right) = (\alpha - 1)\,\overrightarrow{PQ} + \frac{\alpha}{2}\,\overrightarrow{QR}\,,$$

whence, after rearranging,

$$\left(\alpha - 1 + \frac{\beta}{2}\right)\overrightarrow{PQ} + \left(\frac{\alpha - \beta}{2}\right)\overrightarrow{QR} = \mathbf{0}\,.$$

Because $\overrightarrow{PQ}$ and $\overrightarrow{QR}$ are linearly independent, the scalar expressions which are coefficients in this equation must be zero:

$$\alpha - 1 + \frac{\beta}{2} = \frac{\alpha - \beta}{2} = 0$$

It follows quickly that $\alpha = \beta = \dfrac{2}{3}$.

This proves that the median QB intersects the median PA two-thirds along PA. By a similar argument the median RC must also intersect the median PA two-thirds along PA. This proves that all three medians intersect at the same point, and the theorem is proved. The common point of intersection is called the *centroid* of the triangle.

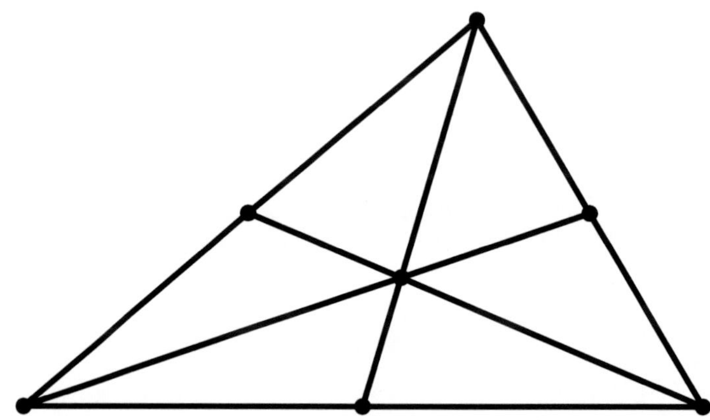

A by-product of the proof is that not only does the centroid exist but it lies exactly two-thirds of the way along any median from a given vertex. If one suspected this in the first place (say, by looking at the diagram and making an intelligent guess) then one could look for a direct proof, without appealing to the definition of linear independence (Exercise 2.7). Of course, if your guess is wrong then looking for a direct proof may lead to considerable frustration!

Chapter 2: Important Ideas and Useful Facts

2.1 If $\mathbf{v}$ is a geometric vector and λ is a scalar then $|\lambda\mathbf{v}| = |\lambda||\mathbf{v}|$.

2.2 A *unit vector* is a vector of length one. There is always a unit vector pointing in any given direction.

2.3 If λ is a nonzero scalar and $\mathbf{v}$ a vector then we write $\dfrac{\mathbf{v}}{\lambda}$ for the scalar multiple $\dfrac{1}{\lambda}\mathbf{v}$.

2.4 Let $\mathbf{v}$ be a nonzero vector. Then $\widehat{\mathbf{v}} = \dfrac{\mathbf{v}}{|\mathbf{v}|}$ is called *the hat of* $\mathbf{v}$, or simply $\mathbf{v}$-*hat*, and is the unit vector pointing in the direction of $\mathbf{v}$. It is designed to satisfy the equation $\mathbf{v} = |\mathbf{v}|\,\widehat{\mathbf{v}}$, which captures precisely the idea of $\mathbf{v}$ being characterised by length and direction.

2.5 Parallel vectors: Nonzero vectors $\mathbf{v}$ and $\mathbf{w}$ are parallel if and only if $\mathbf{v} = \lambda\mathbf{w}$ for some nonzero scalar λ.

2.6 Cartesian form: The unit vectors $\mathbf{i}$, $\mathbf{j}$ and $\mathbf{k}$ point in the directions of the positive x-, y- and z-axes respectively. If $\mathbf{v}$ is the position vector of the point $P(a,b,c)$ then the expression

$$\mathbf{v} = a\,\mathbf{i} + b\,\mathbf{j} + c\,\mathbf{k}$$

is called the *Cartesian form* of $\mathbf{v}$. In this case, a, b, c are called the *components* of $\mathbf{v}$.

2.7 Component-wise operations: To add, subtract or negate vectors, simply add, subtract or negate respective components. To multiply a vector by a scalar, simply multiply the components by the scalar.

2.8 If $\mathbf{v} = a\,\mathbf{i} + b\,\mathbf{j} + c\,\mathbf{k}$ then $|\mathbf{v}| = \sqrt{a^2+b^2+c^2}$.

2.9 If $P(a_1,b_1,c_1)$ and $Q(a_2,b_2,c_2)$ are points in space then

$$\overrightarrow{PQ} = (a_2-a_1)\mathbf{i} + (b_2-b_1)\mathbf{j} + (c_2-c_1)\mathbf{k}\,.$$

2.10 Linear independence of two vectors: Two vectors $\mathbf{v}$ and $\mathbf{w}$ are *linearly independent* if they are not parallel, which is equivalent to saying that the vector equation $a\,\mathbf{v} + b\,\mathbf{w} = \mathbf{0}$ implies that the scalars a and b must both be zero.

2.11 Two vectors are linearly independent if and only if they are not parallel.

2.12 The medians of a triangle intersect in a point.

Exercise 2.1 Consider the vectors

$$\mathbf{u} = 5\mathbf{i} + 12\mathbf{j}, \quad \mathbf{v} = -3\mathbf{j} + 4\mathbf{k}, \quad \mathbf{w} = \mathbf{i} + 2\mathbf{j} - 2\mathbf{k}.$$

Find and simplify the following:

(i) $\mathbf{u} + \mathbf{v}$ (ii) $-\mathbf{v}$ (iii) $\mathbf{u} - \mathbf{v}$ (iv) $3\mathbf{w}$ (v) $\frac{1}{2}(\mathbf{v} - \mathbf{u}) + 3\mathbf{w}$

(vi) $|\mathbf{u}|$ (vii) $|\mathbf{v}|$ (viii) $|\mathbf{w}|$ (ix) $|3\mathbf{w}|$ (x) $|-7\mathbf{w}|$

(xi) $\widehat{\mathbf{u}}$ (xii) $\widehat{\mathbf{v}}$ (xiii) $\widehat{\mathbf{w}}$ (xiv) $|\mathbf{v} + \mathbf{w}|$ (xv) $|\mathbf{v} - \mathbf{w}|$

Exercise 2.2 Consider the following points in space:

$$O(0,0,0), \quad P(1,1,1), \quad Q(-1,-1,0), \quad R(0,1,2), \quad S(2,3,3)$$

(i) Find the Cartesian forms of the following vectors:

$$\overrightarrow{OP}, \ \overrightarrow{OQ}, \ \overrightarrow{OR}, \ \overrightarrow{OS}, \ \overrightarrow{PQ}, \ \overrightarrow{QP}, \ \overrightarrow{QR}, \ \overrightarrow{RS}, \ \overrightarrow{SP}$$

(ii) Verify that the figure $PQRS$ is a rhombus, that is, a parallelogram in which side lengths are equal. How can you tell that it is not a square?

Exercise 2.3 Use properties listed in Section 1.4 to verify the coordinate-wise rules for subtraction of vectors, taking the negative of a vector and multiplying a vector by a scalar.

Exercise 2.4 Consider the following points in space:

$$P(1,-1,2), \quad Q(2,3,0), \quad R(-2,5,0), \quad S(0,1,\lambda)$$

where λ is some real number.

(i) Find the values of λ for which $|\overrightarrow{PQ}| = |\overrightarrow{PS}|$.

(ii) Find the value of λ for which $\overrightarrow{PR}$ is parallel to $\overrightarrow{RS}$.

Exercise 2.5 Draw an appropriate diagram and use two applications of the Theorem of Pythagoras to verify the formula

$$|a\mathbf{i} + b\mathbf{j} + c\mathbf{k}| = \sqrt{a^2 + b^2 + c^2}.$$

Exercise 2.6* Verify that if R divides a line PQ in the ratio $\lambda : \mu$ and O is any point in space then

$$\overrightarrow{OR} = \frac{\mu\,\overrightarrow{OP} + \lambda\,\overrightarrow{OQ}}{\lambda + \mu}.$$

Exercise 2.7* This exercise gives a direct proof (with the benefit of knowing in advance, or guessing from a good diagram, what the ratios ought to be) that the medians of a triangle intersect. Let PQR be a triangle with medians PA, QB and RC, where A, B, C are midpoints of the sides opposite P, Q, R respectively.

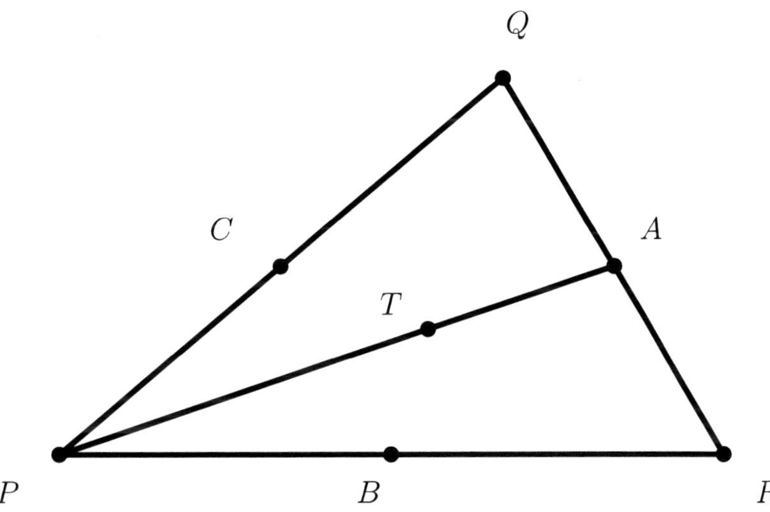

Let T be the point on PA which divides it in the ratio $2:1$. Put

$$\mathbf{u} = \overrightarrow{PQ} \quad \text{and} \quad \mathbf{v} = \overrightarrow{PR}.$$

Express $\overrightarrow{QT}$ and $\overrightarrow{QB}$ in terms of $\mathbf{u}$ and $\mathbf{v}$ and observe that they are parallel. This proves T lies on QB, and similarly on RC, so that the medians intersect.

Exercise 2.8* Suppose that $\mathbf{v}$ and $\mathbf{w}$ are vectors that are not parallel (so are linearly independent) and the following vector equation holds for some scalars α and β:

$$\mathbf{v} + \alpha(\mathbf{w} - \mathbf{v}) = \beta\big(\mathbf{v} + \frac{1}{2}\mathbf{w}\big)$$

Find α and β.

Exercise 2.9* The line joining the vertex of a parallelogram to the midpoint of an opposite side divides one of the diagonals into two pieces of unequal lengths. Find the ratio of the lengths of these pieces.

Exercise 2.10* A *linear combination* of vectors is an expression of the form

$$\lambda_1 \mathbf{v}_1 + \lambda_2 \mathbf{v}_2 + \cdots + \lambda_n \mathbf{v}_n .$$

A sequence of vectors $\mathbf{v}_1, \ldots, \mathbf{v}_n$ is *linearly independent* if

$$\lambda_1 \mathbf{v}_1 + \cdots + \lambda_n \mathbf{v}_n = \mathbf{0} \qquad \text{implies} \qquad \lambda_1 = \cdots = \lambda_n = 0 ,$$

and *linearly dependent* if this implication fails. Prove that a sequence of vectors is linearly dependent if and only if at least one of the vectors can be expressed as a linear combination of the others.

Exercise 2.11* Use diagrams to explain why it is impossible to find three vectors in the plane that are linearly independent, and impossible to find four vectors in space that are linearly independent. (This might suggest that the definition of linear independence in the previous exercise is only meaningful for $n \leq 3$. However, look at the next exercise!)

Exercise 2.12** Define *addition* and *scalar multiplication* of real-valued functions of a real variable as follows. Let $f,\, g : \mathbb{R} \to \mathbb{R}$ and $\lambda \in \mathbb{R}$. Define $f + g$ and λf by the following rules:

$$(f + g)(x) = f(x) + g(x) , \qquad (\lambda f)(x) = \lambda f(x)$$

It is routine to verify that all of the properties listed in Section 1.4 hold, with the zero constant function taking the role of the zero vector. The definitions of linear combinations, linear dependence and independence in Exercise 2.10 make sense with functions in place of vectors. For an integer $n \geq 0$ define $f_n : \mathbb{R} \to \mathbb{R}$ by the following rule:

$$f_n(x) = x^n$$

(i) There is a common name for functions that are linear combinations of $f_0, \ldots, f_n$. What is it?

(ii) Prove that $f_0, \ldots, f_n$ are linearly independent for all nonnegative integers n. (When you have learnt some abstract theory, you will see that this implies that the *vector space* of real-valued functions of a real variable is *infinite dimensional*.)

3 Dot Products and Projections

- **Products of vectors** (Sections 3.1, 3.2)
  Time: 10.28
- **Algebraic and geometric dot products** (Section 3.2)
  Time: 7.16
- **Angle between vectors** (Section 3.3)
  Time: 5.06

3 Dot Products and Projections

In this chapter and the next we discuss two further binary operations on vectors, the first of which is crucial for the development of the theory of *projection*, probably the single most important underyling idea in theories of approximation. So far we have discussed, for vectors $\mathbf{v}$, $\mathbf{w}$ and scalars λ:

- vector addition: forming the vector $\mathbf{v} + \mathbf{w}$
- scalar multiplication: forming the vector $\lambda\mathbf{v}$
- the zero vector $\mathbf{0}$, which behaves as a zero should: $\mathbf{0} + \mathbf{v} = \mathbf{v}$
- the negative vector $-\mathbf{v}$, which takes us back to zero: $-\mathbf{v} + \mathbf{v} = \mathbf{0}$
- vector subtraction, a combined operation: $\mathbf{v} - \mathbf{w} = \mathbf{v} + (-\mathbf{w})$.

One may ask if there is any meaningful and useful binary operation which we should call *multiplication of two vectors* $\mathbf{v}$ and $\mathbf{w}$. In fact, there are several, but we will only discuss two in this book, and they are very different (though closely related):

- the *scalar* or *dot product* $\mathbf{v} \cdot \mathbf{w}$
- the *vector* or *cross product* $\mathbf{v} \times \mathbf{w}$.

We emphasise from the beginning:

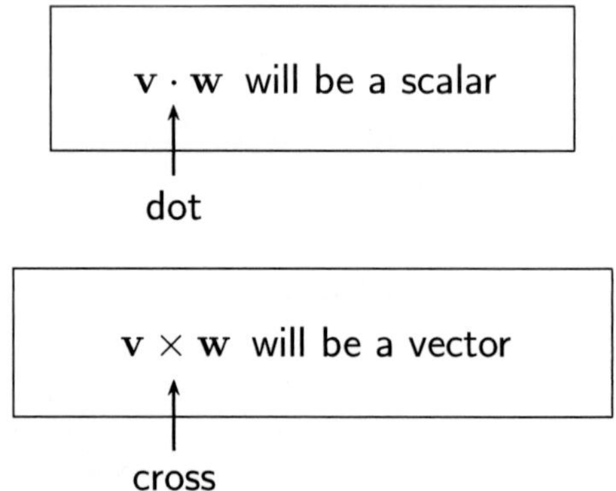

The dot or cross should never be left off:

Simply writing $\mathbf{v}\,\mathbf{w}$ will have no meaning in this book!

3.1 Geometric definition of dot product

Let $\mathbf{v}$ and $\mathbf{w}$ be geometric vectors in the plane or in space, and draw them together so that their tails meet:

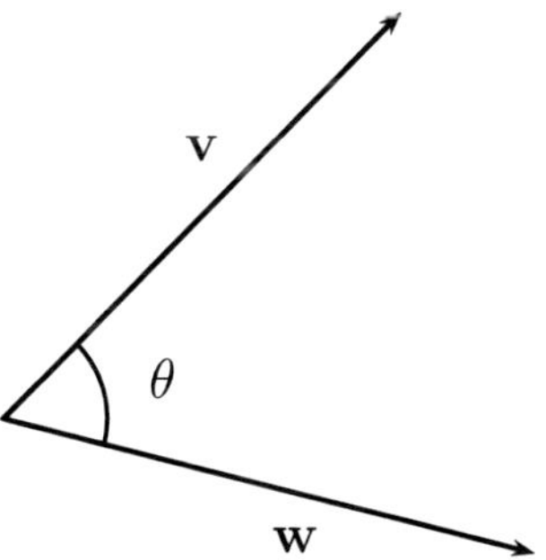

There will be an angle θ between them. It doesn't matter for the purposes of the following definition whether θ is measured clockwise or anticlockwise, since $\cos(-\theta) = \cos\theta$.

> **Geometric definition of dot product:**
>
> $$\mathbf{v} \cdot \mathbf{w} \;=\; |\mathbf{v}||\mathbf{w}|\cos\theta$$

The right-hand side of this formula is closely related to "projection" of the tip of one vector onto the closest point along the other vector. By elementary trigonometry, the lengths of these "projections" are $|\mathbf{v}|\cos\theta$ for $\mathbf{v}$ projected along $\mathbf{w}$:

$$\frac{\mathbf{v} \cdot \mathbf{w}}{|\mathbf{w}|} = |\mathbf{v}|\cos\theta$$

v

θ

w

and $|\mathbf{w}|\cos\theta$ for $\mathbf{w}$ projected along $\mathbf{v}$:

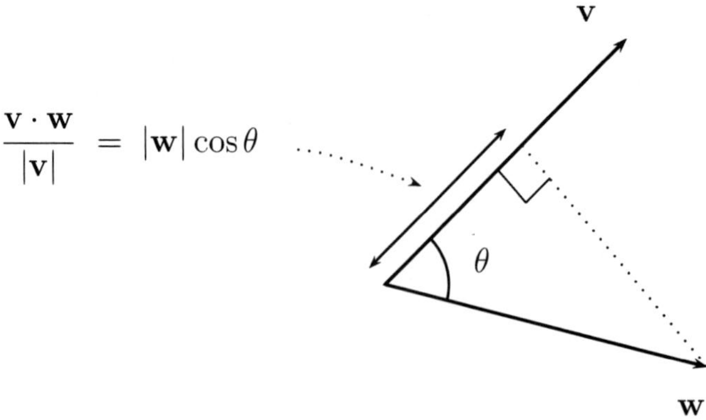

These relationships are formalised and explored in Section 3.4. Notice that, always,

$$-1 \le \cos\theta \le 1$$

so that

$$-|\mathbf{v}||\mathbf{w}| \le |\mathbf{v}||\mathbf{w}|\cos\theta \le |\mathbf{v}||\mathbf{w}|,$$

that is,

$$-|\mathbf{v}||\mathbf{w}| \le \mathbf{v}\cdot\mathbf{w} \le |\mathbf{v}||\mathbf{w}|,$$

or, more succinctly, the following:

Cauchy-Schwarz Inequality:

$$|\mathbf{v}\cdot\mathbf{w}| \le |\mathbf{v}||\mathbf{w}|$$

This is a fancy name for an inequality, but it is worth getting used to, because it is an important elementary result in the theory of *inner product spaces* (again in any course following on from this book). Our dot product turns out to be a special case of an *inner product*, though we won't use this terminology here.

As we saw, the Cauchy-Schwarz Inequality follows immediately from our geometric definition, but if one starts from the algebraic definition of the next section, then there is something of substance to prove. We will in fact show that the algebraic and geometric definitions of the dot product coincide.

We finish this section by noting three extreme cases, depending on whether $\theta = 0°$, $\theta = 90°$ or $\theta = 180°$:

Vectors $\mathbf{v}$ and $\mathbf{w}$ point in the same direction if and only if

$$\mathbf{v} \cdot \mathbf{w} = |\mathbf{v}||\mathbf{w}| .$$

Vectors $\mathbf{v}$ and $\mathbf{w}$ point in opposite directions if and only if

$$\mathbf{v} \cdot \mathbf{w} = -|\mathbf{v}||\mathbf{w}| .$$

Orthogonality Criterion: Vectors $\mathbf{v}$ and $\mathbf{w}$ are mutually perpendicular if and only if

$$\mathbf{v} \cdot \mathbf{w} = 0 .$$

3.2 Algebraic definition of dot product

The following definition is very easy to use when we have components of vectors at our fingertips:

Algebraic definition of dot product:

If $\mathbf{v} = a\,\mathbf{i} + b\,\mathbf{j} + c\,\mathbf{k}$ and $\mathbf{w} = d\,\mathbf{i} + e\,\mathbf{j} + f\,\mathbf{k}$ then

$$\mathbf{v} \cdot \mathbf{w} = ad + be + cf .$$

In other words,

multiply respective components and add everything up.

Later in the book we shall see that *matrix multiplication* is precisely cascades of applications of this rule, but using entries from rows and columns of matrices rather than components of geometric vectors.

To illustrate the algebraic rule, let

$$\mathbf{u} = 2\mathbf{i} - 3\mathbf{j} + \mathbf{k}\,, \qquad \mathbf{v} = 3\mathbf{i} + \mathbf{j} - \mathbf{k}\,, \qquad \mathbf{w} = -4\mathbf{i} - \mathbf{j} + 5\mathbf{k}\,.$$

Then

$$\begin{aligned} \mathbf{u}\cdot\mathbf{v} &= 2(3) + (-3)(1) + (1)(-1) = 2\,, \\ \mathbf{v}\cdot\mathbf{w} &= -12 - 1 - 5 = -18\,, \\ \mathbf{u}\cdot\mathbf{w} &= -8 + 3 + 5 = 0\,. \end{aligned}$$

Since the geometric and algebraic dot products coincide (to be proved shortly), this tells us for example that, if these vectors were brought together, the angle between $\mathbf{u}$ and $\mathbf{v}$ would be acute (since its cosine is positive), the angle between $\mathbf{v}$ and $\mathbf{w}$ would be obtuse (since its cosine is negative), and $\mathbf{u}$ and $\mathbf{w}$ would be mutually perpendicular. This is not at all obvious from just staring at the components, so something nontrivial must be going on!

At this point, we list some easy consequences of the algebraic definition. For all vectors $\mathbf{u}$, $\mathbf{v}$, $\mathbf{w}$ and scalars λ:

(1) $\mathbf{u}\cdot\mathbf{v} = \mathbf{v}\cdot\mathbf{u}$

(2) $(\mathbf{u}+\mathbf{v})\cdot\mathbf{w} = \mathbf{u}\cdot\mathbf{w} + \mathbf{v}\cdot\mathbf{w}$

(3) $(\lambda\mathbf{u})\cdot\mathbf{v} = \lambda(\mathbf{u}\cdot\mathbf{v})$

(4) $\mathbf{v}\cdot\mathbf{v} = |\mathbf{v}|^2$ so $|\mathbf{v}| = \sqrt{\mathbf{v}\cdot\mathbf{v}}$

We illustrate the method of proof for property (2), which is also called

distributivity of dot over plus,

and leave the other verifications for the reader. Suppose

$$\mathbf{u} = u_1\mathbf{i} + u_2\mathbf{j} + u_3\mathbf{k}\,, \quad \mathbf{v} = v_1\mathbf{i} + v_2\mathbf{j} + v_3\mathbf{k}\,, \quad \mathbf{w} = w_1\mathbf{i} + w_2\mathbf{j} + w_3\mathbf{k}\,.$$

We use the template: expand – apply something – contract. First "expand" the left-hand side of (2) by unravelling the definitions, then "apply" properties of real numbers, and finally "contract" by reconstructing the information in reverse to obtain the right-hand side of (2):

$$
\begin{aligned}
(\mathbf{u}+\mathbf{v})\cdot\mathbf{w} &= \big((u_1+v_1)\mathbf{i}+(u_2+v_2)\mathbf{j}+(u_3+v_3)\mathbf{k}\big)\cdot\mathbf{w} \\
&= (u_1+v_1)w_1+(u_2+v_2)w_2+(u_3+v_3)w_3 \\
&= u_1w_1+v_1w_1+u_2w_2+v_2w_2+u_3w_3+v_3w_3 \\
&= (u_1w_1+u_2w_2+u_3w_3)+(v_1w_1+v_2w_2+v_3w_3) \\
&= \mathbf{u}\cdot\mathbf{w}+\mathbf{v}\cdot\mathbf{w}
\end{aligned}
$$

We are now close to proving that the algebraic and geometric definitions of the dot product coincide. First recall, from trigonometry, the *Cosine Rule* for triangles (which follows from the Theorem of Pythagoras):

The Cosine Rule:

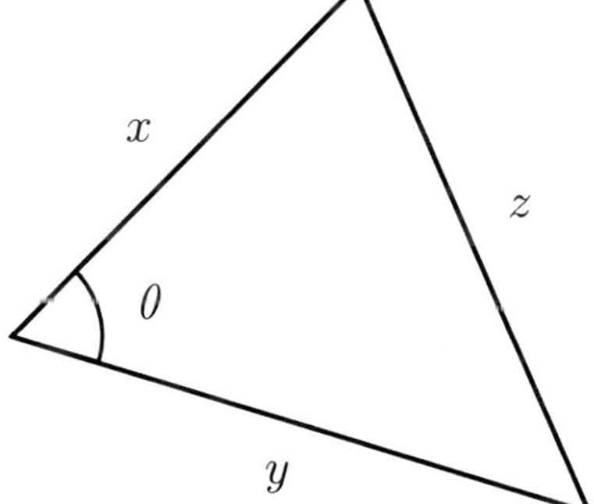

If x, y, z are side lengths of a triangle, and the side for z is opposite the angle θ, then

$$z^2 = x^2+y^2-2xy\cos\theta\,.$$

We apply this to the underlying triangle of the following vector diagram:

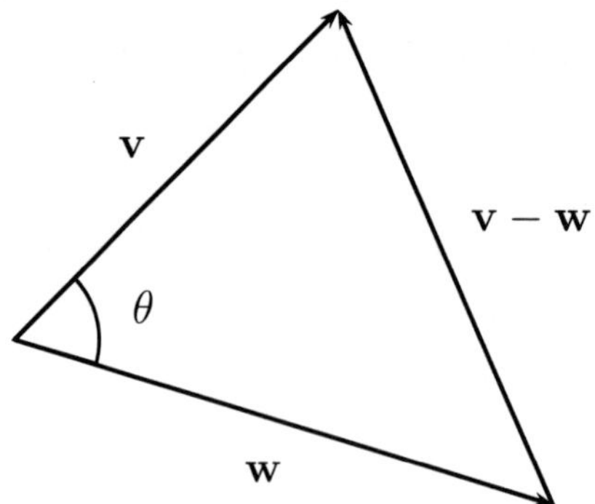

The reader can check that each step in the following is justified by properties (1), (2), (3), (4) listed earlier:

$$|\mathbf{v}-\mathbf{w}|^2 \;=\; |\mathbf{v}|^2+|\mathbf{w}|^2-2|\mathbf{v}||\mathbf{w}|\cos\theta\,,$$

that is, rewriting everything in terms of dot products,

$$\begin{aligned}\mathbf{v}\cdot\mathbf{v}+\mathbf{w}\cdot\mathbf{w}-2\,|\mathbf{v}||\mathbf{w}|\cos\theta \;&=\; (\mathbf{v}-\mathbf{w})\cdot(\mathbf{v}-\mathbf{w})\\ &=\; \mathbf{v}\cdot\mathbf{v}-\mathbf{v}\cdot\mathbf{w}-\mathbf{w}\cdot\mathbf{v}+\mathbf{w}\cdot\mathbf{w}\\ &=\; \mathbf{v}\cdot\mathbf{v}-\mathbf{v}\cdot\mathbf{w}-\mathbf{v}\cdot\mathbf{w}+\mathbf{w}\cdot\mathbf{w}\\ &=\; \mathbf{v}\cdot\mathbf{v}-2\,\mathbf{v}\cdot\mathbf{w}+\mathbf{w}\cdot\mathbf{w}\,,\end{aligned}$$

so that, after some cancelling,

$$-2\,|\mathbf{v}||\mathbf{w}|\cos\theta \;=\; -2\,\mathbf{v}\cdot\mathbf{w}\,,$$

and, finally,

$$\boxed{\mathbf{v}\cdot\mathbf{w} \;=\; |\mathbf{v}||\mathbf{w}|\cos\theta\,.}$$

This completes the proof of the following:

Theorem: The geometric and algebraic definitions of the dot product coincide.

3.3 The angle between two vectors

If we know the components of two vectors then we can calculate their dot product algebraically, and then use the formula for the geometric dot product to deduce the angle between them. If $\mathbf{v}$ and $\mathbf{w}$ are nonzero vectors making an angle θ then $|\mathbf{v}|$ and $|\mathbf{w}|$ are nonzero real numbers, so we can divide through and obtain the following:

$$\cos\theta \;=\; \frac{\mathbf{v}\cdot\mathbf{w}}{|\mathbf{v}||\mathbf{w}|}$$

For example, if

$$\mathbf{v} \;=\; \mathbf{i}+2\mathbf{j}-\mathbf{k} \qquad \text{and} \qquad \mathbf{w} \;=\; 2\mathbf{i}-2\mathbf{j}+4\mathbf{k}$$

then

$$\cos\theta \;=\; \frac{1(2)+2(-2)+(-1)(4)}{\sqrt{1^2+2^2+(-1)^2}\sqrt{2^2+(-2)^2+4^2}}$$

$$=\; \frac{-6}{\sqrt{6}\sqrt{24}} \;=\; -\frac{1}{2}\,,$$

so

$$\theta \;=\; \frac{2\pi}{3} \;=\; 120^\circ\,.$$

The dot product $\mathbf{v}\cdot\mathbf{w}$ can be positive, zero or negative, but as noted near the beginning of this chapter is bounded between $|\mathbf{v}||\mathbf{w}|$ and $-|\mathbf{v}||\mathbf{w}|$.

If θ is the angle between nonzero vectors $\mathbf{v}$ and $\mathbf{w}$ then

(1) θ is acute $\iff$ $0 < \mathbf{v}\cdot\mathbf{w} < |\mathbf{v}||\mathbf{w}|$

(2) θ is obtuse $\iff$ $-|\mathbf{v}||\mathbf{w}| < \mathbf{v}\cdot\mathbf{w} < 0$

(3) $\theta = \frac{\pi}{2} = 90^\circ$ $\iff$ $\mathbf{v}\cdot\mathbf{w} = 0$

3.4 Projections and orthogonal components

This is a difficult section, but worth the effort. The technique described below is the base case of the *Gram-Schmidt Orthogonalisation Process*, which leads, for example, to the theory of *orthogonal polynomials* and *Fourier series*.

For us, the idea in its simplest form is to take two vectors and find the scalar multiple of one vector which is "as close as possible" to the other vector. More precisely, given vectors $\mathbf{v}$ and $\mathbf{w}$, we want to decompose $\mathbf{v}$ as the sum of a vector parallel to $\mathbf{w}$ and a vector perpendicular (orthogonal) to $\mathbf{w}$ as in the following diagram:

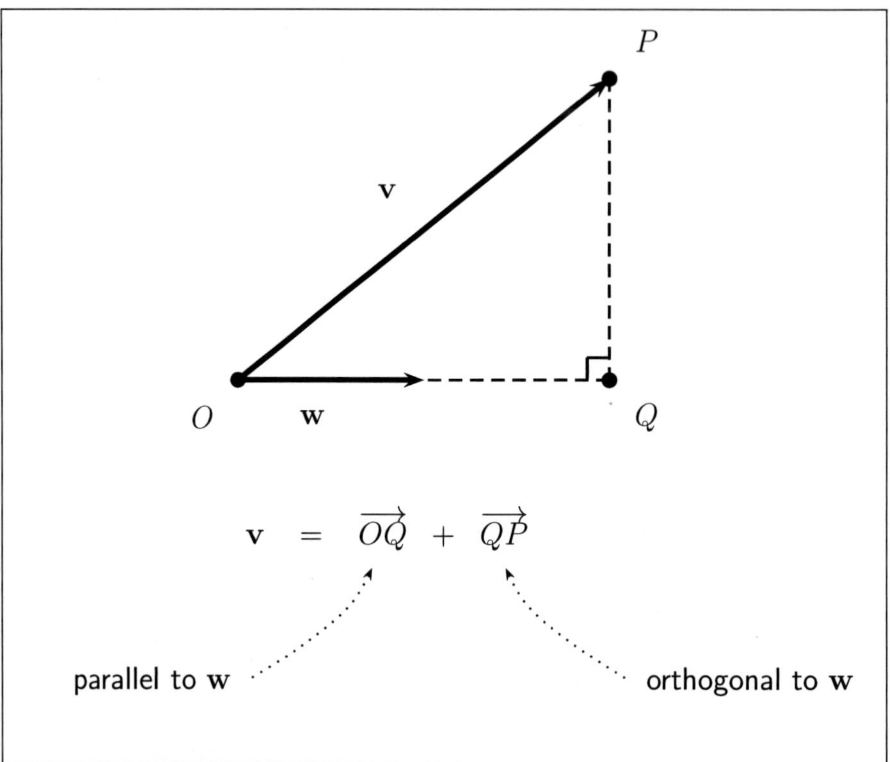

The theory of dot products tells us how to do this, and the hat notation is particularly useful. Recall that

$$\widehat{\mathbf{w}} \;=\; \frac{\mathbf{w}}{|\mathbf{w}|}$$

is the unit vector pointing in the direction of $\mathbf{w}$. We replace $\mathbf{w}$ by $\widehat{\mathbf{w}}$ in the previous diagram, add an angle θ and observe that

$$|\overrightarrow{OQ}| \;=\; |\mathbf{v}|\cos\theta\,.$$

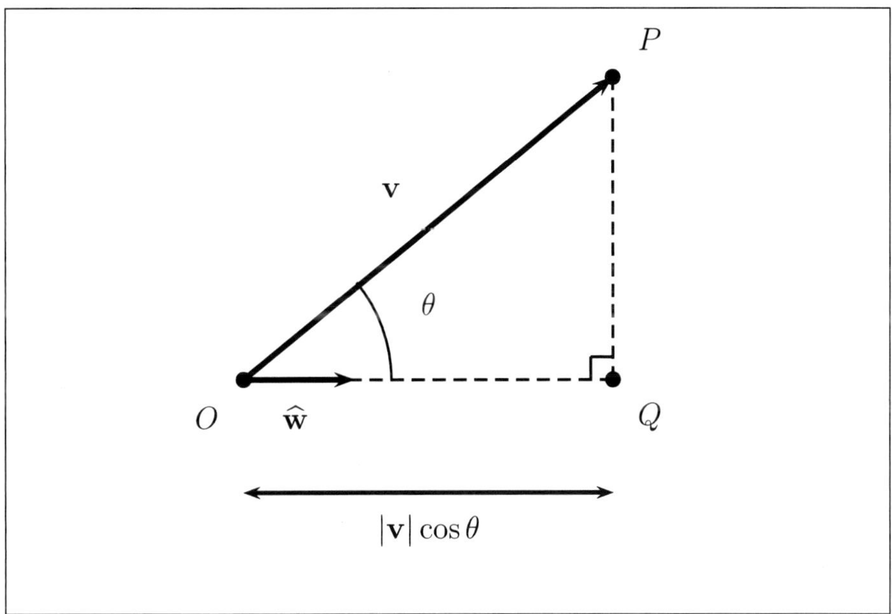

Thus, since $\overrightarrow{OQ}$ points in the direction of $\widehat{\mathbf{w}}$ and has length $|\overrightarrow{OQ}|$, we are able to rewrite it in terms of the dot product:

$$\overrightarrow{OQ} \;=\; \Big(|\mathbf{v}|\cos\theta\Big)\;\widehat{\mathbf{w}} \;=\; |\mathbf{v}|\cos\theta\;\frac{\mathbf{w}}{|\mathbf{w}|} \;=\; \frac{|\mathbf{v}|\,|\mathbf{w}|\,\cos\theta}{|\mathbf{w}|^2}\;\mathbf{w}\,,$$

which becomes

$$\boxed{\overrightarrow{OQ} \;=\; \frac{\mathbf{v}\cdot\mathbf{w}}{|\mathbf{w}|^2}\;\mathbf{w}}$$

and is called

> the (vector) projection of $\mathbf{v}$ in the direction of $\mathbf{w}$.

A careful reader might have noticed that this derivation was linked to a diagram in which θ is acute, so that $|\mathbf{v}|\cos\theta$ is positive. However, if θ turns out to be obtuse, then $\overrightarrow{OQ}$ will point in the opposite direction to $\widehat{\mathbf{w}}$, but now $|\mathbf{v}|\cos\theta$ will be negative. The derivation still holds because, in this case also,

$$\overrightarrow{OQ} \;=\; \Big(|\mathbf{v}|\cos\theta\Big)\;\widehat{\mathbf{w}}\,.$$

The real number,

$$|\mathbf{v}|\cos\theta = \frac{\mathbf{v}\cdot\mathbf{w}}{|\mathbf{w}|},$$

positive if θ is acute, negative if θ is obtuse, is called

the (scalar) component of $\mathbf{v}$ in the direction of $\mathbf{w}$.

It is useful to think of this vector projection

$$\frac{\mathbf{v}\cdot\mathbf{w}}{|\mathbf{w}|^2}\ \mathbf{w}$$

as the

"best approximation" of $\mathbf{v}$ using a scalar multiple of $\mathbf{w}$.

One then wants to find the "error in the approximation", in other words, the difference between $\mathbf{v}$ and its projection. But differences make sense in vector arithmetic! Recall our diagram from before:

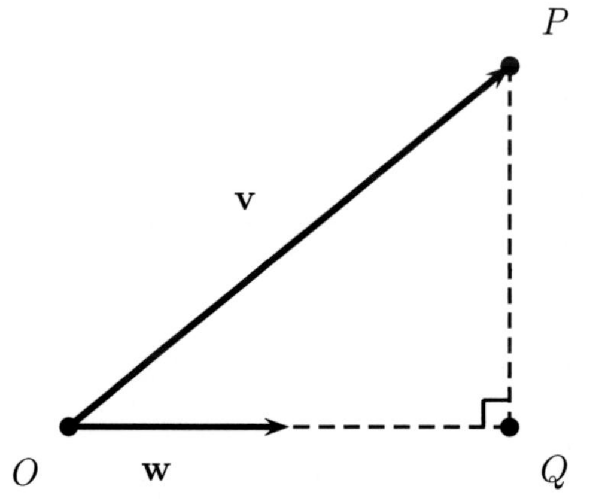

$$\mathbf{v} = \overrightarrow{OQ} + \overrightarrow{QP} = \frac{\mathbf{v}\cdot\mathbf{w}}{|\mathbf{w}|^2}\ \mathbf{w} + \overrightarrow{QP}$$

By subtraction we get the vector $\overrightarrow{QP}$, perpendicular to $\mathbf{w}$,

$$\boxed{\mathbf{v} - \frac{\mathbf{v}\cdot\mathbf{w}}{|\mathbf{w}|^2}\,\mathbf{w}}$$

called

the (vector) component of $\mathbf{v}$ orthogonal to $\mathbf{w}$.

Quick check: If this really is perpendicular to $\mathbf{w}$ then its dot product with $\mathbf{w}$ should evaluate to zero. Yes, all is well:

$$\mathbf{w}\cdot\left(\mathbf{v} - \frac{\mathbf{v}\cdot\mathbf{w}}{|\mathbf{w}|^2}\,\mathbf{w}\right)$$

$$= \mathbf{w}\cdot\mathbf{v} - \mathbf{w}\cdot\left(\frac{\mathbf{v}\cdot\mathbf{w}}{|\mathbf{w}|^2}\,\mathbf{w}\right) = \mathbf{w}\cdot\mathbf{v} - \left(\frac{\mathbf{v}\cdot\mathbf{w}}{|\mathbf{w}|^2}\,\mathbf{w}\cdot\mathbf{w}\right)$$

$$= \mathbf{v}\cdot\mathbf{w} - \left(\frac{\mathbf{v}\cdot\mathbf{w}}{\mathbf{w}\cdot\mathbf{w}}\,\mathbf{w}\cdot\mathbf{w}\right) = \mathbf{v}\cdot\mathbf{w} - \mathbf{v}\cdot\mathbf{w} = 0 \qquad \checkmark$$

Example in the plane: Consider vectors $\mathbf{v} = \mathbf{i} + 2\mathbf{j}$ and $\mathbf{w} = -\mathbf{i} + \mathbf{j}$. The scalar component of $\mathbf{v}$ in the direction of $\mathbf{w}$ is

$$\frac{\mathbf{v}\cdot\mathbf{w}}{|\mathbf{w}|} = \frac{-1+2}{\sqrt{2}} = \frac{1}{\sqrt{2}},$$

the vector projection of $\mathbf{v}$ in the direction of $\mathbf{w}$ is

$$\frac{\mathbf{v}\cdot\mathbf{w}}{|\mathbf{w}|^2}\mathbf{w} = \frac{-1+2}{2}\mathbf{w} = -\frac{\mathbf{i}}{2} + \frac{\mathbf{j}}{2}$$

and the vector component of $\mathbf{v}$ orthogonal to $\mathbf{w}$ is

$$\mathbf{v} - \frac{\mathbf{v}\cdot\mathbf{w}}{|\mathbf{w}|^2}\,\mathbf{w} = \mathbf{i} + 2\mathbf{j} - \left(-\frac{\mathbf{i}}{2} + \frac{\mathbf{j}}{2}\right) = \frac{3}{2}\mathbf{i} + \frac{3}{2}\mathbf{j}.$$

[Check: $\left(\frac{3}{2}\mathbf{i} + \frac{3}{2}\mathbf{j}\right)\cdot\mathbf{w} = -\frac{3}{2} + \frac{3}{2} = 0 \quad \checkmark$]

Thus, we get the following decomposition of $\mathbf{v}$:

$$\mathbf{v} = \left(-\frac{\mathbf{i}}{2}+\frac{\mathbf{j}}{2}\right) + \left(\frac{3}{2}\mathbf{i}+\frac{3}{2}\mathbf{j}\right)$$

parallel to $\mathbf{w}$ orthogonal to $\mathbf{w}$

which is easy to confirm with a diagram (because these vectors lie in the plane):

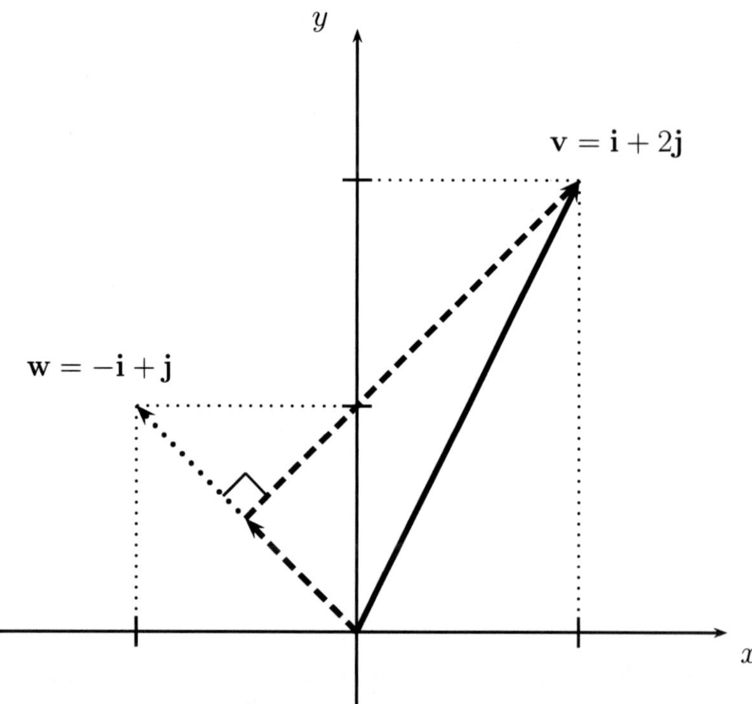

Example in space: Consider vectors $\mathbf{v} = \mathbf{i}-\mathbf{j}-\mathbf{k}$ and $\mathbf{w} = \mathbf{i}+3\mathbf{j}-\mathbf{k}$. The vector projection of $\mathbf{v}$ in the direction of $\mathbf{w}$ is

$$\frac{\mathbf{v}\cdot\mathbf{w}}{|\mathbf{w}|^2}\mathbf{w} = \frac{1-3+1}{1+9+1}\mathbf{w} = -\frac{\mathbf{w}}{11} = \frac{1}{11}(-\mathbf{i}-3\mathbf{j}+\mathbf{k})$$

and the vector component of $\mathbf{v}$ orthogonal to $\mathbf{w}$ is

$$\mathbf{v} - \frac{\mathbf{v}\cdot\mathbf{w}}{|\mathbf{w}|^2}\mathbf{w} = \mathbf{i}-\mathbf{j}-\mathbf{k}-\left(\frac{1}{11}(-\mathbf{i}-3\mathbf{j}+\mathbf{k})\right) = \frac{4}{11}(3\mathbf{i}-2\mathbf{j}-3\mathbf{k}).$$

$$\left[\ \text{Check:}\quad (3\mathbf{i}-2\mathbf{j}-3\mathbf{k})\cdot\mathbf{w} = 3-6+3 = 0 \quad \checkmark\ \right]$$

3.5 Another application to geometry in the plane

In Section 2.5 we used linear independence to discover that the medians of a triangle intersect in a single point. In this section, we use the same general approach, with different details, using the dot product, to prove that the altitudes of a triangle also intersect in a single point, known as the *orthocentre*.

> **Theorem:** The altitudes of a triangle intersect in a point.

Exploration leading to a proof: Consider a triangle PQR with altitudes PA and QB, where A and B lie on sides opposite P and Q respectively. Let D be the point of intersection of PA and QB.

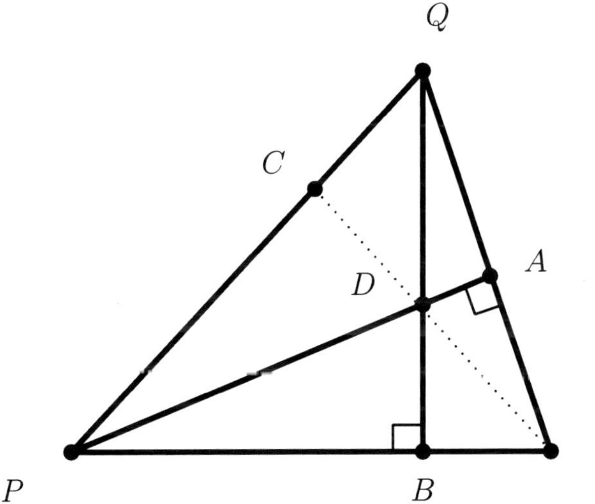

Imagine a line from R passing through D and intersecting the opposite side at C. We would like RC to be the third altitude of the triangle. Since R, D and C lie on the same line, this will be the case if we can show that RD is perpendicular to PQ. It suffices therefore to show

$$\overrightarrow{RD} \cdot \overrightarrow{PQ} = 0 .$$

But PD and QD are perpendicular to RQ and PR respectively, so we can exploit the following assumptions:

$$\overrightarrow{PD} \cdot \overrightarrow{RQ} = \overrightarrow{QD} \cdot \overrightarrow{PR} = 0 .$$

So, our strategy is to **expand** the dot product $\overrightarrow{RD} \cdot \overrightarrow{PQ}$ to incorporate vectors joining points to which we may **apply these assumptions**, and hopefully **contract** everything down to zero:

$$\begin{aligned}
\overrightarrow{RD} \cdot \overrightarrow{PQ} &= (\overrightarrow{RP} + \overrightarrow{PD}) \cdot (\overrightarrow{PR} + \overrightarrow{RQ}) \\
&= \overrightarrow{RP} \cdot \overrightarrow{PR} + \overrightarrow{RP} \cdot \overrightarrow{RQ} + \overrightarrow{PD} \cdot \overrightarrow{PR} + \overrightarrow{PD} \cdot \overrightarrow{RQ} \\
&= \overrightarrow{PR} \cdot \overrightarrow{RP} + \overrightarrow{PR} \cdot \overrightarrow{QR} + \overrightarrow{PR} \cdot \overrightarrow{PD} + 0 \\
&= \overrightarrow{PR} \cdot (\overrightarrow{RP} + \overrightarrow{QR} + \overrightarrow{PD}) \\
&= (\overrightarrow{QR} + \overrightarrow{RP} + \overrightarrow{PD}) \cdot \overrightarrow{PR} \\
&= \overrightarrow{QD} \cdot \overrightarrow{PR} = 0\,,
\end{aligned}$$

which indeed turns out to be as we hoped, and the theorem is proved.

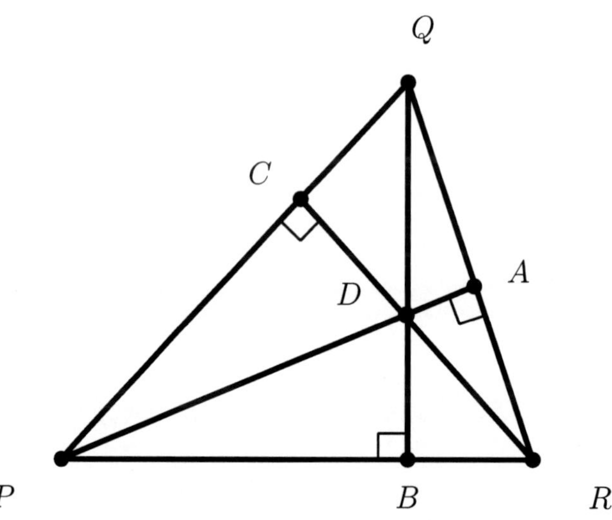

The pathway in the preceding sequence of equalities is by no means unique, and the reader is encouraged to try finding such a pathway him or herself, without looking at the preceding proof again, having grasped the main idea. The same method can be used to show the perpendicular bisectors of the sides of a triangle intersect in a point, known as the *circumcentre*, which is left as a starred exercise.

Chapter 3: Important Ideas and Useful Facts

3.1 Geometric definition of dot product: If $\mathbf{v}$ and $\mathbf{w}$ are vectors and θ is the angle between them, then

$$\mathbf{v}\cdot\mathbf{w} \;=\; |\mathbf{v}||\mathbf{w}|\cos\theta\,,$$

so that

$$\cos\theta \;=\; \frac{\mathbf{v}\cdot\mathbf{w}}{|\mathbf{v}||\mathbf{w}|}\,.$$

3.2 Algebraic definition of dot product: If $\mathbf{v} = a\,\mathbf{i} + b\,\mathbf{j} + c\,\mathbf{k}$ and $\mathbf{w} = d\,\mathbf{i} + e\,\mathbf{j} + f\,\mathbf{k}$ then

$$\mathbf{v}\cdot\mathbf{w} \;=\; ad + be + cf\,.$$

3.3 The angle between two vectors is zero or acute if their dot product is positive. The angle is obtuse or 180° if the dot product is negative. Two vectors are mutually perpendicular if the dot product is zero.

3.4 Cauchy-Schwarz Inequality: $|\mathbf{v}\cdot\mathbf{w}| \;\le\; |\mathbf{v}||\mathbf{w}|$

3.5 Commutativity of dot product: $\mathbf{v}\cdot\mathbf{w} \;=\; \mathbf{w}\cdot\mathbf{v}$

3.6 Distributivity of dot over plus: $(\mathbf{u}+\mathbf{v})\cdot\mathbf{w} \;=\; \mathbf{u}\cdot\mathbf{w} + \mathbf{v}\cdot\mathbf{w}$

3.7 If $\mathbf{v}$ is any vector then $\mathbf{v}\cdot\mathbf{v} = |\mathbf{v}|^2$, so $|\mathbf{v}| = \sqrt{\mathbf{v}\cdot\mathbf{v}}$.

3.8 If $\mathbf{v}$ and $\mathbf{w}$ are vectors and λ is a scalar then $(\lambda\mathbf{v})\cdot\mathbf{w} = \lambda(\mathbf{v}\cdot\mathbf{w}) = \mathbf{v}\cdot(\lambda\mathbf{w})$.

3.9 The *vector projection* of $\mathbf{v}$ in the direction of $\mathbf{w}$ is $\dfrac{\mathbf{v}\cdot\mathbf{w}}{|\mathbf{w}|^2}\mathbf{w}$, which is the best approximation of $\mathbf{v}$ using a scalar multiple of $\mathbf{w}$.

3.10 The *scalar component* of $\mathbf{v}$ in the direction of $\mathbf{w}$ is $\dfrac{\mathbf{v}\cdot\mathbf{w}}{|\mathbf{w}|}$, which is plus or minus the magnitude of the vector projection (minus in the case that the angle is obtuse or 180°).

3.11 The *vector component of* $\mathbf{v}$ *orthogonal to* $\mathbf{w}$ is the difference between $\mathbf{v}$ and its vector projection, which is

$$\mathbf{v} - \frac{\mathbf{v}\cdot\mathbf{w}}{|\mathbf{w}|^2}\mathbf{w}\,.$$

3.12 The altitudes of a triangle intersect in a point.

Exercise 3.1 Given $P = (8, 4, -1)$, $Q = (6, 3, -4)$, $R = (7, 5, -5)$, find

$$\overrightarrow{QP}, \quad |\overrightarrow{QP}|, \quad \overrightarrow{QR}, \quad |\overrightarrow{QR}|, \quad \overrightarrow{QP} \cdot \overrightarrow{QR}$$

and the cosine of $\angle PQR$.

Exercise 3.2 Given that $\mathbf{u} = \mathbf{i} - 2\mathbf{j}$ and $\mathbf{v} = -2\mathbf{i} + \mathbf{j}$, find

(i) $\mathbf{u} \cdot \mathbf{v}$ (ii) $\widehat{\mathbf{u}}$ (iii) $\widehat{\mathbf{v}}$ (iv) $\dfrac{\mathbf{u} \cdot \mathbf{v}}{|\mathbf{u}|}$ (v) $\dfrac{\mathbf{u} \cdot \mathbf{v}}{|\mathbf{v}|}$ (vi) $\dfrac{\mathbf{u} \cdot \mathbf{v}}{|\mathbf{u}||\mathbf{v}|}$

(vii) $\dfrac{\mathbf{u} \cdot \mathbf{v}}{|\mathbf{u}|^2}\,\mathbf{u}$ (viii) $\dfrac{\mathbf{u} \cdot \mathbf{v}}{|\mathbf{v}|^2}\,\mathbf{v}$ (ix) $\mathbf{v} - \dfrac{\mathbf{u} \cdot \mathbf{v}}{|\mathbf{u}|^2}\,\mathbf{u}$ (x) $\mathbf{u} - \dfrac{\mathbf{u} \cdot \mathbf{v}}{|\mathbf{v}|^2}\,\mathbf{v}$

(xi) the cosine of the angle between $\mathbf{u}$ and $\mathbf{v}$

(xii) the scalar component of $\mathbf{u}$ in the direction of $\mathbf{v}$

(xiii) the scalar component of $\mathbf{v}$ in the direction of $\mathbf{u}$

(xiv) the vector projection of $\mathbf{u}$ in the direction of $\mathbf{v}$

(xv) the vector projection of $\mathbf{v}$ in the direction of $\mathbf{u}$

(xvi) the vector component of $\mathbf{u}$ orthogonal to $\mathbf{v}$

(xvii) the vector component of $\mathbf{v}$ orthogonal to $\mathbf{u}$

Exercise 3.3 Consider the following vectors:

$$\mathbf{u} = 5\mathbf{i} + 12\mathbf{j}, \quad \mathbf{v} = -3\mathbf{j} + 4\mathbf{k}, \quad \mathbf{w} = \mathbf{i} + 2\mathbf{j} - 2\mathbf{k}$$

(i) Find $\mathbf{u} \cdot \mathbf{v}$, $\mathbf{u} \cdot \mathbf{w}$, $\mathbf{v} \cdot \mathbf{w}$, $(\mathbf{u} + \mathbf{v}) \cdot \mathbf{w}$ and $(2\mathbf{u} - 3\mathbf{v}) \cdot \mathbf{w}$.

(ii) Determine whether the angles between $\mathbf{u}$ and $\mathbf{v}$, between $\mathbf{v}$ and $\mathbf{w}$, and between $\mathbf{u}$ and $\mathbf{w}$ are acute or obtuse.

(iii) Find the scalar and vector components of $\mathbf{u}$ in the direction of $\mathbf{v}$, and the vector component of $\mathbf{u}$ orthogonal to $\mathbf{v}$.

(iv) Find the vector components of $\mathbf{u} + \mathbf{v}$ in the direction of and orthogonal to $\mathbf{w}$. Find the vector components of $\mathbf{w}$ in the direction of and orthogonal to $\mathbf{u} + \mathbf{v}$.

Exercise 3.4 Consider the following points in space:

$$P(1,1,1)\,,\quad Q(-1,-1,0)\,,\quad R(0,1,2)\,,\quad S(2,3,3)$$

(i) Recall (from Exercise 2.2) that $PQRS$ is a rhombus. Determine whether the angles

$$\angle PQR \qquad \text{and} \qquad \angle QRS$$

are acute or obtuse (which will be further confirmation that this rhombus is not a square).

(ii) Find the midpoint of PR and the midpoint of QS. Conclude that the diagonals of $PQRS$ bisect each other.

(iii) Use a dot product to verify that the diagonals PR and QS are mutually perpendicular.

Exercise 3.5 Verify that if $\mathbf{a}$ and $\mathbf{b}$ have the same length then the vectors $\mathbf{a}+\mathbf{b}$ and $\mathbf{a}-\mathbf{b}$ are mutually perpendicular. Deduce that the diagonals of a rhombus are mutually perpendicular.

Exercise 3.6 Use vectors to show that any angle inscribed in a semicircle is a right angle.

Exercise 3.7 Verify that if $\mathbf{a}$ and $\mathbf{b}$ are mutually perpendicular vectors then

$$|\mathbf{a}+\mathbf{b}|^2 = |\mathbf{a}|^2 + |\mathbf{b}|^2\,.$$

Interpret this in terms of a well-known fact about triangles.

Exercise 3.8 Use dot products to show that

(i) if $\mathbf{v}$ is orthogonal to $\mathbf{x}$ and $\mathbf{y}$, then $\mathbf{v}$ is orthogonal to any linear combination $a\mathbf{x}+b\mathbf{y}$ where a and b are scalars;

(ii) conversely, if $\mathbf{v}$ is orthogonal to $\mathbf{x}$ and $a\mathbf{x}+b\mathbf{y}$ where a and b are scalars such that b is nonzero, then $\mathbf{v}$ is orthogonal to $\mathbf{y}$.

Exercise 3.9 (Uniqueness of the dot product) Suppose that we have a dot operation on vectors such that

$$\mathbf{i}\cdot\mathbf{j} = \mathbf{i}\cdot\mathbf{k} = \mathbf{j}\cdot\mathbf{k} = 0, \qquad \mathbf{i}\cdot\mathbf{i} = \mathbf{j}\cdot\mathbf{j} = \mathbf{k}\cdot\mathbf{k} = 1$$

and properties (1), (2), (3) of Section 3.2 hold. Deduce the usual algebraic rule for dot products:

$$(a\,\mathbf{i} + b\,\mathbf{j} + c\,\mathbf{k})\cdot(\,d\,\mathbf{i} + e\,\mathbf{j} + f\,\mathbf{k}) = ad + be + cf$$

Exercise 3.10 Use the Theorem of Pythagoras to verify the Cosine Rule.

Exercise 3.11* Imagine that you don't know the algebraic formula for the dot product. Starting with the geometric definition for the dot product, apply the Cosine Rule to "discover" the algebraic formula.

Exercise 3.12* Use vectors to find the angles between major diagonals of a cube, between a major diagonal and an edge, and between a major diagonal and a face diagonal.

Exercise 3.13* Verify the following identity for geometric vectors and use it to give an alternative proof that the three altitudes of a triangle intersect in a common point:

$$(\mathbf{a}-\mathbf{b})\cdot(\mathbf{d}-\mathbf{c}) + (\mathbf{b}-\mathbf{c})\cdot(\mathbf{d}-\mathbf{a}) + (\mathbf{c}-\mathbf{a})\cdot(\mathbf{d}-\mathbf{b}) = 0$$

Exercise 3.14* Use the method that showed the orthocentre exists to also prove that the perpendicular bisectors of the sides of a triangle intersect in a common point (known as the *circumcentre*).

Exercise 3.15* Prove that the circumcentre of a triangle is the same distance from each vertex (which explains its name).

Exercise 3.16** This exercise is yet another way of establishing the link between the algebraic and geometric definitions of the dot product. Use trigonometry to prove that the geometric dot product distributes over vector addition so that property (2) holds in Section 3.2. Clearly, for the geometric dot product, (1) and (3) also hold, and

$$\mathbf{i}\cdot\mathbf{j} = \mathbf{i}\cdot\mathbf{k} = \mathbf{j}\cdot\mathbf{k} = 0, \qquad \mathbf{i}\cdot\mathbf{i} = \mathbf{j}\cdot\mathbf{j} = \mathbf{k}\cdot\mathbf{k} = 1.$$

Now apply Exercise 3.9.

4 Cross Products

WATCH IT NOW on DVD VIDEO

- **Cross products** (Section 4.1)
  Time: 4.37
- **Cross products revisited** (Section 4.3)
  Time: 5.13

4 Cross Products

In this chapter we introduce the cross product of two vectors, which will be a vector, in contrast to the last chapter where the dot product of two vectors is a scalar. The cross product always produces a vector perpendicular to the original two vectors, and is perfectly tailored for manipulating directions and planes in space (Chapter 6).

4.1 Definition of cross product

The following definition looks complicated, but there is no need to worry about remembering it. The reader will soon find that it leads to an easy and satisfying technique for manipulating vectors and creating perpendicular directions.

Let $\mathbf{v}$ and $\mathbf{w}$ be geometric vectors in space having Cartesian forms

$$\mathbf{v} = v_1\mathbf{i} + v_2\mathbf{j} + v_3\mathbf{k}, \qquad \mathbf{w} = w_1\mathbf{i} + w_2\mathbf{j} + w_3\mathbf{k}.$$

Define the *cross product* $\mathbf{v} \times \mathbf{w}$ by the following formula:

$$\mathbf{v} \times \mathbf{w} = (v_2w_3 - v_3w_2)\,\mathbf{i} + (v_3w_1 - v_1w_3)\,\mathbf{j} + (v_1w_2 - v_2w_1)\,\mathbf{k}$$

The combinations of products of coefficients on the right-hand side are carefully designed so that

$\mathbf{v} \times \mathbf{w}$ is always perpendicular to both $\mathbf{v}$ and $\mathbf{w}$.

But we have a criterion for perpendicularity, using the dot product:

$\mathbf{v} \times \mathbf{w}$ has the property that

$$(\mathbf{v} \times \mathbf{w}) \cdot \mathbf{v} = 0 \quad \text{and} \quad (\mathbf{v} \times \mathbf{w}) \cdot \mathbf{w} = 0.$$

The reader should take a moment to check that indeed these dot products evaluate to zero, thereby appreciating the delicate balance and symmetry in the definition.

We will learn several methods for finding the cross product, and all of them are easy with a little practice. The first method is to write down the following array:

$$\begin{array}{ccc:cc} \mathbf{i} & \mathbf{j} & \mathbf{k} & \mathbf{i} & \mathbf{j} \\ v_1 & v_2 & v_3 & v_1 & v_2 \\ w_1 & w_2 & w_3 & w_1 & w_2 \end{array}$$

and then imagine forming products up and down diagonals as one moves from left to right:

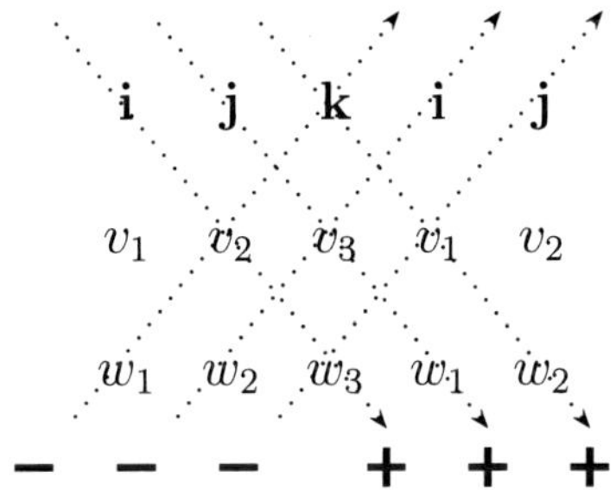

except that

> **up** diagonals are multiplied by -1.

Adding together all the diagonal products produces the cross product $\mathbf{v} \times \mathbf{w}$.

Example: Consider the vectors

$$\mathbf{v} = \mathbf{i} + 2\mathbf{j} + 3\mathbf{k}, \qquad \mathbf{w} = 4\mathbf{i} + 5\mathbf{j} + 6\mathbf{k}.$$

First form an array:

$$\begin{array}{ccc:cc} \mathbf{i} & \mathbf{j} & \mathbf{k} & \mathbf{i} & \mathbf{j} \\ 1 & 2 & 3 & 1 & 2 \\ 4 & 5 & 6 & 4 & 5 \end{array}$$

Then add diagonals, with minus signs for the up diagonals:

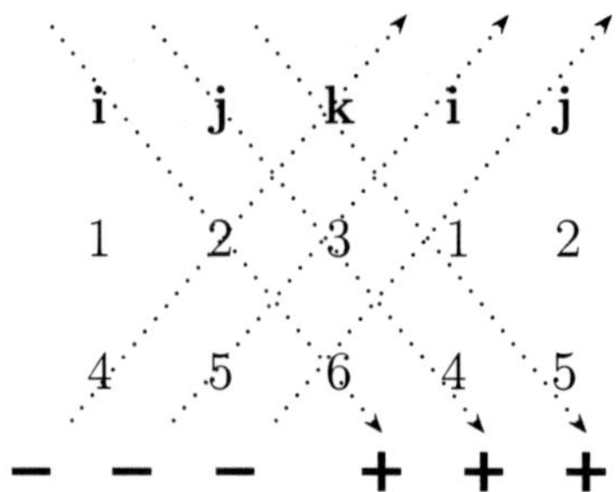

Collect together the down products $12\mathbf{i}$, $12\mathbf{j}$, $5\mathbf{k}$ and the negative up products $-8\mathbf{k}$, $-15\mathbf{i}$, $-6\mathbf{j}$ to get the cross product:

$$\mathbf{v}\times\mathbf{w} \;=\; 12\mathbf{i}+12\mathbf{j}+5\mathbf{k}-8\mathbf{k}-15\mathbf{i}-6\mathbf{j} \;=\; -3\mathbf{i}+6\mathbf{j}-3\mathbf{k}$$

[Check: $\big(-3\mathbf{i}+6\mathbf{j}-3\mathbf{k}\big)\cdot\mathbf{v} \;=\; -3+12-9 \;=\; 0$

$\big(-3\mathbf{i}+6\mathbf{j}-3\mathbf{k}\big)\cdot\mathbf{w} \;=\; -12+30-18 \;=\; 0 \quad\checkmark$]

4.2 List of useful properties

All of the following can be verified easily from the definition:

(1)	$\mathbf{v}\times\mathbf{w}$ is a vector
(2)	$(\mathbf{v}\times\mathbf{w})\cdot\mathbf{v} \;=\; (\mathbf{v}\times\mathbf{w})\cdot\mathbf{w} \;=\; 0$
(3)	$\mathbf{v}\times\mathbf{w} \;=\; -(\mathbf{w}\times\mathbf{v})$
(4)	$\mathbf{v}\times\mathbf{v} \;=\; \mathbf{0}$
(5)	$(\lambda\mathbf{v})\times\mathbf{w} \;=\; \lambda(\mathbf{v}\times\mathbf{w}) \;=\; \mathbf{v}\times(\lambda\mathbf{w})$
(6)	$(\mathbf{u}+\mathbf{v})\times\mathbf{w} \;=\; \mathbf{u}\times\mathbf{w}+\mathbf{v}\times\mathbf{w}$
(7)	$\mathbf{u}\times(\mathbf{v}+\mathbf{w}) \;=\; \mathbf{u}\times\mathbf{v}+\mathbf{u}\times\mathbf{w}$

Property (2) is the *raison d'être* for the cross product. It is good fortune that so many algebraic properties hold, enabling one to manipulate cross products almost as in ordinary arithmetic.

A very important proviso is that **order matters**,

because, by (3), reversing a cross product makes the resultant vector point in the opposite direction! (Property (3) is referred to as *anti-commutativity*.) Property (5) says that scalars can move freely in and about expressions, and

(6) and (7) are **distributivity** laws,

which enable one to expand brackets in the usual way.

Cross products involving unit vectors $\mathbf{i}$, $\mathbf{j}$ and $\mathbf{k}$ are so important that they should be committed to memory quickly:

$$\mathbf{i}\times\mathbf{j} = \mathbf{k}, \qquad \mathbf{j}\times\mathbf{k} = \mathbf{i}, \qquad \mathbf{k}\times\mathbf{i} = \mathbf{j}$$

Remember that reversing the order multiplies the answer by -1:

$$\mathbf{j}\times\mathbf{i} = -\mathbf{k}, \qquad \mathbf{k}\times\mathbf{j} = -\mathbf{i}, \qquad \mathbf{i}\times\mathbf{k} = -\mathbf{j}$$

A useful memory device is to think of $\mathbf{i}$, $\mathbf{j}$, $\mathbf{k}$ as a triple in alphabetical order which repeats itself *ad infinitum*:

$$\ldots, \mathbf{i}, \mathbf{j}, \mathbf{k}, \mathbf{i}, \mathbf{j}, \mathbf{k}, \mathbf{i}, \mathbf{j}, \mathbf{k}, \ldots$$

As you pass from left to right, each step performs the cross product of the previous two consecutive vectors. As you pass from right to left, the rule is the same, but multiply by -1. Finally, some even easier ones to remember:

$$\mathbf{i}\times\mathbf{i} = \mathbf{0}, \qquad \mathbf{j}\times\mathbf{j} = \mathbf{0}, \qquad \mathbf{k}\times\mathbf{k} = \mathbf{0}$$

4.3 Method of expanding brackets

The properties listed in the previous section yield another method for finding cross products, namely, expand brackets and use facts about $\mathbf{i}$, $\mathbf{j}$ and $\mathbf{k}$.

For example, as before, let

$$\mathbf{v} = \mathbf{i} + 2\mathbf{j} + 3\mathbf{k}, \qquad \mathbf{w} = 4\mathbf{i} + 5\mathbf{j} + 6\mathbf{k}.$$

Then

$$\begin{aligned}
\mathbf{v} \times \mathbf{w} &= (\mathbf{i} + 2\mathbf{j} + 3\mathbf{k}) \times (4\mathbf{i} + 5\mathbf{j} + 6\mathbf{k}) \\
&= \mathbf{0} + 5\,(\mathbf{i} \times \mathbf{j}) + 6\,(\mathbf{i} \times \mathbf{k}) \\
&\qquad + 8\,(\mathbf{j} \times \mathbf{i}) + \mathbf{0} + 12\,(\mathbf{j} \times \mathbf{k}) \\
&\qquad\qquad + 12\,(\mathbf{k} \times \mathbf{i}) + 15\,(\mathbf{k} \times \mathbf{j}) + \mathbf{0} \\
&= 5\mathbf{k} - 6\mathbf{j} - 8\mathbf{k} + 12\mathbf{i} + 12\mathbf{j} - 15\mathbf{i} \\
&= -3\mathbf{i} + 6\mathbf{j} - 3\mathbf{k},
\end{aligned}$$

which of course agrees with the earlier answer. With practice one usually goes straight to the third step in a calculation such as this one.

It is easy to make mistakes, often by neglecting or introducing a minus sign. Nevertheless, the method is robust, because one detects (and then corrects) errors almost immediately by checking that both dot products of the cross product with the original vectors are zero.

4.4 Geometric interpretation

The definition of the cross product in Section 4.1 is opaque to say the least! It turns out to have a beautiful geometric interpretation, which we explain now. The following fact relating the cross and dot products is straightforward but takes a while to check, so is left as a starred exercise:

$$\boxed{|\mathbf{v} \times \mathbf{w}|^2 + |\mathbf{v} \cdot \mathbf{w}|^2 = |\mathbf{v}|^2 |\mathbf{w}|^2}$$

Squaring both sides of the Cauchy-Schwarz Inequality (Section 3.1) yields

$$|\mathbf{v} \cdot \mathbf{w}|^2 \;\le\; |\mathbf{v}|^2 |\mathbf{w}|^2 ,$$

so one can think of $|\mathbf{v} \times \mathbf{w}|^2$ in the previous equation as a "correction" that turns inequality into equality. For example, if $\mathbf{v}$ and $\mathbf{w}$ are parallel then the Cauchy-Schwarz Inequality can be replaced by equality, and of course $\mathbf{v} \times \mathbf{w} = \mathbf{0}$ (an easy consequence of properties (4) and (5) above), so that indeed $|\mathbf{v} \times \mathbf{w}|^2 = 0$.

The Cauchy-Schwarz Inequality fails to be equality most spectacularly when $\mathbf{v}$ and $\mathbf{w}$ are orthogonal, which turns out to maximise $|\mathbf{v} \times \mathbf{w}|^2$. As we shall see, this has something to do with maximising areas of triangles.

Manipulating the above equation, using the geometric formula for the dot product (Section 3.1), yields

$$\begin{aligned} |\mathbf{v} \times \mathbf{w}|^2 &= |\mathbf{v}|^2 |\mathbf{w}|^2 \;-\; |\mathbf{v} \cdot \mathbf{w}|^2 \\ &= |\mathbf{v}|^2 |\mathbf{w}|^2 \;-\; |\mathbf{v}|^2 |\mathbf{w}|^2 \cos^2\theta \\ &= |\mathbf{v}|^2 |\mathbf{w}|^2 (1 - \cos^2\theta) \\ &= |\mathbf{v}|^2 |\mathbf{w}|^2 \sin^2\theta \end{aligned}$$

so that, taking nonnegative square roots,

$$\boxed{|\mathbf{v} \times \mathbf{w}| \;=\; |\mathbf{v}||\mathbf{w}| \sin\theta}$$

where we are careful to measure the angle θ between $\mathbf{v}$ and $\mathbf{w}$ to lie between $0°$ and $180°$ (so that $\sin\theta \ge 0$).

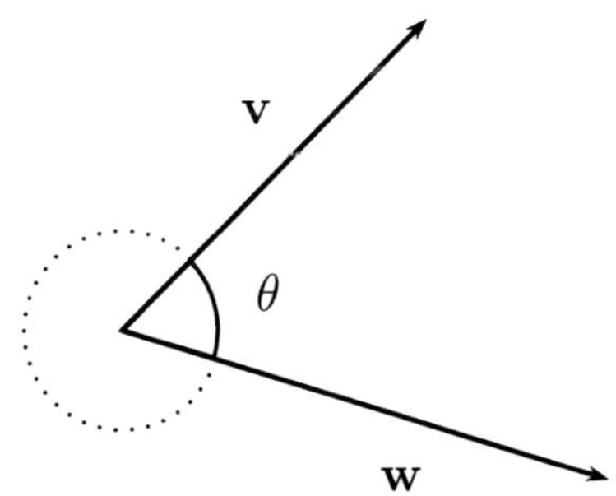

In this diagram we have drawn vectors so that θ is acute, but the observations below hold for all choices of θ.

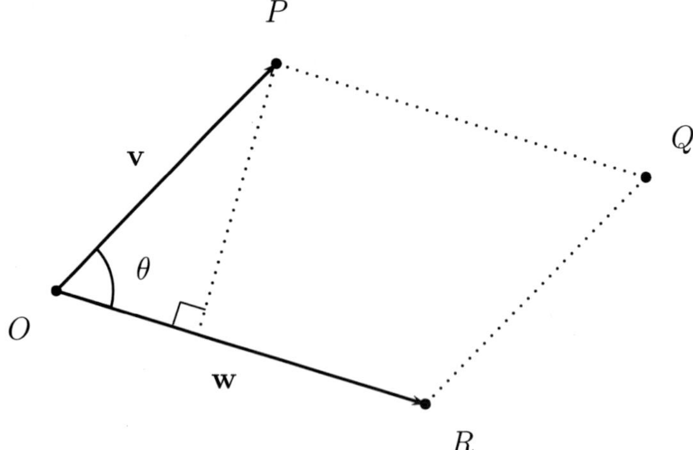

The area of the parallelogram $OPQR$ is the product of the length $|\mathbf{w}|$ of the base OR with the altitude $|\mathbf{v}|\sin\theta$, which is $|\mathbf{v}||\mathbf{w}|\sin\theta$. This proves

> the area of the parallelogram inscribed by vectors $\mathbf{v}$ and $\mathbf{w}$ is
>
> $$|\mathbf{v} \times \mathbf{w}|.$$

Equivalently, taking half the area of the parallelogram,

> the area of the triangle inscribed by vectors $\mathbf{v}$ and $\mathbf{w}$ is
>
> $$\frac{|\mathbf{v} \times \mathbf{w}|}{2}.$$

Thus, to maximise the length of the cross product, one has to maximise the area of the associated triangle (or parallelogram), so that the altitude becomes one of the sides and the triangle is right-angled (or the parallelogram a rectangle).

Example: Find the area of the triangle in space with vertices

$$P(1,-2,1)\,, \qquad Q(4,0,2)\,, \qquad R(-5,-2,3)\,.$$

Solution: Apart from a rough shape, we may have no idea what the triangle actually looks like in space, but it serves our purposes in prompting us towards a solution to draw any old triangle on the page:

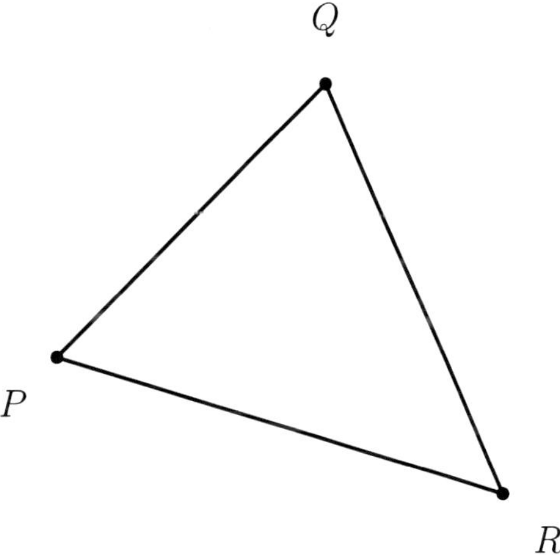

The cross product in the calculation below is found using the "expanding the brackets" method. The area of the triangle is

$$\frac{|\,\overrightarrow{PQ}\times\overrightarrow{PR}\,|}{2} = \frac{|\,(3\mathbf{i}+2\mathbf{j}+\mathbf{k})\times(-6\mathbf{i}+2\mathbf{k})\,|}{2}$$

$$= \frac{|\,-6\mathbf{j}+12\mathbf{k}+4\mathbf{i}-6\mathbf{j}\,|}{2}$$

$$= \frac{|\,4\mathbf{i}-12\mathbf{j}+12\mathbf{k}\,|}{2} = 2|\,\mathbf{i}-3\mathbf{j}+3\mathbf{k}\,| = 2\sqrt{19}\,.$$

To minimise the length of the cross product, one has to minimise the area of the associated triangle (or parallelogram), so that it becomes degenerate, that is, all the vertices lie on a line. For an angle θ between $\mathbf{v}$ and $\mathbf{w}$:

$$\mathbf{v}\times\mathbf{w} = \mathbf{0} \iff |\mathbf{v}\times\mathbf{w}| = 0 \iff |\mathbf{v}||\mathbf{w}|\sin\theta = 0$$

$$\iff \mathbf{v}=\mathbf{0}\,,\quad \mathbf{w}=\mathbf{0}\,,\quad \theta=0 \quad \text{or} \quad \theta=\pi$$

$$\iff \mathbf{v} \text{ and } \mathbf{w} \text{ are parallel}$$

So, what happens if $\mathbf{v}$ and $\mathbf{w}$ are not parallel?

In which direction in space will $\mathbf{v} \times \mathbf{w}$ point?

The direction must certainly be perpendicular to both $\mathbf{v}$ and $\mathbf{w}$. But there are two choices.

Answer: Thumbs up on your right hand!

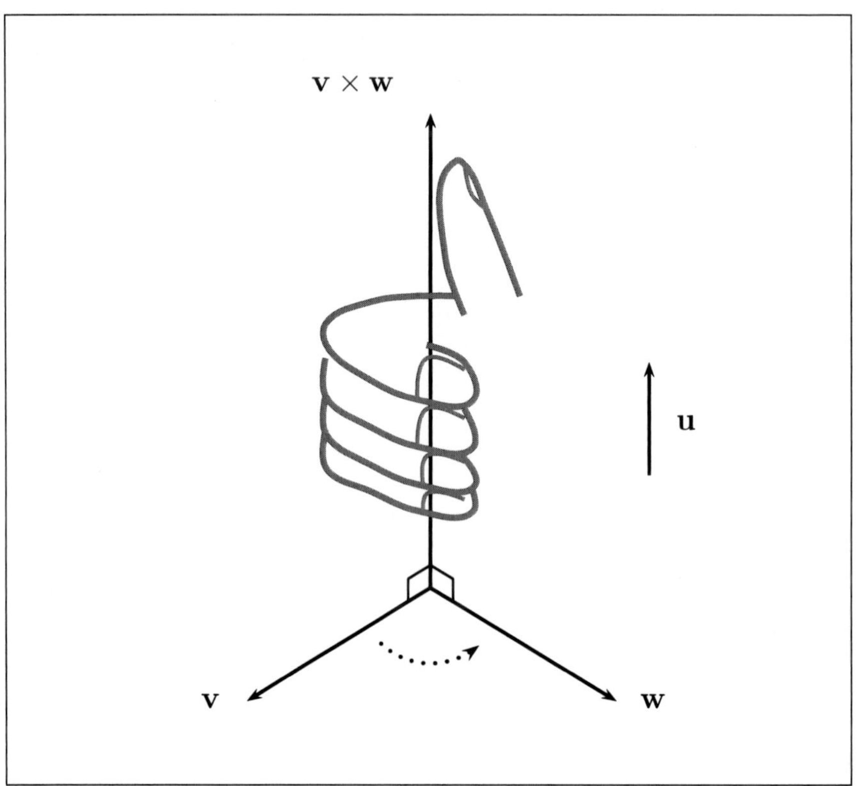

If the fingers of your right hand are curled in the direction of rotation of $\mathbf{v}$ until it points in the direction of $\mathbf{w}$, then the outstretched thumb points in the direction of $\mathbf{v} \times \mathbf{w}$.

This is called the **Right-Hand Rule**.

Thus, if $\mathbf{u}$ is a unit vector perpendicular to both $\mathbf{v}$ and $\mathbf{w}$, and the triple $\mathbf{v}$, $\mathbf{w}$, $\mathbf{u}$ is right-handed, then we have the

geometric formula for the cross product:

$$\mathbf{v} \times \mathbf{w} \quad = \quad |\mathbf{v}|\;|\mathbf{w}|\;\sin\theta\;\;\mathbf{u}$$

4.5 Continuity and the right-hand orientation

But we haven't explained why the Right-Hand Rule holds? It is conceivable that the cross product could follow a Left-Hand Rule, or even oscillate between Right- and Left-Hand Rules, as we vary the choice of vectors fed into the cross product?!

This is a subtle question, and we use some theory of continuity to explain now why the cross product can only follow one such rule.

Suppose $\mathbf{v}$ and $\mathbf{w}$ are any pair of nonparallel vectors in space, making an angle θ strictly between 0 and π.

We claim that the triple $\mathbf{v}$, $\mathbf{w}$, $\mathbf{v} \times \mathbf{w}$ is right-handed.

The first thing to observe is that $\mathbf{i} \times \mathbf{j} = \mathbf{k}$ and the triple $\mathbf{i}$, $\mathbf{j}$, $\mathbf{k}$ is right-handed (because of our convention of choosing the positive directions of our coordinate axes so that this will be the case).

The second thing to note, and it is nontrivial, is that the formulae for the components of $\mathbf{v} \times \mathbf{w}$ are continuous functions of the components of $\mathbf{v}$ and $\mathbf{w}$. The reason is that only multiplications and subtractions are used. The component functions are special cases of polynomial functions of several variables, and it is an important theorem in calculus that these are continuous. Consequently, if we vary $\mathbf{v}$ and $\mathbf{w}$ continously then $\mathbf{v} \times \mathbf{w}$ will also vary continuously.

We are going to transform $\mathbf{v}$, $\mathbf{w}$ into $\mathbf{i}$, $\mathbf{j}$ by a three-stage continuous process.

Firstly, we stretch/shrink $\mathbf{v}$ and $\mathbf{w}$ continuously, up or down as the case may be, until they both have length 1. They are transformed into unit vectors $\widehat{\mathbf{v}}$ and $\widehat{\mathbf{w}}$ respectively. During this process, θ is constant and the length

$|\mathbf{v}||\mathbf{w}|\sin\theta$ remains bounded away from zero.

Imagine all of these vectors have their tails fixed at the origin. The tip of $\mathbf{v}\times\mathbf{w}$ remains always on the line through the origin perpendicular to both $\mathbf{v}$ and $\mathbf{w}$. Hence, if the orientation flipped over during the process then, by the Intermediate Value Theorem from calculus, there must have been a moment when the tip of $\mathbf{v}\times\mathbf{w}$ touched the origin. But this would contradict that $\mathbf{v}\times\mathbf{w}$ remains throughout of nonzero length. This proves that

> the triples
>
> $\mathbf{v}\,,\ \mathbf{w}\,,\ \mathbf{v}\times\mathbf{w}$ and $\widehat{\mathbf{v}}\,,\ \widehat{\mathbf{w}}\,,\ \widehat{\mathbf{v}}\times\widehat{\mathbf{w}}$
>
> both have the same orientation.

For the second stage, we now keep $\widehat{\mathbf{v}}$ fixed and rotate $\widehat{\mathbf{w}}$ continuously in the same plane until θ has moved (up or down) until it equals $\pi/2$. This will have the effect of transforming the length of $\widehat{\mathbf{v}}\times\widehat{\mathbf{w}}$ into 1 without ever becoming zero. Again, by the Intermediate Value Theorem, the orientation of the new triple $\widehat{\mathbf{v}}\,,\ \widehat{\mathbf{w}}\,,\ \widehat{\mathbf{v}}\times\widehat{\mathbf{w}}$ remains unchanged throughout.

For the third and final stage, we rotate the pair of now perpendicular unit vectors $\widehat{\mathbf{v}}$ and $\widehat{\mathbf{w}}$ simultaneously and continuously into $\mathbf{i}$ and $\mathbf{j}$ respectively. The length of the cross product never becomes zero (in fact, remains fixed at 1), so again by continuity the orientation does not change. However, at the end of this process the cross product is equal to $\mathbf{i}\times\mathbf{j}=\mathbf{k}$.

But the triple $\mathbf{i}\,,\ \mathbf{j}\,,\ \mathbf{k}$ is right-handed. Thus, tracing backwards through these three continuous transformations, for which orientation remains invariant, we see that the original triple $\mathbf{v}\,,\ \mathbf{w}\,,\ \mathbf{v}\times\mathbf{w}$ must also be right-handed. This proves our claim, and explains the universality of the Right-Hand Rule.

There is another approach to the invariant nature of the Right-Hand Rule, which starts with the geometric formula as a definition of the cross product, and deduces the algebraic definition. The (considerable) difficulty lies in proving the distributive law, and this is left as a starred exercise.

Chapter 4: Important Ideas and Useful Facts

4.1 Algebraic definition of cross product: If $\mathbf{v} = v_1\,\mathbf{i} + v_2\,\mathbf{j} + v_3\,\mathbf{k}$ and $\mathbf{w} = w_1\,\mathbf{i} + w_2\,\mathbf{j} + w_3\,\mathbf{k}$ then

$$\mathbf{v}\times\mathbf{w} \;=\; (v_2w_3 - v_3w_2)\,\mathbf{i} + (v_3w_1 - v_1w_3)\,\mathbf{j} + (v_1w_2 - v_2w_1)\,\mathbf{k}\,,$$

which can be evaluated by

(a) using the "up-and-down-diagonal" method;

(b) using the "expanding brackets" method and the facts that

$$\mathbf{i}\times\mathbf{j} \;=\; \mathbf{k} \;=\; -(\mathbf{j}\times\mathbf{i})\,,\quad \mathbf{j}\times\mathbf{k} \;=\; \mathbf{i} \;=\; -(\mathbf{k}\times\mathbf{j})\,,$$
$$\mathbf{k}\times\mathbf{i} \;=\; \mathbf{j} \;=\; -(\mathbf{i}\times\mathbf{k})\,,\quad \mathbf{i}\times\mathbf{i} \;=\; \mathbf{j}\times\mathbf{j} \;=\; \mathbf{k}\times\mathbf{k} \;=\; 0\,;$$

(c) evaluating a 3×3 determinant (see Chapter 10):

$$\mathbf{v}\times\mathbf{w} \;=\; \begin{vmatrix} \mathbf{i} & \mathbf{j} & \mathbf{k} \\ v_1 & v_2 & v_3 \\ w_1 & w_2 & w_3 \end{vmatrix}\,.$$

4.2 The cross product $\mathbf{v}\times\mathbf{w}$ is always perpendicular to both $\mathbf{v}$ and $\mathbf{w}$ so that

$$(\mathbf{v}\times\mathbf{w})\cdot\mathbf{v} \;=\; (\mathbf{v}\times\mathbf{w})\cdot\mathbf{w} \;=\; 0\,.$$

4.3 Anti-commutativity of cross product: $\mathbf{v}\times\mathbf{w} \;=\; -(\mathbf{w}\times\mathbf{v})$

4.4 Distributivity of cross over plus: $(\mathbf{u}+\mathbf{v})\times\mathbf{w} \;=\; \mathbf{u}\times\mathbf{w} + \mathbf{v}\times\mathbf{w}$

4.5 If $\mathbf{v}$ and $\mathbf{w}$ are vectors and λ is a scalar then

$$(\lambda\mathbf{v})\times\mathbf{w} = \lambda(\mathbf{v}\times\mathbf{w}) = \mathbf{v}\times(\lambda\mathbf{w}) \qquad \text{and} \qquad \mathbf{v}\times\mathbf{v} = \mathbf{0}\,.$$

4.6 The area of the parallelogram inscribed by $\mathbf{v}$ and $\mathbf{w}$ is $|\mathbf{v}\times\mathbf{w}|$.

4.7 The area of the triangle inscribed by $\mathbf{v}$ and $\mathbf{w}$ is $\dfrac{|\mathbf{v}\times\mathbf{w}|}{2}$.

4.8 Geometric formula for cross product: if θ is the angle between vectors $\mathbf{v}$ and $\mathbf{w}$ chosen so that $0\le\theta\le\pi$ then

$$\mathbf{v}\times\mathbf{w} \;=\; |\mathbf{v}||\mathbf{w}|\sin\theta\;\mathbf{u}\,,$$

where $\mathbf{u}$ is the unit vector perpendicular to both $\mathbf{v}$ and $\mathbf{w}$ such that the triple $\mathbf{u}$, $\mathbf{v}$, $\mathbf{w}$ is right-handed. In particular,

$$|\mathbf{v}\times\mathbf{w}| \;=\; |\mathbf{v}||\mathbf{w}|\sin\theta\,.$$

Exercise 4.1 Write down

(i) $\mathbf{i}\times\mathbf{j}$ (ii) $2\mathbf{i}\times 3\mathbf{j}$ (iii) $\mathbf{i}\times(-4\mathbf{j})$ (iv) $\mathbf{j}\times\mathbf{i}$ (v) $\mathbf{j}\times(-4\mathbf{i})$

(vi) $\mathbf{k}\times\mathbf{k}$ (vii) $\mathbf{k}\times(-\mathbf{k})$ (viii) $(-\mathbf{k})\times\mathbf{i}$ (ix) $(-\mathbf{k})\times(-\mathbf{j})$

(x) $\mathbf{k}\times(\mathbf{i}+\mathbf{k})$ (xi) $(3\mathbf{j}-\mathbf{k})\times 2\mathbf{j}$ (xii) $(\mathbf{j}-\mathbf{k})\times(\mathbf{k}+\mathbf{j})$

Exercise 4.2 Consider the following vectors:

$$\mathbf{u} = 5\mathbf{i}+12\mathbf{j}\,, \quad \mathbf{v} = -3\mathbf{j}+4\mathbf{k}\,, \quad \mathbf{w} = \mathbf{i}+2\mathbf{j}-2\mathbf{k}$$

Find

(i) $\mathbf{u}\times\mathbf{v}$ (ii) $\mathbf{u}\times\mathbf{w}$ (iii) $\mathbf{v}\times\mathbf{w}$

(iv) $(\mathbf{u}+\mathbf{v})\times\mathbf{w}$ (v) $(\mathbf{u}\times\mathbf{v})\times\mathbf{w}$ (vi) $\mathbf{u}\cdot(\mathbf{v}\times\mathbf{w})$

(vii) $(\mathbf{u}\times\mathbf{v})\cdot\mathbf{w}$ (viii) $(\mathbf{u}\times\mathbf{v})\times(\mathbf{v}\times\mathbf{w})$

Exercise 4.3 Given that

$$\mathbf{a} = 2\mathbf{i}-\mathbf{j}+2\mathbf{k}\,, \qquad \mathbf{b} = \mathbf{i}+\mathbf{j}-\mathbf{k}\,,$$

find

(i) $|\mathbf{a}|$ (ii) $|\mathbf{b}|$ (iii) $\mathbf{a}\times\mathbf{b}$ (iv) $|\mathbf{a}\times\mathbf{b}|$

(v) the sine of the angle between $\mathbf{a}$ and $\mathbf{b}$.

Exercise 4.4 Let $P=(5,3,-1)$, $Q=(7,-3,0)$ and $R=(6,5,-6)$. Find

$$\overrightarrow{QP}\times\overrightarrow{QR}$$

and the area of $\triangle PQR$.

Exercise 4.5 Calculate $|\mathbf{a}\times\mathbf{b}|$ given that $|\mathbf{a}|=5$, $|\mathbf{b}|=12$ and $\mathbf{a}\cdot\mathbf{b}=40$.

Exercise 4.6 Suppose $\mathbf{v}$ and $\mathbf{w}$ are the following vectors lying in this page:

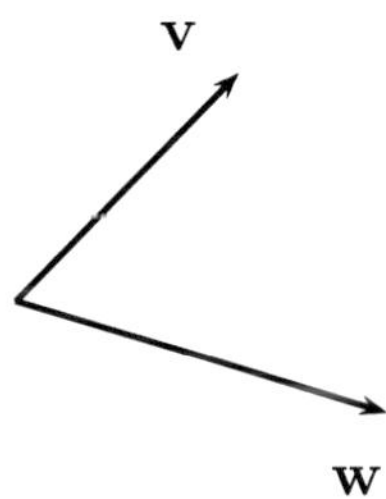

Decide which of

(i) $\mathbf{v} \times (\mathbf{v} - \mathbf{w})$ (ii) $(\mathbf{v} - \mathbf{w}) \times \mathbf{w}$ (iii) $(\mathbf{v} - \mathbf{w}) \times (\mathbf{w} - \mathbf{v})$

(iv) $(\mathbf{v} \times \mathbf{w}) \times \mathbf{w}$ (v) $\mathbf{v} \times (\mathbf{v} \times \mathbf{w})$

(a) is perpendicular to $\mathbf{v}$ but not to $\mathbf{w}$

(b) is perpendicular to $\mathbf{w}$ but not to $\mathbf{v}$

(c) points upwards, away from the page, towards the ceiling

(d) points downwards, away from the page, towards the floor

(e) is the zero vector.

Exercise 4.7 Use the cross product to find:

(i) a unit vector perpendicular to both $\mathbf{i} + \mathbf{j} - \mathbf{k}$ and $-\mathbf{i} + \mathbf{j} + \mathbf{k}$

(ii) a unit vector which points in a direction which is perpendicular to the triangle with vertices

$$P(1,0,2)\,, \quad Q(-1,2,4)\,, \quad R(0,-1,3)$$

and such that looking backwards along the vector (from tip to tail) towards the triangle, the vertices P, Q, R rotate clockwise (in that order).

Exercise 4.8 Check the following properties for any vectors $\mathbf{v}$ and $\mathbf{w}$, using the original definition of cross product:

(i) $(\mathbf{v} \times \mathbf{w}) \cdot \mathbf{v} = 0$ (ii) $(\mathbf{v} \times \mathbf{w}) \cdot \mathbf{w} = 0$

(iii) $\mathbf{v} \times \mathbf{w} = -(\mathbf{w} \times \mathbf{v})$ (iv) $\mathbf{v} \times \mathbf{v} = \mathbf{0}$

Exercise 4.9 Consider the following points in space:

$$P(1,1,1)\,,\quad Q(-1,-1,0)\,,\quad R(0,1,2)\,,\quad S(2,3,3)$$

(i) Use cross products to find the areas of the triangles $\triangle PQR$ and $\triangle QRS$. Are you surprised?

(ii) Find the distance d_1 from P to R, the distance d_2 from Q to S, and evaluate

$$\frac{d_1 d_2}{4}\,.$$

Are you surprised? If not, explain why not.

Exercise 4.10* Let $\mathbf{a}$, $\mathbf{b}$, $\mathbf{c}$, $\mathbf{d}$ be vectors perpendicular to the four faces of a tetrahedron, pointing outwards, of length equal to the respective areas of the faces. Verify the following equation:

$$\mathbf{a}+\mathbf{b}+\mathbf{c}+\mathbf{d} \;=\; \mathbf{0}$$

Exercise 4.11* Carefully verify one of the distributivity laws from the definition, say the law

$$(\mathbf{u}+\mathbf{v})\times\mathbf{w} \;=\; \mathbf{u}\times\mathbf{w}+\mathbf{v}\times\mathbf{w}\,,$$

and deduce the other distributivity law using anti-commutativity.

Exercise 4.12* Verify the following formula, for any vectors $\mathbf{v}$ and $\mathbf{w}$ in space:

$$|\mathbf{v}\times\mathbf{w}|^2 \;+\; |\mathbf{v}\cdot\mathbf{w}|^2 \;=\; |\mathbf{v}|^2|\mathbf{w}|^2$$

Exercise 4.13* Verify the following formula, for any vectors $\mathbf{u}$, $\mathbf{v}$ and $\mathbf{w}$ in space:

$$\mathbf{u}\cdot(\mathbf{v}\times\mathbf{w}) \;=\; (\mathbf{u}\times\mathbf{v})\cdot\mathbf{w}$$

If you found an algebraic proof, now find a geometric proof, and vice versa. The expression $\mathbf{u}\cdot(\mathbf{v}\times\mathbf{w})$ is referred to as a *triple product* and its magnitude is the volume of the parallopiped spanned by the vectors when placed tail-to-tail in space. Verify that $\mathbf{u}\cdot(\mathbf{v}\times\mathbf{w})$ is positive if and only if $\mathbf{u}$, $\mathbf{v}$, $\mathbf{w}$ form a right-handed triple.

Exercise 4.14* Suppose that $\mathbf{u}$, $\mathbf{v}$, $\mathbf{w}$ are vectors in space such that $\mathbf{u} \neq \mathbf{0}$,

$$\mathbf{u} \cdot \mathbf{v} = \mathbf{u} \cdot \mathbf{w} \qquad \text{and} \qquad \mathbf{u} \times \mathbf{v} = \mathbf{u} \times \mathbf{w} .$$

Prove that $\mathbf{v} = \mathbf{w}$.

Exercise 4.15* Let $\mathbf{u}$, $\mathbf{v}$, $\mathbf{w}$ be vectors in space. Prove that

$$(\mathbf{u} \times \mathbf{v}) \times \mathbf{w} = (\mathbf{u} \cdot \mathbf{w})\, \mathbf{v} - (\mathbf{v} \cdot \mathbf{w})\, \mathbf{u} .$$

Use anti-commutativity to deduce that

$$\mathbf{u} \times (\mathbf{v} \times \mathbf{w}) = (\mathbf{u} \cdot \mathbf{w})\, \mathbf{v} - (\mathbf{u} \cdot \mathbf{v})\, \mathbf{w} .$$

Deduce further that

$$(\mathbf{u} \times \mathbf{v}) \times \mathbf{w} \neq \mathbf{u} \times (\mathbf{v} \times \mathbf{w})$$

whenever $\mathbf{u}$ and $\mathbf{w}$ are not parallel and $\mathbf{v}$ is not perpendicular to at least one of $\mathbf{u}$ or $\mathbf{w}$. Thus, associativity of the cross product fails spectacularly! Now verify the *Jacobi identity*:

$$(\mathbf{u} \times \mathbf{v}) \times \mathbf{w} + (\mathbf{v} \times \mathbf{w}) \times \mathbf{u} + (\mathbf{w} \times \mathbf{u}) \times \mathbf{v} = \mathbf{0}$$

Exercise 4.16** Start with the geometric formula for the cross product as a definition, that is, firstly, define the cross product of parallel vectors to be the zero vector, and, secondly,

$$\mathbf{v} \times \mathbf{w} = |\mathbf{v}||\mathbf{w}| \sin\theta \;\, \mathbf{u}$$

when $\mathbf{v}$ and $\mathbf{w}$ are not parallel, where θ is the angle between $\mathbf{v}$ and $\mathbf{w}$, $\mathbf{u}$ is a unit vector and the triple $\mathbf{v}$, $\mathbf{w}$, $\mathbf{u}$ is right-handed. Prove the distributive law:

$$(\mathbf{a} + \mathbf{b}) \times \mathbf{c} = \mathbf{a} \times \mathbf{c} + \mathbf{b} \times \mathbf{c}$$

Now deduce the algebraic definition of the cross product. This yields an alternative proof of the universality of the Right-Hand Rule.

5 Lines in Space

- **Lines in space** (Sections 5.1, 5.2) Time: 7.41

5 Lines in Space

In this chapter and the next we study lines and planes in space and show how to find and manipulate them using vectors and Cartesian equations. We begin with lines.

A line $\mathcal{L}$ in space is determined either by

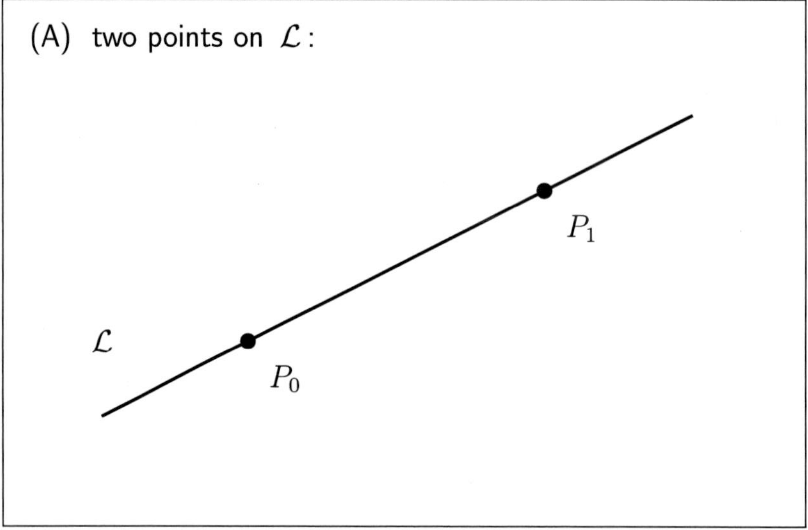

or by

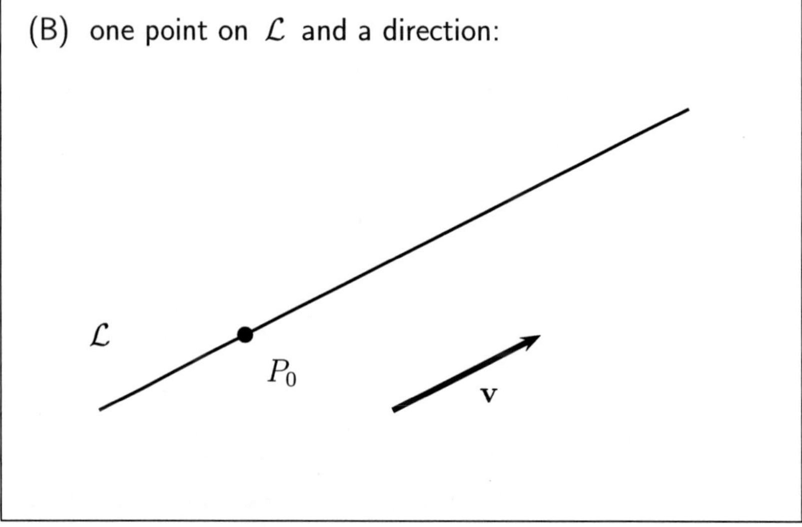

It is easier to begin the analysis of $\mathcal{L}$ with (B), and later reduce (A) to (B).

5.1 Parametric vector and scalar equations of a line

We consider case (B) and give names to all of the coordinates. Suppose

$$P_0(x_0, y_0, z_0)$$

lies on $\mathcal{L}$, and that $\mathcal{L}$ points in the direction of

$$\mathbf{v} = a\mathbf{i} + b\mathbf{j} + c\mathbf{k}\,.$$

It is convenient to add the origin $O(0,0,0)$ somewhere in the diagram and a typical general point

$$P(x, y, z)$$

which moves about on $\mathcal{L}$:

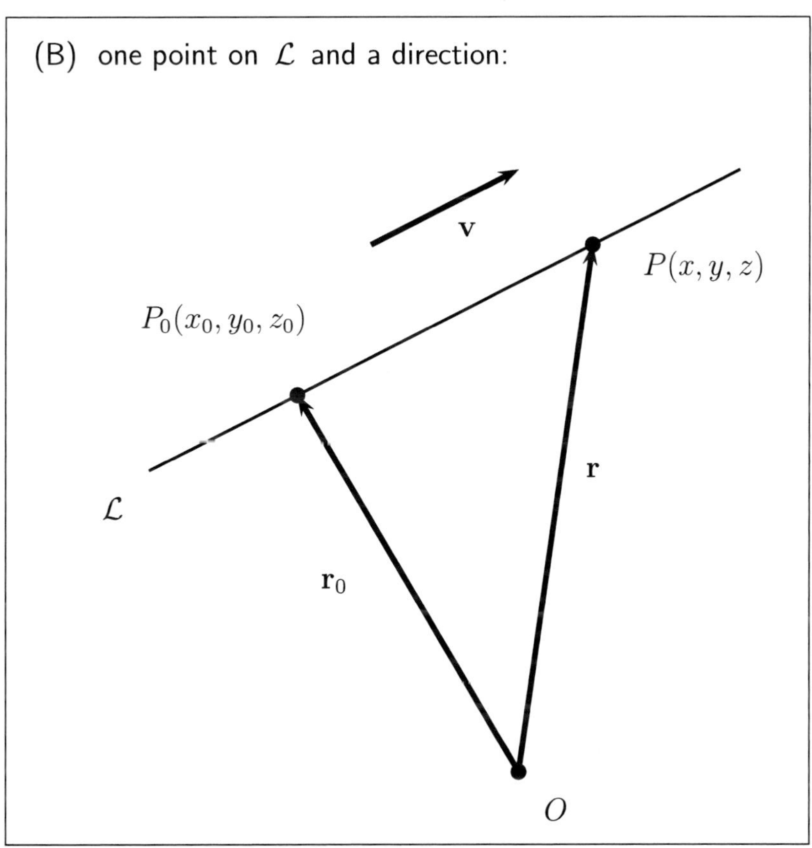

We are looking for formulae for the coordinates x, y, z of our typical point P. To this end we have added to the diagram the position vectors of P and P_0 respectively:

$$\mathbf{r} = \overrightarrow{OP} \quad \text{and} \quad \mathbf{r}_0 = \overrightarrow{OP_0}$$

By vector arithmetic,

$$\begin{aligned}\mathbf{r} - \mathbf{r}_0 &= \overrightarrow{OP} - \overrightarrow{OP_0} = \overrightarrow{OP} + \overrightarrow{P_0O} = \overrightarrow{P_0O} + \overrightarrow{OP} \\ &= \overrightarrow{P_0P},\end{aligned}$$

which is parallel to $\mathbf{v}$. Hence,

$$\boxed{\mathbf{r} - \mathbf{r}_0 = t\,\mathbf{v}}$$

for some scalar t, called the *parametric vector equation of* $\mathcal{L}$. The adjective "parametric" is used because

$t \in \mathbb{R}$ is called the *parameter*,

which is allowed to vary freely. Rearranging this equation yields

$$\boxed{\mathbf{r} = \mathbf{r}_0 + t\,\mathbf{v}\,.}$$

It is easy to interpret these position vectors physically, if one thinks of the origin O as a launching point. To move along the line,

"hop on at P_0 and move forwards and backwards parallel to $\mathbf{v}$".

We can rewrite this last equation and collect components:

$$\begin{aligned} x\,\mathbf{i} + y\,\mathbf{j} + z\,\mathbf{k} &= x_0\,\mathbf{i} + y_0\,\mathbf{j} + z_0\,\mathbf{k} + t\,(a\,\mathbf{i} + b\,\mathbf{j} + c\,\mathbf{k}) \\ &= (x_0 + ta)\,\mathbf{i} + (y_0 + tb)\,\mathbf{j} + (z_0 + tc)\,\mathbf{k}\end{aligned}$$

Equating coefficients of $\mathbf{i}$, $\mathbf{j}$ and $\mathbf{k}$ yields

$$\left.\begin{aligned} x &= x_0 + ta \\ y &= y_0 + tb \\ z &= z_0 + tc \end{aligned}\right\} \quad t \in \mathbb{R}$$

called the *parametric scalar equations of* $\mathcal{L}$.

5.2 Cartesian equations of a line

We can go further and eliminate t from the parametric scalar equations. If $a, b, c \neq 0$ then

$$x = x_0 + ta \iff t = \frac{x - x_0}{a}$$

$$y = y_0 + tb \iff t = \frac{y - y_0}{b}$$

$$z = z_0 + tc \iff t = \frac{z - z_0}{c}$$

so that, making these all equal, t disappears and we get

$$\boxed{\frac{x - x_0}{a} = \frac{y - y_0}{b} = \frac{z - z_0}{c}}$$

called the *Cartesian equations of* $\mathcal{L}$.

There are various other possibilities if one or more of a, b or c is zero, but there is no need to list them all. It is enough to see, for example, that if $a = 0$ then $x = x_0$ is constant and, provided b, $c \neq 0$, we only have to eliminate the parameter from two equations and obtain the following:

$$\boxed{x = x_0\,, \qquad \frac{y - y_0}{b} = \frac{z - z_0}{c}}$$

One soon develops fluency passing between the parametric and Cartesian equations. With a little practice one writes the Cartesian equations immediately from the data, noting that the denominators can be taken to be the coefficients of any vector pointing in the direction of the line.

Example: The line passing through $(1, 2, 3)$ parallel to

$$\mathbf{v} = -\mathbf{i} + 4\mathbf{j} - 3\mathbf{k}$$

has Cartesian equations

$$\frac{x - 1}{-1} = \frac{y - 2}{4} = \frac{z - 3}{-3}\,,$$

parametric scalar equations

$$\left.\begin{aligned} x &= 1 - t \\ y &= 2 + 4t \\ z &= 3 - 3t \end{aligned}\right\} \qquad t \in \mathbb{R}$$

and a parametric vector equation

$$\mathbf{r} \;=\; \mathbf{i} + 2\,\mathbf{j} + 3\,\mathbf{k} \;+\; t\,(-\mathbf{i} + 4\,\mathbf{j} - 3\,\mathbf{k})\,.$$

These equations are not unique. If a different reference point is chosen on the line, or we use a nontrivial scalar multiple of the vector pointing in the direction of the line, then the natural choice of equations will change. In this example, $(2, -2, 6)$ also lies on the line (putting $t = -1$ in the above parametric scalar equations) and $\mathbf{i} - 4\,\mathbf{j} + 3\,\mathbf{k}$ points in the opposite direction of the same line. Using these we get Cartesian equations

$$x - 2 \;=\; \frac{y+2}{-4} \;=\; \frac{z-6}{3}\,,$$

parametric scalar equations

$$\left.\begin{aligned} x &= 2 + s \\ y &= -2 - 4s \\ z &= 6 + 3s \end{aligned}\right\} \qquad s \in \mathbb{R}$$

and a parametric vector equation

$$\mathbf{r} \;=\; 2\,\mathbf{i} - 2\,\mathbf{j} + 6\,\mathbf{k} \;+\; s\,(\mathbf{i} - 4\,\mathbf{j} + 3\,\mathbf{k})\,.$$

Notice that we used s instead of t as the parameter. We can use any convenient name we like for the parameter, even t twice, provided there is no danger of confusing the values of t between the different sets of equations.

We illustrate a case where the parameter disappears from one of the parametric scalar equations.

Example: The line passing through $(-2, 1, 0)$ parallel to

$$\mathbf{v} \;=\; 3\mathbf{i} - 2\,\mathbf{k}$$

has a parametric vector equation

$$\mathbf{r} \;=\; -2\,\mathbf{i}+\mathbf{j}+\; t\,(3\mathbf{i}-2\,\mathbf{k})\,,$$

parametric scalar equations

$$\left.\begin{array}{c} x=-2+3t \\[2ex] y=1 \\[2ex] z=-2t \end{array}\right\} \qquad t\in\mathbb{R}$$

and Cartesian equations

$$\frac{x+2}{3} \;=\; \frac{z}{-2}\,, \quad y\;=\;1\,.$$

5.3 Finding a line using two points

The most natural way to specify a line $\mathcal{L}$ is to give two distinct points on it, say,

$$P_0(x_0, y_0, z_0)\,, \quad P_1(x_1, y_1, z_1)\,.$$

This is case (A) of the introduction, and we reduce to case (B) by forming the vector

$$\overrightarrow{P_0P_1} \;=\; (x_1-x_0)\,\mathbf{i}+(y_1-y_0)\,\mathbf{j}+(z_1-z_0)\,\mathbf{k}\,,$$

which points in the direction of $\mathcal{L}$, and then proceed as before.

Example: Find parametric scalar and Cartesian equations for the line passing through the points

$$P(4, 2, -6)\,, \quad Q(1, 0, 5)\,.$$

Solution: A vector in the direction of the line is

$$\overrightarrow{PQ} \;=\; -3\,\mathbf{i}-2\,\mathbf{j}+11\,\mathbf{k}\,,$$

so, using P as a reference point on the line, parametric equations are

$$\left.\begin{array}{c} x=4-3t \\[2ex] y=2-2t \\[2ex] z=-6+11t \end{array}\right\} \qquad t\in\mathbb{R}$$

and Cartesian equations

$$\frac{x-4}{-3} = \frac{y-2}{-2} = \frac{z+6}{11} \,.$$

If instead we use Q as a reference point on the line, parametric equations are

$$\left.\begin{aligned} x &= 1 - 3s \\ y &= -2s \\ z &= 5 + 11s \end{aligned}\right\} \qquad s \in \mathbb{R}$$

and Cartesian equations

$$\frac{x-1}{-3} = \frac{y}{-2} = \frac{z-5}{11} \,.$$

Of course, both sets of answers are correct.

We can go backwards from an equational description of a line, if we need to, as in the following illustration.

Example: Given Cartesian equations

$$\frac{x+2}{4} = \frac{y-2}{3} = \frac{z-5}{-1}$$

for a line $\mathcal{L}$, find two different points on $\mathcal{L}$ and a vector pointing in the direction of $\mathcal{L}$.

Solution: The most obvious point on $\mathcal{L}$ is $P(-2,2,5)$ which makes the numerators of the Cartesian equations all equal to zero. Using the denominators gives us a vector

$$4\,\mathbf{i} + 3\,\mathbf{j} - \mathbf{k}$$

pointing in the direction of $\mathcal{L}$. We could now write down parametric scalar equations, and feed in values of the parameter to get other points on the line. To get another point quickly, just take Q, say, such that

$$\begin{aligned} \overrightarrow{OQ} &= \overrightarrow{OP} + 4\,\mathbf{i} + 3\,\mathbf{j} - \mathbf{k} \;=\; -2\,\mathbf{i} + 2\,\mathbf{j} + 5\,\mathbf{k} + 4\,\mathbf{i} + 3\,\mathbf{j} - \mathbf{k} \\ &= 2\,\mathbf{i} + 5\,\mathbf{j} + 4\mathbf{k} \,. \end{aligned}$$

Thus, $Q(2,5,4)$ is a second point on $\mathcal{L}$.

5.4 Distance from a point to a line

In this section we discuss the problem of finding the shortest distance d from a point P to a line $\mathcal{L}$, and also the closest point R on $\mathcal{L}$ to P. We apply ideas from the previous two chapters, concerning the dot and cross products of vectors. To get started we need a point Q on $\mathcal{L}$ and a vector $\mathbf{v}$ pointing in the direction of $\mathcal{L}$.

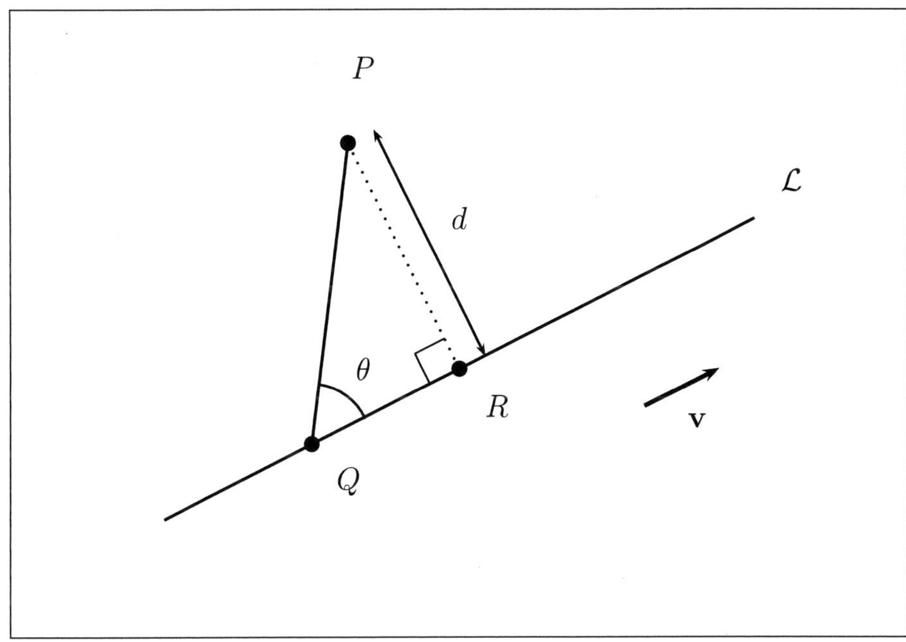

Let θ denote the angle made between the line $\mathcal{L}$ and the line segment joining P to Q. Then the distance we need is

$$d = |\overrightarrow{PQ}| \sin\theta .$$

The appearance of $\sin\theta$ reminds us of something we have recently seen before (in Section 4.4), namely, the formula for the length of a cross product:

$$|\mathbf{v} \times \mathbf{w}| = |\mathbf{v}|\,|\mathbf{w}| \sin\theta$$

where θ is the angle between $\mathbf{v}$ and $\mathbf{w}$.

If we take $\mathbf{w}$ to be $\overrightarrow{PQ}$ then we can rewrite d in terms of $\mathbf{v} \times \overrightarrow{PQ}$:

$$d = |\overrightarrow{PQ}| \sin\theta = \frac{|\mathbf{v}|\,|\overrightarrow{PQ}| \sin\theta}{|\mathbf{v}|} = \frac{|\mathbf{v} \times \overrightarrow{PQ}|}{|\mathbf{v}|}$$

We summarise our findings:

> **Distance from a point to a line:** The shortest distance d from a point P to a line containing the point Q and pointing in the direction of $\mathbf{v}$ is
>
> $$d = \frac{|\mathbf{v} \times \overrightarrow{PQ}|}{|\mathbf{v}|} .$$

Example: Find the distance from the point $P(1,0,1)$ to the line $\mathcal{L}$ given by the parametric scalar equations

$$\left.\begin{aligned} x &= 1 - t \\ y &= 3 + 2t \\ z &= -2 + 5t \end{aligned}\right\} \qquad t \in \mathbb{R} .$$

Solution: We need a point Q on $\mathcal{L}$ and a vector $\mathbf{v}$ in the direction of $\mathcal{L}$. The obvious (but certainly not unique) choices are

$$Q(1, 3, -2) ,$$

setting the parameter $t = 0$, and

$$\mathbf{v} = -\mathbf{i} + 2\,\mathbf{j} + 5\,\mathbf{k} ,$$

where the coefficients match those of t in the equations for x, y and z respectively. Then

$$\overrightarrow{PQ} = 3\,\mathbf{j} - 3\,\mathbf{k}$$

so

$$\begin{aligned} \mathbf{v} \times \overrightarrow{PQ} &= (-\mathbf{i} + 2\,\mathbf{j} + 5\,\mathbf{k}) \times (3\,\mathbf{j} - 3\,\mathbf{k}) \\ &= -3\,\mathbf{k} - 3\,\mathbf{j} - 6\,\mathbf{i} - 15\,\mathbf{i} \\ &= -21\,\mathbf{i} - 3\,\mathbf{j} - 3\,\mathbf{k} = -3\,(7\,\mathbf{i} + \mathbf{j} + \mathbf{k}) , \end{aligned}$$

whence the required distance is

$$d = \frac{|\mathbf{v} \times \overrightarrow{PQ}|}{|\mathbf{v}|} = \frac{|-3\,(7\,\mathbf{i} + \mathbf{j} + \mathbf{k})|}{|-\mathbf{i} + 2\,\mathbf{j} + 5\,\mathbf{k}|} = 3\frac{\sqrt{49 + 1 + 1}}{\sqrt{1 + 4 + 25}} = \frac{\sqrt{1530}}{10} .$$

Observe that the previous method calculated the shortest distance d but did not tell us the closest point R on the line. In fact, R is the point of projection of P on the line $\mathcal{L}$, so we should be able to exploit dot products and their role in the theory of projections (Section 3.4).

Here is our diagram again, but to be viewed from the point of view of vector components of $\overrightarrow{QP}$, one in the direction of $\mathbf{v}$, and the other orthogonal to $\mathbf{v}$. The length of the component orthogonal to $\mathbf{v}$ will of course be the distance d we found previously.

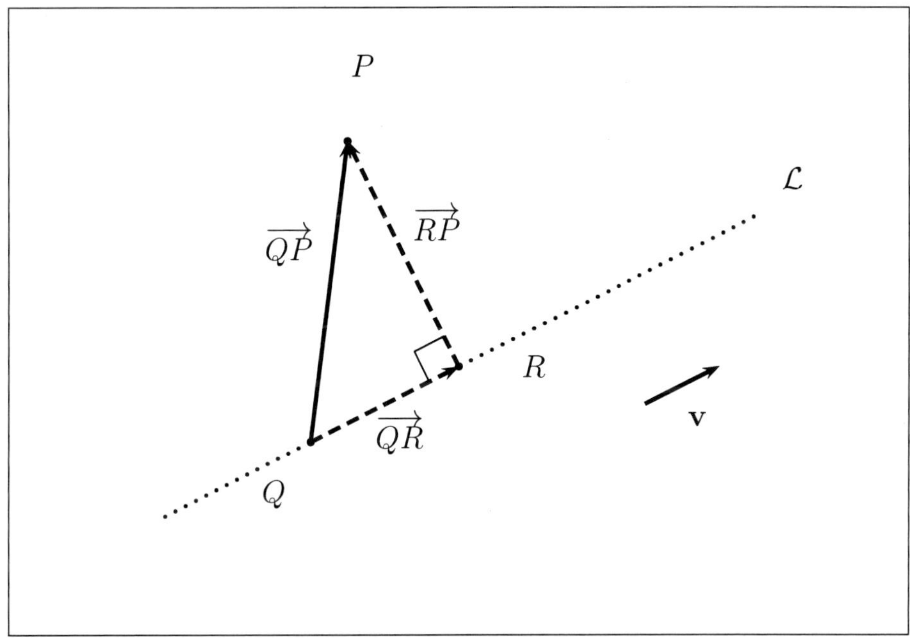

Then

$$|\overrightarrow{QR}| = |\overrightarrow{QP}\cdot\hat{\mathbf{v}}| = \left|\overrightarrow{QP}\cdot\frac{\mathbf{v}}{|\mathbf{v}|}\right| = \frac{|\overrightarrow{QP}\cdot\mathbf{v}|}{|\mathbf{v}|}$$

$$= \frac{|(-3\,\mathbf{j}+3\,\mathbf{k})\cdot(-\mathbf{i}+2\,\mathbf{j}+5\,\mathbf{k})|}{|-\mathbf{i}+2\,\mathbf{j}+5\,\mathbf{k}|}$$

$$= \frac{-6+15}{\sqrt{1+4+25}} = \frac{9}{\sqrt{30}}\,.$$

We can check our arithmetic so far against our earlier work, because

$$|\overrightarrow{QP}| = \sqrt{9+9} = 3\sqrt{2}\,,$$

so, by Pythagoras,

$$d = |\overrightarrow{RP}| = \sqrt{|\overrightarrow{QP}|^2 - |\overrightarrow{QR}|^2} = \sqrt{18 - \frac{81}{30}} = \frac{\sqrt{1530}}{10},$$

as before. $\checkmark$

Now, we can locate the point R indirectly, because

$$\begin{aligned} \overrightarrow{OR} &= \overrightarrow{OQ} + \overrightarrow{QR} = \overrightarrow{OQ} + |\overrightarrow{QR}|\,\hat{\mathbf{v}} \\ &= \mathbf{i} + 3\mathbf{j} - 2\mathbf{k} + \frac{9}{\sqrt{30}}\,\frac{-\mathbf{i} + 2\mathbf{j} + 5\mathbf{k}}{\sqrt{30}} \\ &= \frac{7}{10}\mathbf{i} + \frac{18}{5}\mathbf{j} - \frac{1}{2}\mathbf{k}, \end{aligned}$$

so that the closest point to P on $\mathcal{L}$ is

$$R = \left(\frac{7}{10}, \frac{18}{5}, -\frac{1}{2}\right).$$

(Notice that if the dot product of $\overrightarrow{QP}$ with $\mathbf{v}$ had been negative then we would have used

$$\overrightarrow{OR} = \overrightarrow{OQ} + \overrightarrow{QR} = \overrightarrow{OQ} - |\overrightarrow{QR}|\,\hat{\mathbf{v}}$$

to get the correct answer.)

Naturally, the next stage would be to apply our ideas to find the shortest distance between two lines in space. This is left as a starred exercise, using a combination of dot and cross products.

Chapter 5: Important Ideas and Useful Facts

5.1 A line in space is determined by two points, or by one point and a direction.

5.2 If the vector $\mathbf{v}$ points in the direction of a line $\mathcal{L}$ containing the point P_0, then the *parametric vector equation* of $\mathcal{L}$ is

$$\mathbf{r} - \mathbf{r}_0 = t\mathbf{v} \qquad \text{or equivalently} \qquad \mathbf{r} = \mathbf{r}_0 + t\mathbf{v}$$

where $\mathbf{r}$ is the position vector of a typical point on $\mathcal{L}$, $\mathbf{r}_0$ is the position vector of P_0 and t is a parameter which varies over all real numbers.

5.3 If the vector $\mathbf{v} = a\,\mathbf{i} + b\,\mathbf{j} + c\,\mathbf{k}$ points in the direction of a line $\mathcal{L}$ containing the point $P_0(x_0, y_0, z_0)$, then the *parametric scalar equations* of $\mathcal{L}$ are

$$\left.\begin{aligned} x &= x_0 + ta \\ y &= y_0 + tb \\ z &= z_0 + tc \end{aligned}\right\} \quad t \in \mathbb{R}$$

and the *Cartesian equations* are

$$\frac{x - x_0}{a} = \frac{y - y_0}{b} = \frac{z - z_0}{c}$$

(in the case that a, b, c are all nonzero).

5.4 The shortest distance d from a point P to a line containing the point Q and pointing in the direction of $\mathbf{v}$ is

$$d = \frac{|\mathbf{v} \times \overrightarrow{PQ}|}{|\mathbf{v}|}\,.$$

Exercise 5.1 Find the parametric vector, parametric scalar and Cartesian equations of the lines passing through P in the direction of $\mathbf{v}$ in each of the following cases:

(i) $P = (3, 2, 1)\,,\ \ \mathbf{v} = \mathbf{i} + 2\,\mathbf{j} + 3\,\mathbf{k}$ (ii) $P = (0, 0, 1)\,,\ \ \mathbf{v} = \mathbf{k}$

(iii) $P = (2, 0, -1)\,,\ \ \mathbf{v} = -\mathbf{i} + 2\,\mathbf{j} - 3\,\mathbf{k}$ (iv) $P = (0, 0, 0)\,,\ \ \mathbf{v} = \mathbf{j} + \mathbf{k}$

(v) $P = (-4, -3, 0)\,,\ \ \mathbf{v} = \mathbf{j} - \mathbf{k}$ (vi) $P = (1, 0, 1)\,,\ \ \mathbf{v} = \mathbf{i} - \mathbf{j}$

Exercise 5.2 Find the parametric vector, parametric scalar and Cartesian equations of the lines passing through P and Q in each of the following cases:

(i) $P = (3, 2, 1)\,,\ \ Q = (1, 2, 3)$ (ii) $P = (0, 0, 1)\,,\ \ Q = (1, 0, 0)$

(iii) $P = (2, 0, 1)\,,\ \ Q = (4, 7, -3)$ (iv) $P = (1, 1, 1)\,,\ \ Q = (-1, -1, 1)$

Exercise 5.3 Verify that the line $\mathcal{L}_1$ described by

$$\frac{x-1}{4} = \frac{y-2}{3} = \frac{z-10}{5}$$

is identical to the line $\mathcal{L}_2$ described by

$$\left.\begin{array}{rcl} x &=& -7-4t \\ y &=& -4-3t \\ z &=& -5t \end{array}\right\} \qquad t \in \mathbb{R}\,.$$

Exercise 5.4 Two lines are called *skew* if they are not parallel and do not intersect. Verify that the lines $\mathcal{L}_1$ described by

$$\frac{x+3}{2} = \frac{y}{3} = \frac{z-4}{5}$$

and $\mathcal{L}_2$ described by

$$\left.\begin{array}{rcl} x &=& 1+3s \\ y &=& 6-2s \\ z &=& 15-s \end{array}\right\} \qquad s \in \mathbb{R}$$

are skew.

Exercise 5.5 The following lines are not parallel. Show that they are not skew, by finding their point of intersection:

$$\mathcal{L}_1 : \mathbf{r} = \mathbf{i} + \mathbf{j} + \mathbf{k} + t(3\mathbf{i} - \mathbf{j} + 4\mathbf{k})\,,\ \ \mathcal{L}_2 : \mathbf{r} = 6\mathbf{i} - 6\mathbf{j} + \mathbf{k} + s(-7\mathbf{i} + 5\mathbf{j} - 6\mathbf{k})$$

Exercise 5.6 Consider the following points in space:

$$P(1,1,1)\,,\quad Q(5,-5,-3)\,,\quad R(6,-3,-1)\,,\quad S(2,3,3)$$

(i) Find the parametric vector, parametric scalar and Cartesian equations of the line $\mathcal{L}_1$ passing through P and R, and also the line $\mathcal{L}_2$ passing through Q and S.

(ii) Find the intersection point T of $\mathcal{L}_1$ and $\mathcal{L}_2$.

(iii) Verify that T is the midpoint of both PR and QS. Are you surprised? If not, explain why not.

Exercise 5.7* Find the distance from $P(2,1,1)$ to the line $\mathcal{L}$ given by the following equations:

$$x-1 \;=\; \frac{y-1}{3} \;=\; \frac{z+4}{-1}$$

Find the closest point to P lying on $\mathcal{L}$.

Exercise 5.8* Describe a general technique for finding the shortest distance between two parallel lines in space. Use your method to find the distance between the lines given by the following sets of Cartesian equations:

$$x-1 \;=\; \frac{y-2}{4} \;=\; \frac{z-3}{-3} \qquad \text{and} \qquad x+1 \;=\; \frac{y-3}{4} \;=\; \frac{z+1}{-3}$$

Exercise 5.9* Let P and Q be fixed points in space. Suppose R is a point in space (which varies) such that

$$\overrightarrow{OR} \;=\; \lambda\,\overrightarrow{OP} + \mu\,\overrightarrow{OQ}$$

for real numbers λ and μ subject to the constraint

$$\lambda + \mu \;=\; 1\,.$$

Prove that R varies over the line that passes through P and Q. Describe the values of λ such that R is located as follows:

(i) somewhere on the line segment joining P to Q

(ii) somewhere on the line beyond P on the side away from Q

(iii) somewhere on the line beyond Q on the side away from P

(iv) twice as far from P as it is from Q

Exercise 5.10* Describe the geometric relationship, in terms of a rotation in space, between the line whose parametric vector form is $\mathbf{r} = \mathbf{r}_0 + t\,\mathbf{v}$ and each of the following lines, where $\mathbf{w}$ and $\mathbf{v}$ are not parallel:

(i) $\mathbf{r} = \mathbf{r}_0 - t\,\mathbf{v}$ (ii) $\mathbf{r} = \mathbf{r}_0 + t\,(\mathbf{v} \times \mathbf{w})$

(iii) $\mathbf{r} = -\mathbf{r}_0 + t\,\mathbf{v}$ (iv) $\mathbf{r} = \mathbf{r}_0 + t\left(\mathbf{v} - \dfrac{\mathbf{v} \cdot \mathbf{w}}{|\mathbf{w}|^2}\mathbf{w}\right)$

Exercise 5.11** Let $\mathcal{L}_1$ and $\mathcal{L}_2$ be two skew lines in space, containing points P_1 and P_2, and pointing in directions $\mathbf{v}_1$ and $\mathbf{v}_2$ respectively. Explain why the shortest distance d between $\mathcal{L}_1$ and $\mathcal{L}_2$ is given by the following formula:

$$d = \frac{\left|\overrightarrow{P_1P_2} \cdot (\mathbf{v}_1 \times \mathbf{v}_2)\right|}{|\mathbf{v}_1 \times \mathbf{v}_2|}$$

Give a general method for finding the closest points Q_1 and Q_2 on $\mathcal{L}_1$ and $\mathcal{L}_2$ respectively. Find d, Q_1 and Q_2 for the following lines:

$$\mathcal{L}_1 : x = 2\,,\ y = 1 - z \qquad \text{and} \qquad \mathcal{L}_2 : x + 1 = \frac{y-3}{4} = \frac{z+1}{-3}$$

Exercise 5.12* A smooth curve $\mathcal{C}$ in space is defined parametrically by the equations

$$\left.\begin{aligned} x &= f(t) \\ y &= g(t) \\ z &= h(t) \end{aligned}\right\} \qquad t \in \mathbb{R}$$

where f, g and h are differentiable functions. Let $\mathbf{r} = \mathbf{r}(t)$ denote the position vector of a point on $\mathcal{C}$ at time t. Define the *derivative* of $\mathbf{r}$ to be

$$\mathbf{r}' = \mathbf{r}'(t) = \lim_{\delta \to 0} \frac{\mathbf{r}(t+\delta) - \mathbf{r}(t)}{\delta}\,.$$

Give a physical interpretation of $\mathbf{r}'$. Prove that $\mathbf{r}'$ is given parametrically by

$$\left.\begin{aligned} x &= f'(t) \\ y &= g'(t) \\ z &= h'(t) \end{aligned}\right\} \qquad t \in \mathbb{R}\,.$$

6 Planes in Space

- **Planes in space** (Sections 6.1, 6.2)
  Time: 7.19
- **Finding a plane** (Section 6.3)
  Time: 3.53
- **Distance from a point to a plane** (Section 6.4)
  Time: 10.02

6 Planes in Space

In this chapter we study planes in 3-space, which one can think of as two-dimensional analogues of lines in the xy-plane. This leads to the useful notion of *hyperplanes* of dimension $n-1$ sitting inside n-space, where n is any positive integer. We will not use this terminology except in passing, but in fact almost everything we do from now on in this book is an exploration of the algebra and geometry of hyperplanes!

A plane $\mathcal{P}$ in space is determined either by three *non-collinear* points, that is, three points which do not lie on a single line,

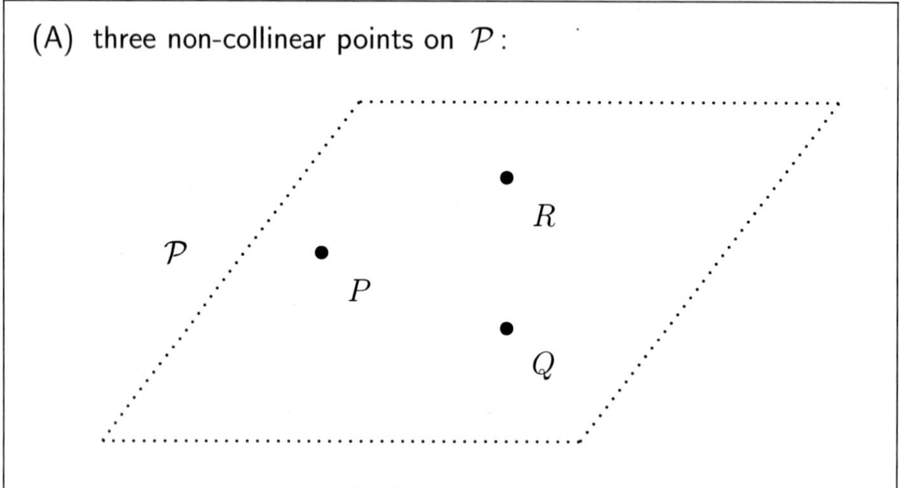

or by one point and a perpendicular direction

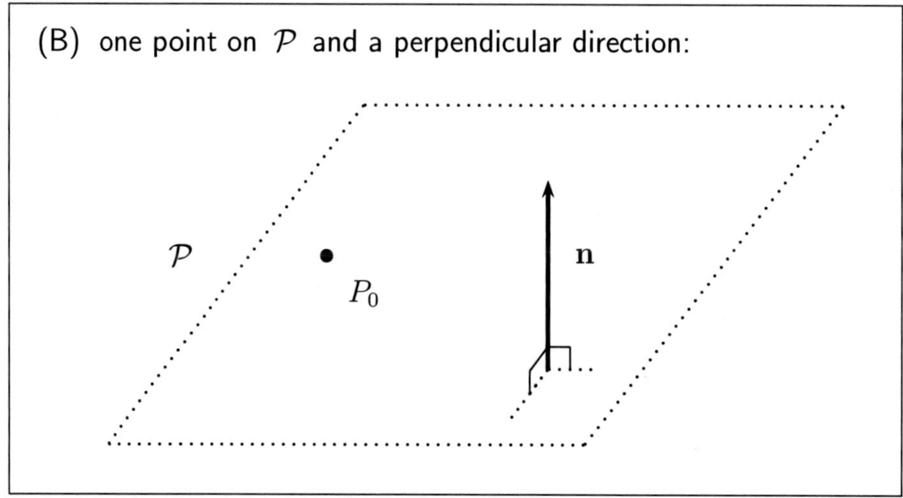

It is typical to label the vector in the above diagram $\mathbf{n}$ for *normal*. In this context *normal* and *perpendicular* have the same meaning. (Unfortunately, the word "normal" is as overused in mathematics as it is in real life.)

At this point the reader might be puzzled that case (B) here looks qualitatively different to case (B) of the previous chapter. But if our line $\mathcal{L}$ of Chapter 5 happened to lie in the xy-plane, then specifying the direction of $\mathcal{L}$ would be equivalent to specifying a direction perpendicular to $\mathcal{L}$. With that heavy restriction, case (B) of Chapter 5 would become an analogue of what we are doing here.

But, for lines in space, a direct natural analogue of case (B) of Chapter 5 would suggest that we should now specify two directions for our plane $\mathcal{P}$ in space. That turns out to be less efficient than specifying just one normal vector. In fact, in higher dimensions, a hyperplane of dimension $n-1$ is specified either by $n-1$ directions, which is a lot of information, or, more compactly, by just a single normal direction. (Of course, we have to explain what *normal* means in n-space, but that is part of the *theory of inner product spaces.*)

We are on the tip of an iceberg, which is obscured only because the dimensions we are dealing with here are small. A line has dimension one, and that was exploited in Chapter 5. A plane in space has codimension one, which means we only need one more (normal) direction to get everything, and that is what is being exploited in this chapter.

As in the last chapter it is easier to begin the analysis with (B), and later reduce (A) to (B).

6.1 Vector equation of a plane

We consider case (B) and again give names to all of the coordinates. Suppose that

$$P_0 = P_0(x_0, y_0, z_0)$$

lies on $\mathcal{P}$, and a normal vector to $\mathcal{P}$ is

$$\mathbf{n} = a\,\mathbf{i} + b\,\mathbf{j} + c\,\mathbf{k}.$$

Shift the vector $\mathbf{n}$ so that it emanates from P_0, and add the origin $O(0, 0, 0)$ somewhere in the diagram:

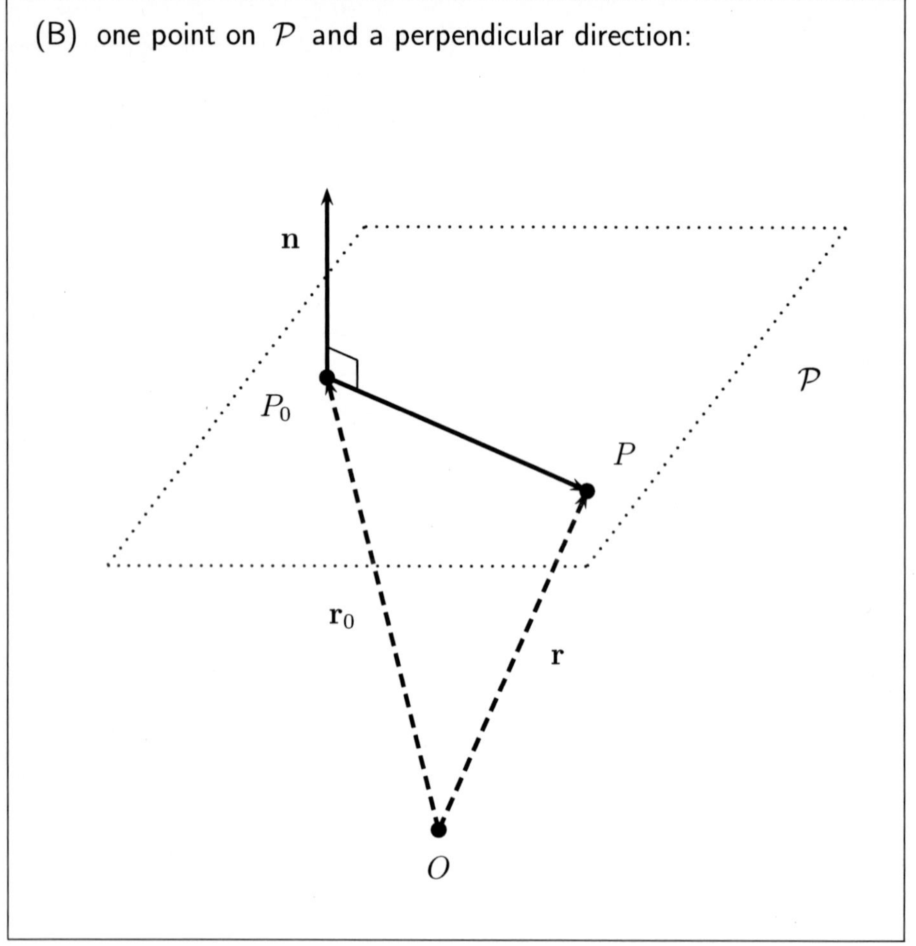

Add a typical point

$$P = P(x, y, z)$$

which moves arbitrarily about on the plane $\mathcal{P}$, and denote the respective position vectors of P, P_0 by

$$\mathbf{r} = \overrightarrow{OP}, \qquad \mathbf{r}_0 = \overrightarrow{OP_0},$$

so again $\mathbf{r} - \mathbf{r}_0 = \overrightarrow{P_0P}$. But this vector is perpendicular to $\mathbf{n}$, so we get the following:

$$\boxed{(\mathbf{r} - \mathbf{r}_0) \cdot \mathbf{n} = 0}$$

This is called the *vector equation* of the plane $\mathcal{P}$.

6.2 Cartesian equation of a plane

We use components now to reformulate the vector equation of $\mathcal{P}$. Observe that

$$\mathbf{r} - \mathbf{r}_0 \;=\; \overrightarrow{P_0P} \;=\; (x - x_0)\mathbf{i} + (y - y_0)\mathbf{j} + (z - z_0)\mathbf{k}\,,$$

so that the vector equation becomes

$$\big((x - x_0)\mathbf{i} + (y - y_0)\mathbf{j} + (z - z_0)\mathbf{k}\big) \cdot (a\,\mathbf{i} + b\,\mathbf{j} + c\,\mathbf{k}) \;=\; 0\,,$$

that is,

$$(x - x_0)a + (y - y_0)b + (z - z_0)c \;=\; 0\,,$$

which rearranges to

$$ax + by + cz \;=\; ax_0 + by_0 + cz_0\,.$$

If we put $d = ax_0 + by_0 + cz_0$, then we can write this more simply as

$$\boxed{ax + by + cz \;=\; d}$$

called the *Cartesian equation* of the plane $\mathcal{P}$.

(This is the direct analogue of the equation $ax + by = c$ of a line in the xy-plane, as explained in the Introduction.)

Example: The plane containing the point $(-2, 4, 7)$ with normal vector

$$\mathbf{n} \;=\; 3\mathbf{i} + 2\,\mathbf{j} + 5\,\mathbf{k}$$

has Cartesian equation

$$3x + 2y + 5z \;=\; d$$

where

$$d \;=\; 3(-2) + 2(4) + 5(7) \;=\; 37\,,$$

that is, the following:

$$\boxed{3x + 2y + 5z \;=\; 37}$$

The important thing to remember is that the coefficients of x, y and z are just the respective coefficients of $\mathbf{i}$, $\mathbf{j}$ and $\mathbf{k}$ in a vector normal to the plane. We have derived the Cartesian equation from geometric considerations, but in fact every such equation arises in this way (Exercise 6.10):

> Provided at least one of a, b or c is nonzero, the equation
>
> $$ax + by + cz = d$$
>
> always describes a plane in space, with normal vector
>
> $$\mathbf{n} = a\,\mathbf{i} + b\,\mathbf{j} + c\,\mathbf{k}\,.$$

There is no need to remember the formula $d = ax_0 + by_0 + cz_0$. The value of d comes out in the wash, when the left-hand side of the Cartesian equation is evaluated at some point.

It is easy to decide when two planes are parallel:

> Two planes are parallel if and only if their normal vectors are parallel.

Example: The planes

$$\mathcal{P}_1 : \qquad 3x + 2y + 5z = 2$$

$$\mathcal{P}_2 : \quad -6x - 4y - 10z = 7$$

have respective normals

$$\mathbf{n}_1 = 3\,\mathbf{i} + 2\,\mathbf{j} + 5\,\mathbf{k}\,, \qquad \mathbf{n}_2 = -6\,\mathbf{i} - 4\,\mathbf{j} - 10\,\mathbf{k}\,.$$

But the vectors $\mathbf{n}_1$ and $\mathbf{n}_2$ are parallel, since $\mathbf{n}_2 = -2\mathbf{n}_1$, so the planes $\mathcal{P}_1$ and $\mathcal{P}_2$ must also be parallel.

These planes are not identical, however, because we can find a point on the plane $\mathcal{P}_1$ that doesn't lie on $\mathcal{P}_2$. To see this, look for a simple triple, say, $(0, 1, 0)$, which satisfies the equation for $\mathcal{P}_1$ and check that this does not satisfy the equation for $\mathcal{P}_2$:

$$-6(0) - 4(1) - 10(0) = -4 \neq 7 \qquad \surd$$

6.3 Finding a plane using three points

What happens if the plane $\mathcal{P}$ is specified by three non-collinear points P_1, P_2, P_3?

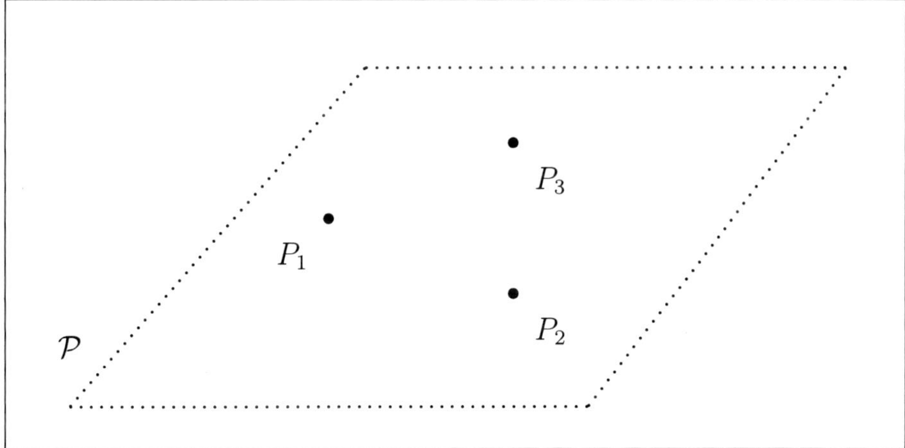

This is case (A) at the beginning of this chapter, and the answer is to reduce to case (B) by joining up one point, say P_1, to the other two points to form vectors

$$\overrightarrow{P_1P_2}\,, \qquad \overrightarrow{P_1P_3}$$

and then take their cross product

$$\mathbf{n} \;=\; \overrightarrow{P_1P_2} \times \overrightarrow{P_1P_3}$$

which is normal to the plane $\mathcal{P}$:

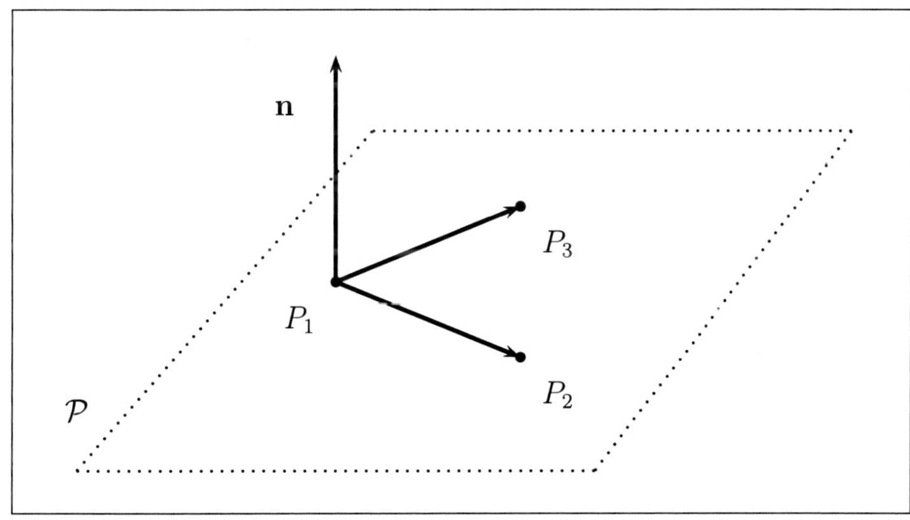

Example: Find the equation of the plane containing

$$P_1(1,2,3)\,, \quad P_2(2,3,4)\,, \quad P_3(-2,0,6)\,.$$

Solution: We form the normal vector

$$\begin{aligned}
\mathbf{n} &= \overrightarrow{P_1P_2} \times \overrightarrow{P_1P_3} \\
&= (\mathbf{i}+\mathbf{j}+\mathbf{k}) \times (-3\,\mathbf{i} - 2\,\mathbf{j} + 3\mathbf{k}) \\
&= -2\,\mathbf{k} - 3\,\mathbf{j} + 3\,\mathbf{k} + 3\,\mathbf{i} - 3\,\mathbf{j} + 2\,\mathbf{i} \\
&= 5\,\mathbf{i} - 6\,\mathbf{j} + \mathbf{k}\,,
\end{aligned}$$

so that the equation of the plane is

$$5\,x - 6\,y + z = d$$

for some d. Using the coordinates of P_1, say, we get

$$d = 5(1) - 6(2) + 3 = -4\,,$$

so the Cartesian equation of the plane is the following:

$$\boxed{5\,x - 6\,y + z = -4}$$

[Check: This equation must be satisfied by the coordinates of the other points:

$$P_2:\ 5(2) - 6(3) + 4 = -4 \quad \checkmark \qquad P_3:\ 5(-2) - 6(0) + 6 = -4 \quad \checkmark$$]

6.4 Distance from a point to a plane

In this final section we discuss the problem of finding the shortest distance d from a point P to a plane, and also the closest point R on the plane to P. Again we exploit the dot product and projection. To get started we need a point Q on the plane and a normal vector $\mathbf{n}$. Both of these are easy to find, for example, if we have the Cartesian equation of the plane.

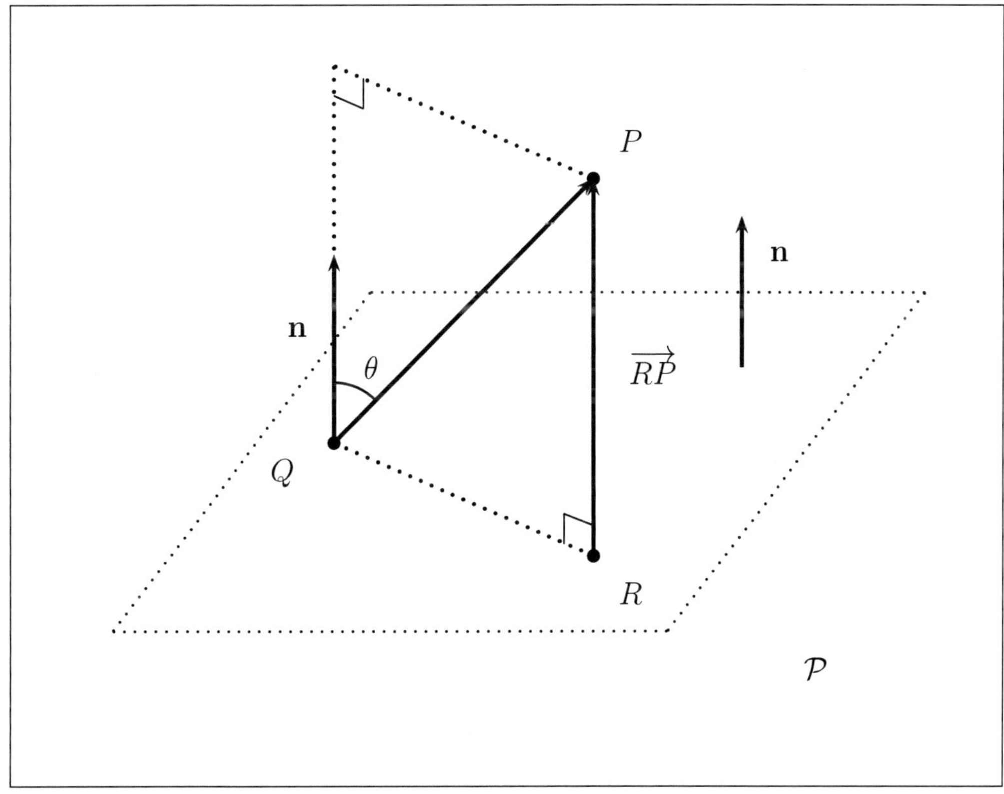

Let θ be the angle between the vectors $\overrightarrow{QP}$ and $\mathbf{n}$. Then the distance from P to the plane $\mathcal{P}$ is

$$d = \left|\overrightarrow{RP}\right| = \left|\overrightarrow{QP}\right| \left|\cos\theta\right| .$$

Note that we have used $\left|\cos\theta\right|$ to account for the possibility that θ might be obtuse (though it happens to be acute in the above diagram). Working towards incorporating a dot product, we get

$$d = \frac{|\mathbf{n}| \left|\overrightarrow{QP}\right| \left|\cos\theta\right|}{|\mathbf{n}|} = \frac{\left| |\mathbf{n}| \left|\overrightarrow{PQ}\right| \cos\theta \right|}{|\mathbf{n}|} = \frac{\left|\mathbf{n} \cdot \overrightarrow{PQ}\right|}{|\mathbf{n}|} .$$

We summarise our findings:

Distance from a point to a plane: The shortest distance d from a point P to a plane containing the point Q with normal vector $\mathbf{n}$ is

$$d = \frac{\left|\mathbf{n} \cdot \overrightarrow{PQ}\right|}{|\mathbf{n}|} .$$

Example: Find the distance from the point $P(2,-1,3)$ to the plane

$$\mathcal{P}: \quad 2x+y-z = 2$$

and the closest point R on the plane $\mathcal{P}$ to the point P.

Solution: We need a point Q on $\mathcal{P}$ and a normal vector $\mathbf{n}$. The most obvious choices are

$$Q(1,0,0), \qquad \mathbf{n} = 2\,\mathbf{i}+\mathbf{j}-\mathbf{k}.$$

The required distance is

$$d = \frac{|\mathbf{n}\cdot\overrightarrow{PQ}|}{|\mathbf{n}|} = \frac{\left|(2\,\mathbf{i}+\mathbf{j}-\mathbf{k})\cdot(-\mathbf{i}+\mathbf{j}-3\,\mathbf{k})\right|}{\sqrt{4+1+1}} = \frac{-2+1+3}{\sqrt{6}} = \frac{2}{\sqrt{6}}.$$

We can now use detective work to find R indirectly since

$$\overrightarrow{OR} = \overrightarrow{OP}+\overrightarrow{PR}.$$

Clearly, $\overrightarrow{OP} = 2\,\mathbf{i}-\mathbf{j}+3\,\mathbf{k}$, but finding $\overrightarrow{PR}$ is tricky. Since R is the closest point on the plane to P, the vector $\overrightarrow{PR}$ is normal to the plane, so is parallel to $\mathbf{n}$. Notice that, in every situation (Exercise 6.11),

$$\boxed{\overrightarrow{PQ}\cdot\overrightarrow{PR} > 0.}$$

But, in this example,

$$\boxed{\overrightarrow{PQ}\cdot\mathbf{n} = 2 > 0.}$$

It follows that

$\overrightarrow{PR}$ **must be a** ***positive scalar multiple*** **of** $\mathbf{n}$.

(The diagram on the previous page would be wrong in this case, since the angle θ would be obtuse!) Hence,

$$\overrightarrow{PR} = |\overrightarrow{PR}|\,\hat{\mathbf{n}} = \frac{2}{\sqrt{6}}\,\frac{2\,\mathbf{i}+\mathbf{j}-\mathbf{k}}{\sqrt{6}} = \frac{2\,\mathbf{i}+\mathbf{j}-\mathbf{k}}{3},$$

so

$$\overrightarrow{OR} = 2\,\mathbf{i}-\mathbf{j}+3\,\mathbf{k}+\frac{2\,\mathbf{i}+\mathbf{j}-\mathbf{k}}{3} = \frac{8}{3}\,\mathbf{i}-\frac{2}{3}\,\mathbf{j}+\frac{8}{3}\,\mathbf{k}.$$

Thus, finally, R is the point

$$(8/3,\ -2/3,\ 8/3).$$

Chapter 6: Important Ideas and Useful Facts

6.1 A plane in space is determined either by three non-collinear points, or by one point and a perpendicular (normal) direction.

6.2 If the vector $\mathbf{n}$ is normal to a plane $\mathcal{P}$ containing the point P_0, then the *vector equation* of $\mathcal{P}$ is

$$(\mathbf{r} - \mathbf{r}_0) \cdot \mathbf{n} = 0 \qquad \text{or equivalently} \qquad \mathbf{r} \cdot \mathbf{n} = \mathbf{r}_0 \cdot \mathbf{n}$$

where $\mathbf{r}$ is the position vector of a typical point and $\mathbf{r}_0$ is the position vector of P_0.

6.3 If the vector $\mathbf{n} = a\,\mathbf{i} + b\,\mathbf{j} + c\,\mathbf{k}$ is normal to the plane $\mathcal{P}$ containing the point $P_0(x_0, y_0, z_0)$, then the *Cartesian equation* of $\mathcal{P}$ is

$$ax + by + cz = d$$

where $d = ax_0 + by_0 + cz_0$.

6.4 If P_1, P_2, P_3 are non-collinear points on a plane, then a normal vector to the plane is

$$\mathbf{n} = \overrightarrow{P_1P_2} \times \overrightarrow{P_1P_3}.$$

6.5 The shortest distance d from a point P to a plane containing the point Q with normal vector $\mathbf{n}$ is

$$d = \frac{|\mathbf{n} \cdot \overrightarrow{PQ}|}{|\mathbf{n}|}.$$

Exercise 6.1 Find the equation of the plane containing the point $(3, 2, -8)$ and perpendicular to the vector $\mathbf{v} = -\mathbf{i} + \mathbf{j} + 3\mathbf{k}$.

Exercise 6.2 Let $P = (2, 1, 3)$, $Q = (-2, -2, -1)$ and $R = (3, -4, 1)$.

(i) Express $\overrightarrow{PQ}$ and $\overrightarrow{PR}$ in Cartesian form.

(ii) Find the cross product $\overrightarrow{PQ} \times \overrightarrow{PR}$.

(iii) Find the Cartesian equation of the plane containing P, Q, R.

Exercise 6.3 Find the equation for the plane containing the three points in each case:

(i) $(0, 1, -1), (3, 0, -3), (2, -2, 0)$ (ii) $(4, -6, 2), (2, 1, -4), (2, 5, 7)$

Exercise 6.4 For each of (i)–(vii), find matching descriptions from (a)–(n):

(i) line containing $(0, 0, 0)$ in the direction of $\mathbf{i} + \mathbf{j} + \mathbf{k}$

(ii) line containing $(-1, 2, -1)$ in the direction of $-\mathbf{i} + 2\mathbf{j} + 2\mathbf{k}$

(iii) line containing $(-1, 2, -1)$ and $(0, 0, -2)$

(iv) plane containing $(0, 0, 0)$ with normal vector $\mathbf{i} + \mathbf{j} + \mathbf{k}$

(v) plane containing $(-1, 2, -1)$ with normal vector $-\mathbf{i} + 2\mathbf{j} + 2\mathbf{k}$

(vi) plane containing $(-1, 2, -1)$, $(0, 0, -2)$ and $(1, 3, 3)$

(vii) plane containing $(-1, 2, -1)$, $(0, 0, -2)$ and $(1, 3, 2)$

(a) $x + y + z = 0$ (b) $x = y = z$ (c) $x + y - z = 2$

(d) $x + 1 = \dfrac{y - 2}{-2} = \dfrac{z + 1}{-2}$ (e) $7x + 6y - 5z = 10$

(f) $x + 1 = \dfrac{y - 2}{-2} = \dfrac{z + 1}{-1}$ (g) $x - 2y - 2z = -3$

(h) $(\mathbf{r} + 2\mathbf{k}) \cdot (7\mathbf{i} + 6\mathbf{j} - 5\mathbf{k}) = 0$ (i) $\mathbf{r} = \mathbf{i} - 2\mathbf{j} - 5\mathbf{k} + t(\mathbf{i} - 2\mathbf{j} - 2\mathbf{k})$

(j) $(\mathbf{r} + 2\mathbf{k}) \cdot (\mathbf{i} + \mathbf{j} - \mathbf{k}) = 0$ (k) $\mathbf{r} \cdot (\mathbf{i} + \mathbf{j} + \mathbf{k}) = 0$

(l) $(\mathbf{r} + 3\mathbf{i}) \cdot (\mathbf{i} - 2\mathbf{j} - 2\mathbf{k}) = 0$ (m) $\mathbf{r} = \mathbf{i} - 2\mathbf{j} - 3\mathbf{k} + t(\mathbf{i} - 2\mathbf{j} - \mathbf{k})$

(n) $\mathbf{r} = \mathbf{i} + \mathbf{j} + \mathbf{k} - t(\mathbf{i} + \mathbf{j} + \mathbf{k})$

Exercise 6.5 Find Cartesian equations of the line passing through $(3,-1,2)$ and perpendicular to the plane $x-3y+2z=4$.

Exercise 6.6 Verify that the line

$$\frac{x+2}{3} = \frac{y+1}{3} = \frac{z-7}{-1}$$

is parallel to the plane $-3x+4y+3z=14$.

Exercise 6.7 The following planes intersect in a line:

$$2x+y-z=6 \qquad \text{and} \qquad -3x+2y+z=-9\,.$$

Find a point on this line of intersection and its direction. Now write down parametric and Cartesian equations for this line.

Exercise 6.8 Find the Cartesian equation of the plane containing $(2,-1,3)$ and the line

$$\frac{x+3}{-3} = y-1 = \frac{z+4}{2}\,.$$

Exercise 6.9 What do we mean by the angle between two planes? Find the cosine of the angle between the two planes given by equations

$$x+y+z=6 \qquad \text{and} \qquad x-2y-z=3\,.$$

Exercise 6.10 Suppose a, b, c and d are real numbers such that at least one of a, b or c is nonzero. (There is no restriction on d.) Find a point Q and a direction $\mathbf{n}$ such that the plane containing Q perpendicular to $\mathbf{n}$ has Cartesian equation

$$ax+by+cz = d\,.$$

What happens when $a=b=c=d=0$ and when $a=b=c=0\neq d$?

Exercise 6.11 Suppose that P is a point in space and $\mathcal{P}$ is a plane not containing P. Let Q be any point on $\mathcal{P}$ and let R be the closest point on $\mathcal{P}$ to P. Explain why the dot product $\overrightarrow{PQ}\cdot\overrightarrow{PR}$ must be positive.

Exercise 6.12* Find the shortest distance from the point $P(-3,2,7)$ to the plane $\mathcal{P}$ given by the equation

$$x-y+5z = -2\,.$$

Find the closest point on $\mathcal{P}$ to P.

Exercise 6.13* Let $z = f(x,y)$ be a real-valued function of two real variables such that the partial derivatives $\partial f/\partial x$ and $\partial f/\partial y$ exist and are continuous. It is a theorem of calculus that the tangent plane is a good approximation to the surface $z = f(x,y)$ at a given point (x_0, y_0, z_0). Explain why the vectors

$$\mathbf{i} + \frac{\partial f}{\partial x}(x_0, y_0)\,\mathbf{k} \qquad \text{and} \qquad \mathbf{j} + \frac{\partial f}{\partial y}(x_0, y_0)\,\mathbf{k}$$

are parallel to the tangent plane. Deduce the following equation for the tangent plane to the surface at (x_0, y_0, z_0):

$$z - z_0 \;=\; \frac{\partial f}{\partial x}(x_0, y_0)\,(x - x_0) + \frac{\partial f}{\partial y}(x_0, y_0)\,(y - y_0)$$

Exercise 6.14* Let r be a fixed positive real number. Describe geometrically the configuration $\mathcal{S}$ in space of points whose position vectors $\mathbf{r}$ satisfy the equation

$$|\mathbf{r}| \;=\; r\,.$$

Let $P(x,y,z)$ be a point on $\mathcal{S}$. Find the Cartesian equation of the tangent plane to $\mathcal{S}$ at P.

Exercise 6.15* You are the painter René Magritte pondering the mysteries of the horizon, standing on the xy-plane, with your eyeball positioned at the point $(10, 0, \sqrt{101})$. You are looking at your canvas, which is the yz-plane. You wish to faithfully depict two infinitely thin roads on your canvas, one to the left and one to the right of your vision, represented by the lines

$$\ell_{\text{left}}: \quad x = t\,, \;\; y = -10\,, \;\; z = 0\,, \qquad (t \le 0)$$

and

$$\ell_{\text{right}}: \quad x = t\,, \;\; y = 10\,, \;\; z = 0\,, \qquad (t \le 0)$$

as t heads off towards negative infinity.

For a fixed value of t, let $\mathcal{L}_{\text{left}}(t)$ be the line joining your eyeball to the point $(t, -10, 0)$ on ℓ_{left} and $\mathcal{L}_{\text{right}}(t)$ be the line joining your eyeball to $(t, 10, 0)$ on ℓ_{right}. Find the point of intersection of $\mathcal{L}_{\text{left}}(t)$ and $\mathcal{L}_{\text{right}}(t)$ with the canvas.

Show that the intersection points form two lines on your canvas which are not parallel, yet do not intersect! What happens to the z-values as $t \to -\infty$? (This is not a proof that $\sqrt{101}$ exists; that is a different story!)

7 Systems of Linear Equations

- **Solving a system of equations** (Section 7.4)
  Time: 6.13
- **An inconsistent system** (Section 7.4)
  Time: 3.07
- **A system with a parametric solution** (Section 7.4)
  Time: 4.48

7 Systems of Linear Equations

Recall from the Introduction that

$$\boxed{ax + by \;=\; c}$$

is the general equation of a line in the xy-plane where a, b, c are constants and x, y are variables, and we call the left-hand side

$$\boxed{ax + by}$$

a *linear combination* of x and y. This terminology characteristically refers to a cascade of multiplications followed by a single addition, and there is no need to stop at two variables. A *linear equation* in n variables

$$x_1\,,\; x_2\,,\; \ldots\,,\; x_n$$

has the form

$$\boxed{a_1x_1 + a_2x_2 + a_3x_3 + \cdots + a_nx_n \;=\; b}$$

where now a_1, a_2, $\ldots$, a_n and b are constants. A *solution* to this equation is a sequence of real numbers

$$(s_1\,,\; s_2\,,\; \ldots\,,\; s_n)$$

(usually tidily enclosed in round brackets) such that if we substitute these for the variables, in the correct order, that is, put

$$x_1 = s_1\,,\quad x_2 = s_2\,,\quad x_3 = s_3\,,\quad \ldots\,,\quad x_n = s_n$$

and evaluate, then we get the following true statement in the real number system $\mathbb{R}$:

$$\boxed{a_1s_1 + a_2s_2 + a_3s_3 + \cdots + a_ns_n \;=\; b}$$

Deciding whether this statement is true is a matter of simple arithmetic.

Example: Consider the linear equation

$$2x - 3y + z \;=\; 6$$

of three variables, describing a plane in space. Here are some solutions, but there are many more:

$$x = 0\,, \quad y = 0\,, \quad z = 6$$

$$x = 3\,, \quad y = 0\,, \quad z = 0$$

$$x = 1\,, \quad y = 1\,, \quad z = 7$$

However,

$$x = 1\,, \quad y = 1\,, \quad z = 6$$

is not a solution, because when we do the sums we get an inequality:

$$2(1) - 3(1) + 6 \;= 5 \;\neq\; 6$$

In fact, there are infinitely many solutions (because there are infinitely many points on a plane!) which can be described, say, by rearranging the equation to express x in terms of y and z:

$$x \;=\; \frac{1}{2}(6 + 3y - z)$$

If we think of y and z as *parameters*, even give them typical "parametric" names

$$y = s \quad \text{and} \quad z = t\,,$$

then we get a general solution which can be expressed compactly as a triple

$$\left(\frac{6 + 3s - t}{2}\,,\; s\,,\; t \right)$$

where s and t are allowed to vary freely over $\mathbb{R}$. Since there are two parameters the solution is called *two-dimensional* (matching our geometric notion of a plane being two-dimensional). In this example, we could equally have expressed y in terms of x and z, or z in terms of x and y, obtaining parametric solutions with a different appearance.

Example: Consider the linear equation

$$x - y + 2z - 3w \ = \ 5$$

of four variables (the equation of a *hyperplane* in 4-space). Here are two solutions, but again there are many more:

$$x = 11\,, \quad y = 1\,, \quad z = 2\,, \quad w = 3$$

$$x = -2\,, \quad y = 2\,, \quad z = 6\,, \quad w = 1$$

We can express x in terms of y, z and w:

$$x \ = \ 5 + y - 2z + 3w$$

and think of y, z, w as taking parametric values:

$$y = r\,, \qquad z = s\,, \qquad w = t$$

Then we get a typical solution expressed as a quadruple

$$\big(5 + r - 2s + 3t\,,\ r\,,\ s\,,\ t\big)$$

where r, s and t are allowed to vary freely over $\mathbb{R}$. Since there are three parameters the set of solutions is called *three-dimensional.*

7.1 Consistent and inconsistent systems

A *system* of linear equations is a collection of equations in the same variables. The variables are usually lined up vertically for the purposes of creating and manipulating matrices of coefficients (explained later). Beware: a common source of errors is to misalign variables due to the presence of zero coefficients. A *solution* to the system is a solution simultaneously to each of the equations in the system.

A system which has at least one solution is called *consistent.*

A system which has no solutions is called *inconsistent.*

Example: Consider the following system of three equations in four variables:

$$\begin{array}{rcrcrcrcr} x_1 & + & x_2 & - & x_3 & + & x_4 & = & 5 \\ 2x_1 & - & x_2 & + & 3x_3 & + & x_4 & = & -2 \\ -x_1 & + & 4x_2 & & & - & x_4 & = & 6 \end{array}$$

Here is a solution:

$$x_1 = 1\,, \quad x_2 = 2\,, \quad x_3 = -1\,, \quad x_4 = 1$$

and here is another:

$$x_1 = 19\,, \quad x_2 = 1\,, \quad x_3 = -6\,, \quad x_4 = -21$$

It is easy to verify that indeed these are solutions (just by substituting into the equations and performing arithmetic), but it is not obvious how to find them in the first place. Already something interesting is going on. The fact that two distinct solutions have been exhibited guarantees that there must be infinitely many solutions, somewhere in the mathematical universe. This is a consequence of a theorem that will be stated and carefully proved later as an elegant application of matrix arithmetic.

In general, a system of m linear equations in n variables looks as follows:

$$\begin{array}{ccccccccccc} a_{11}x_1 & + & a_{12}x_2 & + & a_{13}x_3 & + & \cdots & + & a_{1n}x_n & = & b_1 \\ a_{21}x_1 & + & a_{22}x_2 & + & a_{23}x_3 & + & \cdots & + & a_{2n}x_n & = & b_2 \\ \vdots & & \vdots & & \vdots & & & & \vdots & & \vdots \\ a_{m1}x_1 & + & a_{m2}x_2 & + & a_{m3}x_3 & + & \cdots & + & a_{mn}x_n & = & b_m \end{array}$$

It is conventional to denote the coefficient of x_j in the ith equation by a_{ij}. This is quite easy to get used to: when we form a *matrix of coefficients* shortly, the first subscript i will refer to the row and the second subscript j to the column containing a_{ij}.

Example: Not all systems have solutions. The following system is inconsistent:

$$\begin{array}{rcrcl} x & + & y & = & 0 \\ -2x & - & 2y & = & 1 \end{array}$$

Clearly, if the first equation holds for some real numbers x and y, then

$$-2x - 2y \;=\; -2(x+y) \;=\; -2(0) \;=\; 0 \neq 1$$

so that the second equation fails. It is impossible therefore for any substitution of values to satisfy both equations simultaneously.

Consider a general system of two equations in two variables:

$$\begin{array}{rcrcl} a_{11}\,x & + & a_{12}\,y & = & b_1 \\ a_{21}\,x & + & a_{22}\,y & = & b_2 \end{array}$$

Their graphs are lines in the xy-plane, say $\mathcal{L}_1$ and $\mathcal{L}_2$ respectively. A solution to the system corresponds to a point of intersection, and there are three possible outcomes:

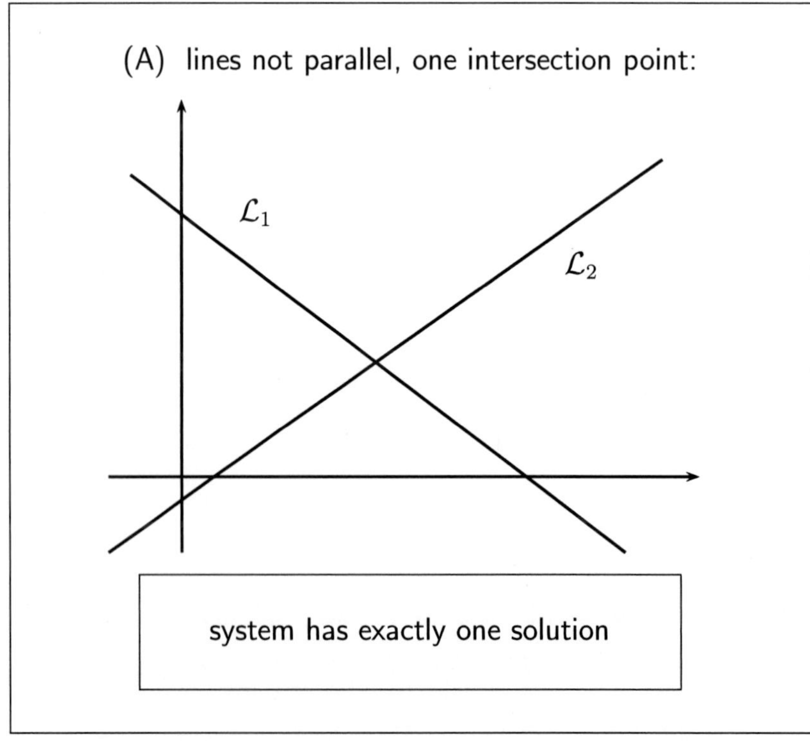

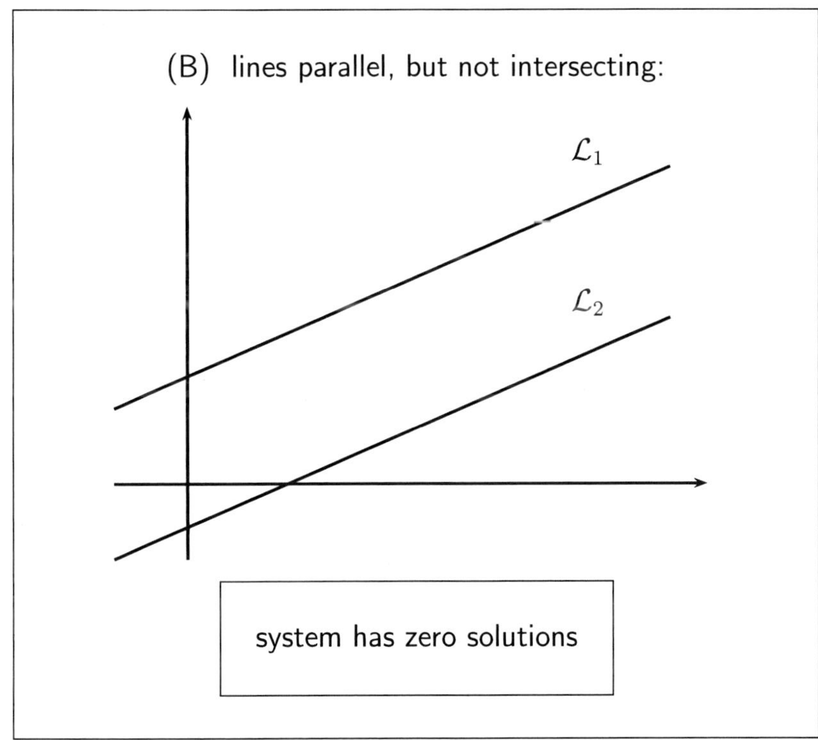
(B) lines parallel, but not intersecting:
$\mathcal{L}_1$
$\mathcal{L}_2$
system has zero solutions

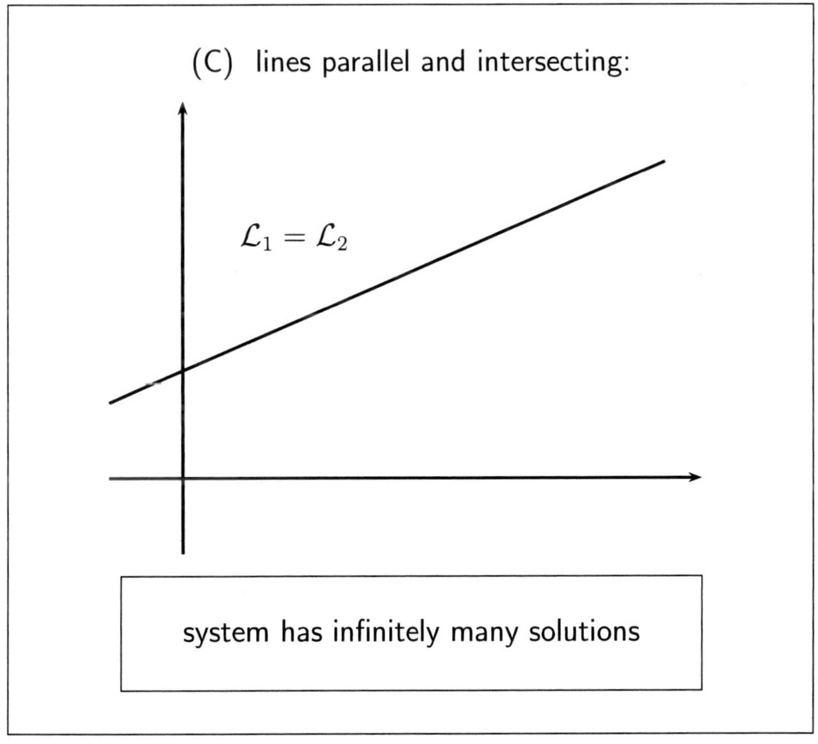
(C) lines parallel and intersecting:
$\mathcal{L}_1 = \mathcal{L}_2$
system has infinitely many solutions

Perhaps surprisingly at first, the same qualitative conclusion holds for all systems regardless of the number of equations or variables:

Theorem: Every system of linear equations has either

(A) no solutions,

(B) exactly one solution, or

(C) infinitely many solutions.

This theorem is saying that if a system of linear equations is consistent, then either the solution is unique, or else the "space" of solutions blows up to infinite size. There is nothing in between. This can be seen as a by-product of the *row echelon form* of the *augmented matrix* of a linear system, which is explained later in this chapter. In the next chapter, we also give an elegant proof using matrix arithmetic.

7.2 Parametric solutions

How might we express infinitely many solutions (case (C) of the previous theorem)? The answer is to use *parameters*, which vary arbitrarily over the infinite set $\mathbb{R}$ of real numbers. We have already seen examples of this with just one equation, but now give some examples of systems with two equations, gradually increasing in complexity.

Example: Consider the following system with three variables:

$$\begin{array}{ccccccc} x & & & + & z & = & 3 \\ & & y & - & 2z & = & 1 \end{array}$$

This has infinitely many solutions, because, for any value of z, the first equation immediately rearranges to give a solution for x:

$$x = 3 - z$$

and the second a solution for y:

$$y = 1 + 2z$$

Putting $z = t$ (the customary symbol for a parameter), the solution can be expressed parametrically as the following:

$$x = 3 - t\,, \quad y = 1 + 2t\,, \quad z = t \qquad (t \in \mathbb{R})$$

Indeed, this is an instance of a set of parametric scalar equations for a line in space (Section 5.1).

Example: Now consider the following system:

$$\begin{array}{ccccccc} x & + & y & + & z & = & 2 \\ & & y & - & 2z & = & 4 \end{array}$$

Again put $z = t$. Rearrange the second equation:

$$y \;=\; 4 + 2t$$

Now we go back to the first equation, and feed in also the solution for y:

$$x \;=\; 2 - y - z \;= 2 - (4 + 2t) - t \;=\; -2 - 3t$$

Finally we have the complete parametric solution:

$$x = -2 - 3t\,, \quad y = 4 + 2t\,, \quad z = t \qquad (t \in \mathbb{R})$$

Notice in this example and the last that the parameter was assigned to the "hidden" variable, which you do not see immediately when you scan the equations from left to right. The first variables you see are x and y, and they are called *leading*. The *nonleading* variable is z, and that gets assigned the role of parameter. Notice also in this example that we were careful to solve the second equation before "going back" to the first equation. This is an instance of a technique called *back substitution*.

Example: The following system will have a "doubly-infinite" solution set:

$$\begin{array}{ccccccccc} x_1 & + & x_2 & + & x_3 & - & x_4 & = & 3 \\ & & x_2 & + & 3x_3 & - & 2x_4 & = & -2 \end{array}$$

The leading variables (the first you see as you scan the equations from left to right) are x_1 and x_2. The nonleading variables x_3 and x_4 are assigned parametric values:

$$x_3 \;=\; s\,, \qquad x_4 \;=\; t$$

We now *back substitute* by using the solution to the second equation:

$$x_2 \;=\; -2 - 3x_3 + 2x_4 \;=\; -2 - 3s + 2t$$

to solve the first:

$$x_1 \;=\; 3 - x_2 - x_3 + x_4 \;=\; 3 - (-2 - 3s + 2t) - s + t \;=\; 5 + 2s - t$$

The entire solution can then be expressed parametrically:

$$\left.\begin{aligned} x_1 &= 5 + 2s - t \\ x_2 &= -2 - 3s + 2t \\ x_3 &= s \\ x_4 &= t \end{aligned}\right\} \qquad s, t \in \mathbb{R}$$

The underlying ideas of these examples, namely assigning parameters to the nonleading variables and using back substitution, generalise to systems of any size in any number of variables. However, the equations have to be in the correct form, namely, such that

> leading variables appear successively to the right as one moves down the list of equations.

In the next sections we explain how this can be achieved for any system.

7.3 Augmented matrix of a system

The idea of this section is to streamline the information encoded in a linear system using an array of coefficients and constants called an *augmented matrix*. We illustrate the idea with actual numbers, and then describe what happens in general.

Example: Consider the following system:

$$\begin{array}{rcrcrcr} 3x & + & 4y & - & 2z & = & 3 \\ x & - & y & + & 5z & = & -2 \\ -x & & & + & z & = & 7 \end{array}$$

All of the important information is contained in the following array of coefficients of the variables:

$$\begin{bmatrix} 3 & 4 & -2 \\ 1 & -1 & 5 \\ -1 & 0 & 1 \end{bmatrix}$$

and in the following column of constants from the right-hand sides of the equations:

$$\begin{bmatrix} 3 \\ -2 \\ 7 \end{bmatrix}$$

An array of numbers is called a *matrix* (plural: *matrices*).

The two matrices in this example are in fact linked by a *matrix equation*, which also incorporates the variables and which will make sense in the next chapter on matrix arithmetic:

$$\begin{bmatrix} 3 & 4 & -2 \\ 1 & -1 & 5 \\ -1 & 0 & 1 \end{bmatrix} \begin{bmatrix} x \\ y \\ z \end{bmatrix} = \begin{bmatrix} 3 \\ -2 \\ 7 \end{bmatrix}$$

The totality of this information can be summarised by a single *augmented matrix*, which is just a matrix in two halves divided by a vertical line:

$$\left[\begin{array}{rrr|r} 3 & 4 & -2 & 3 \\ 1 & -1 & 5 & -2 \\ -1 & 0 & 1 & 7 \end{array}\right]$$

We say that the augmented matrix is "the matrix obtained by augmenting the matrix of coefficients by the column of constants". In the next section we will see how to make simple changes to the rows, using *elementary row operations*, which lead us to a solution to the original system of equations.

The reader can probably guess what happens in general. For the general system

$$\begin{array}{ccccccccccc} a_{11}\,x_1 & + & a_{12}\,x_2 & + & a_{13}\,x_3 & + & \cdots & + & a_{1n}\,x_n & = & b_1 \\ a_{21}\,x_1 & + & a_{22}\,x_2 & + & a_{23}\,x_3 & + & \cdots & + & a_{2n}\,x_n & = & b_2 \\ \vdots & & \vdots & & \vdots & & & & \vdots & & \vdots \\ a_{m1}\,x_1 & + & a_{m2}\,x_2 & + & a_{m3}\,x_3 & + & \cdots & + & a_{mn}\,x_n & = & b_m \end{array}$$

we use the following augmented matrix:

$$\left[\begin{array}{ccccc|c} a_{11} & a_{12} & a_{13} & \cdots & a_{1n} & b_1 \\ a_{21} & a_{22} & a_{23} & \cdots & a_{2n} & b_2 \\ \vdots & \vdots & \vdots & \ddots & \vdots & \vdots \\ a_{m1} & a_{m2} & a_{m3} & \cdots & a_{mn} & b_m \end{array}\right]$$

Warning: Carefully line up the variables before writing down the augmented matrix!

Example: Even though the variables appear in correct order in each equation in the following system, the layout is sloppy from the point of view of matrix analysis:

$$\begin{aligned} x_1 - x_2 + x_3 - x_4 + x_5 &= -1 \\ 3x_2 - x_3 + x_5 &= 4 \\ 2x_1 + 4x_4 &= 0 \end{aligned}$$

and the system should be rewritten:

$$\begin{array}{ccccccccccc} x_1 & - & x_2 & + & x_3 & - & x_4 & + & x_5 & = & -1 \\ & & 3x_2 & - & x_3 & & & + & x_5 & = & 4 \\ 2x_1 & & & & & + & 4x_4 & & & = & 0 \end{array}$$

yielding the following augmented matrix:

$$\left[\begin{array}{ccccc|c} 1 & -1 & 1 & -1 & 1 & -1 \\ 0 & 3 & -1 & 0 & 1 & 4 \\ 2 & 0 & 0 & 4 & 0 & 0 \end{array}\right]$$

7.4 Gaussian elimination

In this section we describe a method attributed to the prolific German mathematician and scientist Karl Friedrich Gauss (1777–1855). His name appears often in attributions of theorems and techniques and he is widely regarded as one of the most influential mathematicians of all time.

The method solves a system of linear equations by progressively replacing it by simpler systems which have exactly the same solutions as the original. We use three kinds of operations:

Elementary operations on systems of equations:

(1) interchanging two equations

(2) multiplying an equation through by a nonzero constant

(3) adding a multiple of one equation to another

It is clear that interchanging two equations has no effect on the solutions of the system. Multiplying an equation through by a nonzero constant can be reversed by multiplying through again by the reciprocal. Adding a multiple of one equation to another is also clearly reversible by subtracting off from the new equation the same multiple of the old equation that was previously added. Whenever operations on a system are reversible, the solution set remains the same, so this verifies the following:

Simple Fact: Each of (1), (2) and (3) does not alter the solutions of the system.

The advantage of these operations is that, by using them in an organised way, the system can be "straightened out" so that solutions can be read off easily. The method becomes streamlined by instead manipulating the augmented matrix of the system. The three operations above correspond exactly to the following, which are given an obvious symbolic representation where R_i denotes the ith row of the matrix:

Elementary row operations on matrices:

(1) interchanging two rows

$$R_i \leftrightarrow R_j$$

(2) multiplying a row through by a nonzero constant

$$R_i \to \lambda R_j \qquad (\lambda \neq 0)$$

(3) adding a multiple of one row to another

$$R_i \to R_i + \lambda R_j \qquad (i \neq j)$$

Each row of the augmented matrix corresponds exactly to one equation of the system, and at any time in the simplification process we can go back and forth between matrices and systems as whim dictates. Typically, we transform an augmented matrix into a particularly simple form before translating back into equations. With some practice, particularly in the case of *reduced row echelon form* (Section 7.5), the reader will develop a facility for writing down solutions even without translating into equations. It is prudent to write everything out carefully when learning the method for the first time.

Before formalising the method, we give a numerical illustration. The elementary row operations that are used at each step are indicated on the right-hand side of each matrix. The tilde symbol $\sim$ placed between two matrices means that they can be transformed into each other using some sequence of elementary row operations, in which case we say the matrices are *row equivalent*. An application of elementary row operations is called *row reduction*.

Care needs to be taken regarding order of operations. **The composition of elementary operations is not commutative!** However, in cases where order does not matter (and even where order does matter, but the order is clear from context), it is common to list applications of several elementary row operations in one step. The second step in the example below is the compression of two elementary row operations, where order does not matter.

In reading the following example, the reader should try to anticipate the way the matrix simplifies. The aim is to produce zeros in obvious positions in the matrix which later enable back substitution (as illustrated in Section 7.2).

Example: Solve the following system:

$$\begin{array}{rcrcrcr} 2x & + & 4y & & & = & 6 \\ -x & & & - & 2z & = & -5 \\ -3x & - & 5y & + & z & = & -4 \end{array}$$

Solution: Form the augmented matrix and apply the following operations:

$$\left[\begin{array}{rrr|r} 2 & 4 & 0 & 6 \\ -1 & 0 & -2 & -5 \\ -3 & -5 & 1 & -4 \end{array}\right] \sim \left[\begin{array}{rrr|r} 1 & 2 & 0 & 3 \\ -1 & 0 & -2 & -5 \\ -3 & -5 & 1 & -4 \end{array}\right] \begin{array}{l} R_1/2 \\ \\ \\ \end{array}$$

$$\sim \left[\begin{array}{ccc|c} 1 & 2 & 0 & 3 \\ 0 & 2 & -2 & -2 \\ 0 & 1 & 1 & 5 \end{array}\right] \begin{array}{l} \\ R_2 + R_1 \\ R_3 + 3R_1 \end{array}$$

$$\sim \left[\begin{array}{ccc|c} 1 & 2 & 0 & 3 \\ 0 & 1 & -1 & -1 \\ 0 & 1 & 1 & 5 \end{array}\right] \begin{array}{l} \\ R_2/2 \\ \\ \end{array}$$

$$\sim \left[\begin{array}{ccc|c} 1 & 2 & 0 & 3 \\ 0 & 1 & -1 & -1 \\ 0 & 0 & 2 & 6 \end{array}\right] \begin{array}{l} \\ \\ R_3 - R_2 \end{array}$$

$$\sim \left[\begin{array}{ccc|c} 1 & 2 & 0 & 3 \\ 0 & 1 & -1 & -1 \\ 0 & 0 & 1 & 3 \end{array}\right] \begin{array}{l} \\ \\ R_3/2 \end{array}$$

which corresponds to the system:

$$\begin{array}{ccccccc} x & + & 2y & & & = & 3 \\ & & y & - & z & = & -1 \\ & & & & z & = & 3 \end{array}$$

which can be solved by back substitution:

$$\begin{aligned} z &= 3 \\ y &= -1 + z = -1 + 3 = 2 \\ x &= 3 - 2y = 3 - 4 = -1 \end{aligned}$$

giving a final solution:

$$\boxed{x = -1, \quad y = 2, \quad z = 3}$$

The method is robust, because even if there are arithmetic slips, one is likely to find them almost immediately, by checking that the solutions really do work in the original equations:

$$\begin{array}{cccccccc} 2(-1) & + & 4(2) & & & = & 6 & \checkmark \\ -(-1) & & & - & 2(3) & = & -5 & \checkmark \\ -3(-1) & - & 5(2) & + & 3 & = & -4 & \checkmark \end{array}$$

In general we use the following:

> **Strategy:** Systematically alter the augmented matrix into *row echelon form*, which means that:
>
> (a) rows of zeros appear at the bottom
>
> (b) first nonzero entries of consecutive rows appear further to the right
>
> (c) first nonzero entries of rows are equal to 1.

At any stage in the process, the first nonzero entry of a row is called a *pivot* or *leading entry*. Pivots are used to "knock out" other entries in the matrix (in the sense of replacing them with zero).

Here are a couple of examples of matrices in row echelon form. The terminology applies whether or not the matrix is augmented, so in these examples the vertical dividing line has been removed. (Indeed, some authors dispense with the dividing line altogether and expect the reader to interpret an augmentated matrix from context.)

$$\begin{bmatrix} 1 & 3 & -5 & 10 \\ 0 & 1 & 2 & 0 \\ 0 & 0 & 0 & 1 \end{bmatrix} \qquad \begin{bmatrix} 1 & 0 & 2 & 0 & 3 & 4 & 5 \\ 0 & 0 & 1 & 3 & 0 & 2 & -4 \\ 0 & 0 & 0 & 1 & -1 & 2 & 0 \\ 0 & 0 & 0 & 0 & 0 & 1 & 1 \\ 0 & 0 & 0 & 0 & 0 & 0 & 0 \end{bmatrix}$$

The process of using elementary row operations (often abbreviated to e.r.o.'s) to put a matrix into row echelon form is called

> *Gaussian elimination*

and it is one of the most widely used procedures in mathematics.

A system having an augmented matrix in row echelon form can be solved by *back substitution*, which means:

(a) the *leading variables* corresponding to leading entries are evaluated one equation at a time from the bottom towards the top

(b) parameters are assigned to each nonleading variable (if any).

Example: Solve the following system:

$$\begin{array}{rcrcrcrcr} 2x_1 & + & 4x_2 & + & 6x_3 & - & 2x_4 & = & 0 \\ x_1 & + & 2x_2 & + & x_3 & + & 2x_4 & = & 1 \\ & & 2x_2 & + & 4x_3 & + & 2x_4 & = & 2 \\ -2x_1 & + & x_2 & & & + & 10x_4 & = & 10 \end{array}$$

Solution: Form the augmented matrix and apply the following operations, which systematically aim for row echelon form:

$$\left[\begin{array}{rrrr|r} 2 & 4 & 6 & -2 & 0 \\ 1 & 2 & 1 & 2 & 1 \\ 0 & 2 & 4 & 2 & 2 \\ -2 & 1 & 0 & 10 & 10 \end{array}\right] \sim \left[\begin{array}{rrrr|r} 1 & 2 & 3 & -1 & 0 \\ 1 & 2 & 1 & 2 & 1 \\ 0 & 1 & 2 & 1 & 1 \\ 0 & 5 & 6 & 8 & 10 \end{array}\right] \begin{array}{l} R_1/2 \\ \\ R_3/2 \\ R_4 + R_1 \end{array}$$

$$\sim \left[\begin{array}{rrrr|r} 1 & 2 & 3 & -1 & 0 \\ 0 & 0 & -2 & 3 & 1 \\ 0 & 1 & 2 & 1 & 1 \\ 0 & 0 & -4 & 3 & 5 \end{array}\right] \begin{array}{l} \\ R_2 - R_1 \\ \\ R_4 - 5R_3 \end{array}$$

$$\sim \left[\begin{array}{rrrr|r} 1 & 2 & 3 & -1 & 0 \\ 0 & 1 & 2 & 1 & 1 \\ 0 & 0 & -2 & 3 & 1 \\ 0 & 0 & -4 & 3 & 5 \end{array}\right] \begin{array}{l} \\ R_2 \leftrightarrow R_3 \\ \\ \\ \end{array}$$

$$\sim \left[\begin{array}{rrrr|r} 1 & 2 & 3 & -1 & 0 \\ 0 & 1 & 2 & 1 & 1 \\ 0 & 0 & 1 & -\frac{3}{2} & -\frac{1}{2} \\ 0 & 0 & 0 & -3 & 3 \end{array}\right] \begin{array}{l} \\ \\ -R_3/2 \\ R_4 - 2R_3 \end{array}$$

$$\sim \left[\begin{array}{rrrr|r} 1 & 2 & 3 & -1 & 0 \\ 0 & 1 & 2 & 1 & 1 \\ 0 & 0 & 1 & -\frac{3}{2} & -\frac{1}{2} \\ 0 & 0 & 0 & 1 & -1 \end{array}\right] \begin{array}{l} \\ \\ \\ -R_4/3 \end{array}$$

which corresponds to the system:

$$\begin{array}{rcrcrcrcr} x_1 & + & 2x_2 & + & 3x_3 & - & x_4 & = & 0 \\ & & x_2 & + & 2x_3 & + & x_4 & = & 1 \\ & & & & x_3 & - & \frac{3}{2}x_4 & = & -\frac{1}{2} \\ & & & & & & x_4 & = & -1 \end{array}$$

which can be solved by back substitution:

$$\begin{aligned} x_4 &= -1 \\ x_3 &= -\tfrac{1}{2} + \tfrac{3}{2}x_4 \;=\; -\tfrac{1}{2} - \tfrac{3}{2} \;=\; -2 \\ x_2 &= 1 - 2x_3 - x_4 \;=\; 1 + 4 + 1 \;=\; 6 \\ x_1 &= -2x_2 - 3x_2 + x_4 \;=\; -12 + 6 - 1 \;=\; -7 \end{aligned}$$

giving, finally, the unique solution:

$$\boxed{x_1 = -7, \quad x_2 = 6, \quad x_3 = -2, \quad x_4 = -1}$$

Example: Solve the following system:

$$\begin{array}{rcrcrcrcrcr} x_1 & + & 2x_2 & & & + & 2x_4 & - & x_5 & = & 4 \\ 2x_1 & + & 4x_2 & + & 2x_3 & + & 4x_4 & - & 6x_5 & = & 6 \\ -x_1 & - & 2x_2 & + & 3x_3 & - & 2x_4 & + & 6x_5 & = & 4 \end{array}$$

Solution: Form the augmented matrix and row reduce:

$$\left[\begin{array}{rrrrr|r} 1 & 2 & 0 & 2 & -1 & 4 \\ 2 & 4 & 2 & 4 & -6 & 6 \\ -1 & -2 & 3 & -2 & 6 & 4 \end{array}\right] \sim \left[\begin{array}{rrrrr|r} 1 & 2 & 0 & 2 & -1 & 4 \\ 0 & 0 & 2 & 0 & -4 & -2 \\ 0 & 0 & 3 & 0 & 5 & 8 \end{array}\right] \begin{array}{l} \\ R_2 - 2R_1 \\ R_3 + R_1 \end{array}$$

$$\sim \left[\begin{array}{rrrrr|r} 1 & 2 & 0 & 2 & -1 & 4 \\ 0 & 0 & 1 & 0 & -2 & -1 \\ 0 & 0 & 3 & 0 & 5 & 8 \end{array}\right] \quad R_2/2$$

$$\sim \left[\begin{array}{rrrrr|r} 1 & 2 & 0 & 2 & -1 & 4 \\ 0 & 0 & 1 & 0 & -2 & -1 \\ 0 & 0 & 0 & 0 & 11 & 11 \end{array}\right] \begin{array}{l} \\ \\ R_3 - 3R_2 \end{array}$$

which corresponds to the system:

$$\begin{array}{rcrcrcrcrcr} x_1 & + & 2x_2 & & & + & 2x_4 & - & x_5 & = & 4 \\ & & & & x_3 & & & - & 2x_5 & = & -1 \\ & & & & & & & & 11x_5 & = & 11 \end{array}$$

which can be solved by back substitution, assigning parameters to the non-leading variables, which in this case are x_2 and x_4 :

$$\begin{aligned} x_5 &= 1 \\ x_4 &= t \\ x_3 &= -1 + 2x_5 \;=\; -1 + 2 \;=\; 1 \\ x_2 &= s \\ x_1 &= 4 - 2x_2 - 2x_4 + x_5 \;=\; 5 - 2s - 2t \end{aligned}$$

giving, finally, the two-dimensional parametric solution:

$$\boxed{x_1 = 5 - 2s - 2t\,,\quad x_2 = s\,,\quad x_3 = 1\,,\quad x_4 = t\,,\quad x_5 = 1 \quad (s, t \in \mathbb{R})}$$

Notice that we did not go completely to row echelon form, before converting to equations, since the leading entry of the last row of the augmented matrix is 11 , not 1 . But clearly the arithmetic is not disadvantaged by doing this. In practice, one takes any obvious shortcuts to streamline the process.

Example: Consider the following system:

$$\begin{array}{ccccccc} x & + & y & + & 2z & = & 1 \\ x & & & + & z & = & 2 \\ & & y & + & z & = & 0 \end{array}$$

Solution: As usual, form the augmented matrix and row reduce:

$$\left[\begin{array}{ccc|c} 1 & 1 & 2 & 1 \\ 1 & 0 & 1 & 2 \\ 0 & 1 & 1 & 0 \end{array}\right] \sim \left[\begin{array}{ccc|c} 1 & 1 & 2 & 1 \\ 0 & -1 & -1 & 1 \\ 0 & 1 & 1 & 0 \end{array}\right] \begin{array}{l} \\ R_2 - R_1 \\ \\ \end{array}$$

$$\sim \left[\begin{array}{ccc|c} 1 & 1 & 2 & 1 \\ 0 & 1 & 1 & -1 \\ 0 & 0 & 0 & 1 \end{array}\right] \begin{array}{l} \\ -R_2 \\ R_3 + R_2 \end{array}$$

This corresponds to an inconsistent system, because the last row corresponds to the following equation:

$$0 \;=\; 0x + 0y + 0z \;=\; 1$$

This, of course, can never be satisfied by any substitution of x, y, z . Thus, the system has no solutions.

This last example is an instance of the following general fact:

> A system of linear equations is inconsistent if and only if at some stage in the process of row reducing the augmented matrix to row echelon form a row of the form
>
> $$\begin{array}{cccc|c} 0 & 0 & \cdots & 0 & k \end{array}$$
>
> is produced for some nonzero real number k.

Clearly, as in the example, if such a row exists then the equation it corresponds to is saying that 0 and k are equal, which is impossible since k is nonzero. Thus, the original system would be equivalent to a system with no solutions, so itself would have no solutions.

But conversely, if a row such as the one in the box above never arises, then from the equations corresponding to row echelon form, solutions are guaranteed to exist by the process of back substitution, so the original system must be consistent.

Now, in the consistent case, in the process of back substitution, if there are no nonleading variables, then the system has a unique solution. If there are nonleading variables, then they are assigned parameters, and so the system has infinitely many solutions, since the parameters vary over $\mathbb{R}$, which is infinite. Indeed, the number of parameters is called the *dimension* of the solution. The case of a unique solution then is zero-dimensional (which accords with common parlance that a single point has dimension zero).

All of this incidentally proves the theorem at the end of Section 7.1, that a system of linear equations always has no solutions, a unique solution, or infinitely many solutions.

7.5 Reduced row echelon form

A row echelon matrix may be simplified further by using the pivots (leading entries) to remove all nonzero entries above them (as well as below). This leads to the *reduced row echelon form.*

A matrix is in *reduced row echelon form* if

(a) rows of zeros appear at the bottom

(b) first nonzero entries of consecutive rows appear further to the right

(c) first nonzero entries of rows are equal to 1

(d) entries above (and below) leading entries are zero.

Here are some examples of (non-augmented) matrices in reduced row echelon form.

$$\begin{bmatrix} 1 & 0 & 0 \\ 0 & 1 & 0 \\ 0 & 0 & 1 \end{bmatrix} \quad \begin{bmatrix} 1 & -1 & 0 & 0 \\ 0 & 0 & 1 & 2 \\ 0 & 0 & 0 & 0 \end{bmatrix} \quad \begin{bmatrix} 1 & 0 & 2 & 0 & 3 & -4 & 0 \\ 0 & 1 & 3 & 0 & 4 & 2 & 0 \\ 0 & 0 & 0 & 1 & 5 & 6 & 0 \\ 0 & 0 & 0 & 0 & 0 & 0 & 1 \end{bmatrix}$$

One advantage of reduced row echelon form is that the corresponding system is as simple as possible and back substitution arrives at the solution immediately when the solution is unique (as in the example below), and almost immediately when parameters are involved. However, for large systems, it is considerably more efficient to apply the method of the previous section, that is, to stop at row echelon form and then apply back substitution. The extra manipulation at the final stage is more than compensated for by fewer elementary row operations on the matrix (Exercise 7.19). Nevertheless, reduced row echelon forms are important (Chapter 9) for inverting matrices and also for theoretical reasons.

Example: Use the reduced row echelon form of the augmented matrix to solve the following system:

$$\begin{array}{rcrcrcr} 2x & + & 2y & - & 4z & = & 0 \\ 2x & + & 2y & - & z & = & 1 \\ 3x & + & 2y & - & 3z & = & 3 \end{array}$$

Solution: Writing down the augmented matrix and row reducing produces

$$\left[\begin{array}{ccc|c} 2 & 2 & -4 & 0 \\ 2 & 2 & -1 & 1 \\ 3 & 2 & -3 & 3 \end{array}\right] \sim \left[\begin{array}{ccc|c} 1 & 1 & -2 & 0 \\ 0 & 0 & 3 & 1 \\ 3 & 2 & -3 & 3 \end{array}\right] \begin{array}{l} R_1/2 \\ R_2 - R_1 \\ \end{array}$$

$$\sim \left[\begin{array}{ccc|c} 1 & 1 & -2 & 0 \\ 0 & 0 & 1 & \frac{1}{3} \\ 0 & -1 & 3 & 3 \end{array}\right] \begin{array}{l} \\ R_2/3 \\ R_3 - 3R_1 \end{array} \sim \left[\begin{array}{ccc|c} 1 & 1 & -2 & 0 \\ 0 & -1 & 3 & 3 \\ 0 & 0 & 1 & \frac{1}{3} \end{array}\right] R_2 \leftrightarrow R_3$$

$$\sim \left[\begin{array}{ccc|c} 1 & 0 & 1 & 3 \\ 0 & 1 & -3 & -3 \\ 0 & 0 & 1 & \frac{1}{3} \end{array}\right] \begin{array}{l} R_1 + R_2 \\ -R_2 \\ \end{array} \sim \left[\begin{array}{ccc|c} 1 & 0 & 0 & \frac{8}{3} \\ 0 & 1 & 0 & -2 \\ 0 & 0 & 1 & \frac{1}{3} \end{array}\right] \begin{array}{l} R_1 - R_3 \\ R_2 + 3R_3 \\ \end{array}$$

corresponding immediately to the unique solution:

$$\boxed{x_1 = \tfrac{8}{3}, \quad x_2 = -2, \quad x_3 = \tfrac{1}{3}}$$

It should be remarked that there are many different ways of getting to the reduced row echelon form using elementary row operations, and the reader will find other pathways to those used in the examples in this chapter. However, the final reduced row echelon form that is reached from a given matrix is always the same, regardless of the pathway. This is a theorem, a special case of which is left as a starred exercise.

> **Theorem:** A given matrix can be row reduced to one and only one matrix in reduced row echelon form.

The process of row reducing all the way to reduced row echelon form is referred to as *Gauss-Jordan elimination*. Gauss we have already mentioned. The other half of the attribution is to Wilhelm Jordan (1842–1899), a German engineer. A source of confusion is that there is another Jordan mentioned frequently in linear algebra, namely the French mathematician Camille Jordan (1838–1922). He is famous for the Jordan Normal Form (mentioned in Chapter 12) and the Jordan Curve Theorem, not directly related to linear algebra, but notorious for being the first extremely difficult theorem in mathematics, which almost everyone considered obvious!

Chapter 7: Important Ideas and Useful Facts

7.1 A *linear equation* in variables x_1, x_2, ..., x_n has the form

$$a_1x_1 + a_2x_2 + \cdots + a_nx_n = b$$

where a_1, a_2, ..., a_n, b are constants.

7.2 A *system* of linear equations has the form

$$\begin{array}{ccccccccc} a_{11}x_1 & + & a_{12}x_2 & + & \cdots & + & a_{1n}x_n & = & b_1 \\ a_{21}x_1 & + & a_{22}x_2 & + & \cdots & + & a_{2n}x_n & = & b_2 \\ \vdots & & \vdots & & & & \vdots & & \vdots \\ a_{m1}x_1 & + & a_{m2}x_2 & + & \cdots & + & a_{mn}x_n & = & b_m \end{array}$$

with the following *augmented matrix*:

$$\left[\begin{array}{cccc|c} a_{11} & a_{12} & \cdots & a_{1n} & b_1 \\ a_{21} & a_{22} & \cdots & a_{2n} & b_2 \\ \vdots & \vdots & \ddots & \vdots & \vdots \\ a_{m1} & a_{m2} & \cdots & a_{mn} & b_m \end{array}\right]$$

7.3 When $b_1 = b_2 = \ldots = b_m = 0$ above, the system is called *homogeneous*, and has at least the *trivial* solution (where the variables all take the value zero).

7.4 Every system of linear equations has either no solutions, one solution, or infinitely many solutions.

7.5 If a system has no solutions then it is called *inconsistent*. If the system has at least one solution then it is called *consistent*.

7.6 There are three types of *elementary row operations* performed on augmented matrices:

(a) interchanging the ith and jth rows (denoted by $R_i \leftrightarrow R_j$)

(b) multiplying the ith row through by a nonzero constant λ (denoted by $R_i \to \lambda R_i$)

(c) adding a multiple of the jth row to the ith row (denoted by $R_i \to R_i + \lambda R_j$).

7.7 A matrix is in *row echelon form* if

(a) rows of zeros appear at the bottom

(b) first nonzero (*leading*) entries of consecutive rows appear further to the right

(c) leading entries of rows are equal to 1

and in *reduced row echelon form* if, in addition,

(d) entries above (and below) leading entries are zero.

7.8 The process of *Gaussian elimination* applies elementary row operations (*row reduction*) to transform the augmented matrix of a system into row echelon form, after which the associated system is solved using *back substitution*:

(a) the *leading variables* corresponding to leading entries are evaluated one equation at a time from the bottom towards the top

(b) parameters are assigned to each nonleading variable (if any).

One may think of the number of parameters as the *dimension* or *degrees of freedom* of the system.

7.9 A system is inconsistent if and only if at some stage in the process of row reduction a row of the form

$$0 \quad 0 \quad \cdots \quad 0 \quad | \quad k$$

is produced for some nonzero real number k.

7.10 The process of *Gauss-Jordan elimination* row reduces the augmented matrix to reduced row echelon form, after which the process of back substitution simplifies. However, Gauss-Jordan elimination is usually less efficient in terms of the overall number of arithmetic operations used than Gaussian elimination.

7.11 A given matrix can be row reduced to one and only one matrix in reduced row echelon form.

Exercise 7.1 Look at the corner of the room. The walls are two planes which meet in a line. Follow the line upwards towards the ceiling. Where it meets the ceiling is the point of intersection of three planes. Find the point (x, y, z) of intersection of the following three planes:

$$\begin{array}{rcrcrcr} 2x & + & 3y & + & 4z & = & -4 \\ 5x & + & 5y & + & 6z & = & -3 \\ 3x & + & y & + & 2z & = & -1 \end{array}$$

Exercise 7.2 Solve the following system of linear equations:

$$\begin{array}{rcrcrcr} x_1 & + & x_2 & + & x_3 & = & 1 \\ x_1 & - & 2x_2 & + & 2x_3 & = & -2 \end{array}$$

Exercise 7.3 Solve the following system:

$$\begin{array}{rcrcrcrcrcr} x_1 & + & x_2 & + & x_3 & + & x_4 & + & x_5 & = & 2 \\ x_1 & - & x_2 & + & x_3 & & & - & x_5 & = & 0 \\ x_1 & + & 5x_2 & + & x_3 & + & 3x_4 & + & 3x_5 & = & 5 \end{array}$$

Exercise 7.4 Solve the following system:

$$\begin{array}{rcrcrcrcrcr} x_1 & - & x_2 & + & x_3 & - & x_4 & + & x_5 & = & -1 \\ 3x_2 & - & x_3 & + & x_5 & & & & & = & 4 \\ 2x_1 & + & 4x_4 & & & & & & & = & 0 \end{array}$$

Exercise 7.5 A system of equations is called *homogeneous* if the constants on the right-hand side are all equal to zero. Solve the following homogeneous system of equations:

$$\begin{array}{rcrcrcr} 2x & - & 2y & + & z & = & 0 \\ x & + & y & + & 2z & = & 0 \\ x & + & 3y & - & 7z & = & 0 \end{array}$$

Exercise 7.6 Solve the following homogeneous system of equations:

$$\begin{array}{rcrcrcrcr} x_1 & + & 3x_2 & + & 2x_3 & - & x_4 & = & 0 \\ x_1 & + & x_2 & + & x_3 & - & 2x_4 & = & 0 \\ x_1 & & & - & x_3 & + & x_4 & = & 0 \\ 4x_1 & + & 4x_2 & + & x_3 & - & x_4 & = & 0 \end{array}$$

Exercise 7.7 Find parametric scalar equations for the line of intersection of the two planes:

$$\begin{array}{rcrcrcr} -3x & + & 2y & + & 7z & = & 1 \\ 5x & - & 3y & - & 2z & = & -2 \end{array}$$

Exercise 7.8 For each of the following augmented matrices, decide whether the system of equations to which it corresponds has (a) no solution, (b) a unique solution, or (c) an infinite solution.

$$\text{(i)} \quad \left[\begin{array}{rrr|r} 1 & 2 & 3 & 3 \\ 1 & 2 & -2 & 0 \\ -1 & -2 & 7 & 3 \end{array}\right] \qquad \text{(ii)} \quad \left[\begin{array}{rrr|r} 1 & 2 & 3 & 3 \\ 1 & 2 & -2 & 0 \\ -1 & -2 & 0 & 1 \end{array}\right]$$

$$\text{(iii)} \quad \left[\begin{array}{rrr|r} 1 & 2 & 3 & 3 \\ 1 & 2 & -2 & 0 \\ -1 & -1 & 7 & 3 \end{array}\right] \qquad \text{(iv)} \quad \left[\begin{array}{rrrrr|r} 1 & 1 & -2 & 3 & 1 & -1 \\ -2 & -2 & 4 & 6 & 2 & 0 \\ 0 & 0 & 0 & -3 & -1 & 4 \end{array}\right]$$

Exercise 7.9 By solving a system of four equations in x, y and z, find the point of intersection of the following two lines in space, expressed in Cartesian form:

$$\frac{x-9}{4} = \frac{y-11}{4} = \frac{z+8}{-5} \quad \text{and} \quad \frac{x-15}{7} = y-5 = \frac{z-8}{3}$$

Exercise 7.10 A cubic polynomial $p(x)$ with derivative $p'(x)$ has the properties that $p(1) = p'(x) = 4$, $p(2) = 14$ and $p'(2) = 17$. Find $p(x)$.

Exercise 7.11 Find A, B, C and D to satisfy the following:

$$\frac{x^3}{(x-1)^4} = \frac{A}{x-1} + \frac{B}{(x-1)^2} + \frac{C}{(x-1)^3} + \frac{D}{(x-1)^4}$$

Exercise 7.12 Find the smallest positive integer values for x, y, z, w such that the chemical reaction

$$x\,C_8H_{18} + y\,O_2 \longrightarrow z\,CO_2 + w\,H_2O$$

describing the combustion of octane is balanced, in the sense that we have the same number of atoms of each type on both sides.

Exercise 7.13* Find the respective values of λ such that the following system is inconsistent, has infinitely many solutions, or has a unique solution:

$$\begin{array}{rcrcrcr} x & & & - & 3z & = & -3 \\ -2x & - & \lambda y & + & z & = & 2 \\ x & + & 2y & + & \lambda z & = & 1 \end{array}$$

Exercise 7.14* Use row reduction on augmented matrices to show that the lines $ax + by = k$ and $cx + dy = \ell$ intersect in a single point if and only if $ad - bc \neq 0$.

Exercise 7.15* Given a system of at least two linear equations, is the operation of simultaneously applying the elementary row operations $R_1 \to R_1 - R_2$ and $R_2 \to R_2 - R_1$ reversible? If not, why not?

Exercise 7.16* Catalogue the possible reduced row echelon forms for coefficient matrices for the following homogeneous system in α, β, γ:

$$\begin{array}{rcrcrcr} u_1\alpha & + & v_1\beta & + & w_1\gamma & = & 0 \\ u_2\alpha & + & v_2\beta & + & w_2\gamma & = & 0 \end{array}$$

Find a simple connection that explains why any three geometric vectors in the plane must be linearly dependent.

Exercise 7.17* Catalogue the possible reduced row echelon forms for coefficient matrices for the following homogeneous system in α, β, γ, δ:

$$\begin{array}{rcrcrcrcr} u_1\alpha & + & v_1\beta & + & w_1\gamma & + & t_1\delta & = & 0 \\ u_2\alpha & + & v_2\beta & + & w_2\gamma & + & t_2\delta & = & 0 \\ u_3\alpha & + & v_3\beta & + & w_3\gamma & + & t_3\delta & = & 0 \end{array}$$

Find a simple connection that explains why any four geometric vectors in space must be linearly dependent.

Exercise 7.18* Let $\mathcal{S}$ denote the collection of systems of m equations in n variables x_1, x_2, ..., x_n where m and n are positive integers. Let $\mathcal{A}$ denote the set of augmented matrices with m rows and $n+1$ columns. Let $f : \mathcal{S} \to \mathcal{A}$ where $f(\sigma)$ is the augmented matrix of a given system σ. Let ρ denote an elementary operation either on a system or on a matrix, read in context. Make sense of the following:

(i) f is one-one and onto (ii) $f(\rho\sigma) = \rho f(\sigma)$ for all ρ and σ

Exercise 7.19* This exercise illustrates the fact that Gaussian elimination is more efficient than Gauss-Jordan elimination. In the following, all leading coefficients are assumed to be nonzero.

(i) Count the number $f(n)$ of arithmetic operations needed to find the values x_1, x_2, …, x_n by back substitution, given that the following equations hold:

$$\begin{array}{ccccccccccc} a_{11}x_1 & + & a_{12}x_2 & + & a_{13}x_3 & + & \cdots & + & a_{1n}x_n & = & b_1 \\ & & a_{22}x_2 & + & a_{23}x_3 & + & \cdots & + & a_{2n}x_n & = & b_2 \\ & & & & & & \ddots & & \vdots & & \vdots \\ & & & & & & & & a_{nn}x_n & = & b_n \end{array}$$

(ii) Count the number $g(n)$ of arithmetic operations needed to bring the following matrix into reduced row echelon form:

$$\left[\begin{array}{cccccc|c} a_{11} & a_{12} & a_{13} & \cdots & a_{1,n-1} & a_{1n} & b_1 \\ 0 & a_{22} & a_{23} & \cdots & a_{2,n-1} & a_{2n} & b_2 \\ 0 & 0 & a_{33} & \cdots & a_{3,n-1} & a_{3n} & b_3 \\ \vdots & \vdots & \vdots & \ddots & \vdots & \vdots & \vdots \\ 0 & 0 & 0 & \cdots & a_{n-1,n-1} & a_{n-1,n} & b_{n-1} \\ 0 & 0 & 0 & \cdots & 0 & a_{nn} & b_n \end{array}\right]$$

(iii) Verify that $\lim\limits_{n\to\infty} \dfrac{g(n)}{f(n)} = \infty$.

Exercise 7.20** Prove that for any 2×3 matrix, the reduced row echelon form is unique, that is, no matter how one applies elementary row operations, it is possible to reach one and only one reduced row echelon matrix. (Now think about matrices of any size.)

8 Matrix Operations

- **Matrix multiplication** (Section 8.2) Time: 10.01

8 Matrix Operations

In the last chapter, we introduced the notion of an *augmented matrix*, as an ingredient for a streamlined process, called *Gaussian elimination*, for determining whether solutions exist for systems of linear equations, and finding them when they do. In Chapter 2, we introduced the binary operations of *addition* of vectors and *multiplication* of a vector *by a scalar*. The algebraic versions of these have exact analogues for operations on matrices. *Multiplication of matrices*, however, which is also introduced in this chapter, is more involved and relates to *composition* of systems of equations and "cascades" of dot products of vectors. When we discuss determinants in Chapter 10, we will also find a strong connection with cross products of vectors.

A *matrix* is a rectangular array of numbers, called *entries*. We have seen and used augmented matrices in the last chapter, but for now we dispense with the vertical dividing line. The *size* of a matrix is specified by the number of rows and number of columns. These numbers are also called the *dimensions* of the matrix.

> If a matrix A has m rows and n columns then we say
>
> "A is m by n" and write "A is $m \times n$".

For example, we say the following:

$$\begin{bmatrix} 1 & 2 & 3 \\ 4 & 5 & 6 \end{bmatrix} \quad \text{is} \quad 2 \times 3$$

$$\begin{bmatrix} -1 & 0 & 77 & -5 & 6 \end{bmatrix} \quad \text{is} \quad 1 \times 5$$

$$\begin{bmatrix} 4 & 0 \\ 0 & 5 \\ 2 & -3 \\ -4 & 1 \end{bmatrix} \quad \text{is} \quad 4 \times 2$$

A *square* matrix is $n \times n$ for some n, and we think of 1×1 matrices as ordinary numbers and usually (but not always) delete brackets:

$$\begin{bmatrix} 1 & 0 & 0 \\ 0 & 1 & 0 \\ 0 & 0 & 1 \end{bmatrix} \quad \text{is} \quad 3 \times 3, \qquad \begin{bmatrix} 6 \end{bmatrix} = 6 \quad \text{is} \quad 1 \times 1$$

Entries of matrices are also referred to as *scalars*, but can be anything we like. In this book, most of the time, entries will be real numbers, but later we will also allow them to be complex numbers (an *algebraically closed* arithmetic, explained in Appendix 3).

A matrix consisting of one row

$$\begin{bmatrix} \lambda_1 & \lambda_2 & \dots & \lambda_n \end{bmatrix}$$

is called a *row matrix* or *row vector*.

A matrix consisting of one column

$$\begin{bmatrix} \lambda_1 \\ \lambda_2 \\ \vdots \\ \lambda_n \end{bmatrix}$$

is called a *column matrix* or *column vector*.

In general, if

$$A = \begin{bmatrix} a_{11} & a_{12} & \cdots & a_{1n} \\ a_{21} & a_{22} & \cdots & a_{2n} \\ \vdots & \vdots & \ddots & \vdots \\ a_{m1} & a_{m2} & \cdots & a_{mn} \end{bmatrix}$$

then A is $m \times n$ and it is typical to write

$$A = \begin{bmatrix} a_{ij} \end{bmatrix}_{m \times n} \qquad \text{or} \qquad A = \begin{bmatrix} a_{ij} \end{bmatrix}$$

if the dimensions are understood. We call a_{ij} the *(i, j)th entry* or *(i, j)-entry*. The first subscript i refers to the ith row and the second subscript j to the jth column in which a_{ij} lies. For example, in the matrix

$$\begin{bmatrix} 0 & -1 & -7 \\ 4 & 0 & 3 \end{bmatrix}$$

the $(2, 1)$-entry is 4, the $(1, 3)$ entry is -7 and both the $(1, 1)$ and $(2, 2)$-entries are 0.

8.1 Addition, subtraction and scalar multiplication

Addition and *subtraction* of matrices are exactly what we would guess, by analogy with operations involving components of geometric vectors. (We have already surreptitiously introduced the terminology *row* and *column vectors*. In a more advanced course on linear algebra you will find that matrices themselves can be regarded as "vectors" regardless of the number of rows and columns.)

Let A and B be matrices of the same size. Their *sum* is called $A+B$ and is obtained by adding corresponding entries. Here is a 2×2 example:

$$\begin{bmatrix} 1 & 2 \\ -3 & 4 \end{bmatrix} + \begin{bmatrix} 0 & 3 \\ 6 & -5 \end{bmatrix} = \begin{bmatrix} 1 & 5 \\ 3 & -1 \end{bmatrix}$$

Their *difference* is called $A-B$ and is obtained by subtracting corresponding entries:

$$\begin{bmatrix} 1 & 2 \\ -3 & 4 \end{bmatrix} - \begin{bmatrix} 0 & 3 \\ 6 & -5 \end{bmatrix} = \begin{bmatrix} 1 & -1 \\ -9 & 9 \end{bmatrix}$$

Adding or subtracting matrices of different sizes does not make sense. For example,

$$\begin{bmatrix} 1 & 2 & -5 \\ -3 & 4 & 0 \end{bmatrix} + \begin{bmatrix} 0 & 3 \\ 6 & -5 \end{bmatrix}$$

is not defined. If λ is any scalar then we can form the *scalar multiple* λA by multiplying every entry of A by λ. Here are contrasting examples:

$$3\begin{bmatrix} 1 & 2 \\ -3 & 4 \end{bmatrix} = \begin{bmatrix} 3 & 6 \\ -9 & 12 \end{bmatrix}, \qquad (-1)\begin{bmatrix} 0 & 3 \\ 6 & -5 \end{bmatrix} = \begin{bmatrix} 0 & -3 \\ -6 & 5 \end{bmatrix}$$

Notice, in this last example, that to multiply by the scalar -1 is the same as taking negatives of the entries. We call $(-1)B$ the *negative* of B and denote it by $-B$. We can use it to reformulate subtraction in terms of addition:

$$\boxed{A - B = A + (-B)}$$

If we subtract a matrix from itself we end up with a matrix of zeros, called the *zero matrix*, denoted by 0, or by $0_{m\times n}$ if we need to emphasise its dimensions. For example, we have the following:

$$\begin{bmatrix} 1 & 2 & 0 & -7 \\ -3 & 4 & 3 & 1 \end{bmatrix} - \begin{bmatrix} 1 & 2 & 0 & -7 \\ -3 & 4 & 3 & 1 \end{bmatrix} = \begin{bmatrix} 0 & 0 & 0 & 0 \\ 0 & 0 & 0 & 0 \end{bmatrix} = 0_{2\times 4} = 0$$

> 0 is the most used symbol in mathematics and has to be read and understood in context!

Below is a compendium of simple properties of matrix arithmetic using the above operations, which are exact analogues of those listed in Section 1.4 for vectors. Here, A, B, C are matrices of the same size and λ, μ are scalars. The proofs are immediate, because corresponding real number arithmetic is taking place at each position of the matrix. (Formal proofs, however, are good exercises in notation, and practice for proving more difficult facts later about matrix multiplication.)

$$A + B = B + A\,, \qquad (A + B) + C = A + (B + C)$$

$$A + 0 = 0 + A = A\,, \qquad -(-A) = A$$

$$A + (-A) = A - A = 0\,, \qquad \lambda(\mu A) = (\lambda\mu)A$$

$$\lambda(A + B) = \lambda A + \lambda B\,, \qquad (\lambda + \mu)A = \lambda A + \mu A$$

$$1A = A\,, \qquad (-\lambda)A = -(\lambda A)\,, \qquad (-1)A = -A$$

8.2 Matrix multiplication

By contrast to the operations of the previous section, it is not obvious at all how to define matrix multiplication in a way that is useful. Many students' first guess is simply to multiply corresponding entries of the matrix. This is certainly not a stupid idea, and leads to the *Hadamard product*, attributed to Jacques Hadamard, who is famous for proving the celebrated Prime Number Theorem. However, we will not use the Hadamard product.

Let A and B be matrices. We will define the *matrix product* AB used in linear algebra. For this to be possible, A and B need to have compatible sizes.

For AB to be defined,

the number of columns of A must equal the number of rows of B.

There is no restriction on the number of rows of A and the number of columns of B.

Suppose then that A is $m \times n$ and B is $n \times p$, so that n is the common number of columns of A and rows of B. Using our earlier notation,

$$A = \left[a_{ij}\right]_{m\times n}, \qquad B = \left[b_{jk}\right]_{n\times p}.$$

Now define AB to be the matrix C where

$$C = \left[c_{ik}\right]_{m\times p}$$

and, for each pair of subscripts i and k,

$$c_{ik} = a_{i1}b_{1k} + a_{i2}b_{2k} + \cdots + a_{in}b_{nk}.$$

Note that C is $m \times p$ so has m rows and p columns. The number n has disappeared!

When defined,
the matrix product AB has the same number of rows as A and the same number of columns as B.

The formula for c_{ik} takes some digesting, but it is the result of adding together certain corresponding products of matrix entries. It is just like the algebraic dot product of two geometric vectors discussed in Chapter 3. Indeed, we say

"the (i,k)-th entry of C is the product of the ith row of A with the kth column of B".

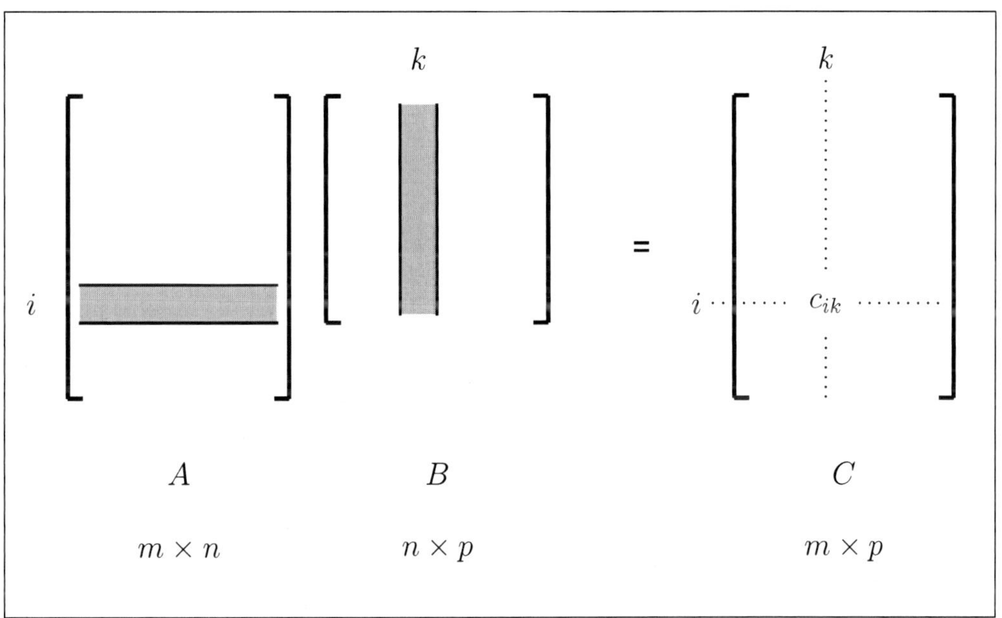

Example: A product of 2×2 matrices is 2×2:

$$\begin{bmatrix} 1 & 2 \\ -3 & 4 \end{bmatrix} \begin{bmatrix} 0 & 3 \\ 6 & -5 \end{bmatrix} = \begin{bmatrix} 1(0)+2(6) & 1(3)+2(-5) \\ -3(0)+4(6) & -3(3)+4(-5) \end{bmatrix}$$

$$= \begin{bmatrix} 12 & -7 \\ 24 & -29 \end{bmatrix}$$

but multiplied in the other order may give a different answer:

$$\begin{bmatrix} 0 & 3 \\ 6 & -5 \end{bmatrix} \begin{bmatrix} 1 & 2 \\ -3 & 4 \end{bmatrix} = \begin{bmatrix} 0(1)+3(-3) & 0(2)+3(4) \\ 6(1)-5(-3) & 6(2)-5(4) \end{bmatrix}$$

$$= \begin{bmatrix} -9 & 12 \\ 21 & -8 \end{bmatrix}$$

Very important: order matters!

Multiplication of matrices is not commutative in general. Most of the time $AB \neq BA$.

In the previous example, AB and BA were different, but of the same dimensions. Matrix products AB and BA may not even have the same dimensions. As an extreme example, if A is $n \times 1$ and B is $1 \times n$, then AB is $n \times n$ and BA is 1×1, so changing the order can mean the difference between a huge square matrix, if n is large, and a single real number!

Example: Multiplying a 1×3 by a 3×1 produces a 1×1 matrix:

$$\begin{bmatrix} a & b & c \end{bmatrix} \begin{bmatrix} d \\ e \\ f \end{bmatrix} = \begin{bmatrix} ad + be + cf \end{bmatrix} = ad + be + cf$$

whereas, multiplying a 3×1 by a 1×3 produces a 3×3 matrix:

$$\begin{bmatrix} d \\ e \\ f \end{bmatrix} \begin{bmatrix} a & b & c \end{bmatrix} = \begin{bmatrix} da & db & dc \\ ea & eb & ec \\ fa & fb & fc \end{bmatrix}$$

Why is the definition of matrix multiplication so complicated? In fact, much of the apparent complication comes from lack of familiarity, and with a little practice matrix multiplication will become almost second nature. However, especially when learning the technique, one has to be very careful with order, because of the failure of the commutative law of multiplication.

In a certain sense, the definition is forced upon us, because of the close relationship with systems of equations. It turns out that a system of linear equations corresponds to a single matrix equation (see Section 8.3). When one "composes" two systems to produce a third system, the two matrix equations are modified exactly by matrix multiplication (see Exercise 8.7).

In a more advanced course in linear algebra, these ideas are taken much further and a close connection is established between matrices and *linear transformations*. In a sense that can be made precise, a matrix "encodes" a linear transformation, in such a way that the latter becomes tractable, and questions about the real world reduce to matrix arithmetic and ultimately to real number arithmetic.

We finish this section by listing some very important properties of matrix multiplication. Here, A, B, C are matrices of dimensions for which the expressions make sense, and λ is a scalar. The proofs are left as exercises. Note that in property (5), the interpretation of 0 alters as one passes from one side of each equation to the other.

(1)	$(AB)C = A(BC)$
(2)	$A(B+C) = AB+AC$
(3)	$(A+B)C = AC+BC$
(4)	$\lambda(BC) = (\lambda B)C = B(\lambda C)$
(5)	$0A = 0 = A0$

8.3 Connections with systems of equations

With our definition of matrix multiplication, it is immediate now that the following system of linear equations

$$\begin{array}{ccccccccccc} a_{11}x_1 & + & a_{12}x_2 & + & a_{13}x_3 & + & \cdots & + & a_{1n}x_n & = & b_1 \\ a_{21}x_1 & + & a_{22}x_2 & + & a_{23}x_3 & + & \cdots & + & a_{2n}x_n & = & b_2 \\ \vdots & & \vdots & & \vdots & & & & \vdots & & \vdots \\ a_{m1}x_1 & + & a_{m2}x_2 & + & a_{m3}x_3 & + & \cdots & + & a_{mn}x_n & = & b_m \end{array}$$

can be expressed succinctly as a single matrix equation

$$A\mathbf{x} = \mathbf{b}$$

where

$$A = \begin{bmatrix} a_{11} & a_{12} & \cdots & a_{1n} \\ a_{21} & a_{22} & \cdots & a_{2n} \\ \vdots & \vdots & \vdots & \vdots \\ a_{m1} & a_{m2} & \cdots & a_{mn} \end{bmatrix}, \quad \mathbf{x} = \begin{bmatrix} x_1 \\ x_2 \\ \vdots \\ x_n \end{bmatrix}, \quad \mathbf{b} = \begin{bmatrix} b_1 \\ b_2 \\ \vdots \\ b_m \end{bmatrix}.$$

Note that $\mathbf{x}$ and $\mathbf{b}$ are column vectors, so multiplication on the left by the matrix A has a "transforming effect", turning $\mathbf{x}$ into $\mathbf{b}$, so to speak. If we can "reverse" this effect, then we should be able to get back from $\mathbf{b}$ to $\mathbf{x}$, and then read off solutions to the original system.

If the symbols in this matrix equation represented ordinary real numbers, then we wouldn't hesitate to "restore" the equation (in the sense of Al-Khawarizmi's ancient treatise on algebra) by "dividing through" by A:

$$A\mathbf{x} = \mathbf{b} \quad \Longrightarrow \quad \mathbf{x} = \frac{\mathbf{b}}{A}$$

For real numbers, we would need at least A to be nonzero. But the situation appears to be much worse. The left-hand side of this "implication" is a **matrix equation**, so it is not immediately clear how to make sense of the quotient on the right-hand side. The trick is to use **multiplicative notation** for division, and to **be careful about the order of multiplication**:

$$A\mathbf{x} = \mathbf{b} \quad \Longrightarrow \quad \mathbf{x} = A^{-1}\mathbf{b}$$

Now the implication makes perfectly good sense for matrices, provided the matrix A^{-1}, called the *inverse* of A, exists. Matrix inverses will be studied in Chapter 9, and a very useful existence criterion will be given in Chapter 10 using determinants.

We will finish this chapter by revisiting the theorem stated at the end of Section 7.1 about the number of solutions of systems of linear equations. A proof was provided indirectly at the end of Chapter 7. The proof offered now is short and direct, intending to be an illustration of the elegance and power of abstract matrix arithmetic. The proof may have the appearance of a *fait accompli*, and this is typical: mathematicians often cover their tracks and present proofs more or less as magicians produce rabbits out of hats. In fact, the proof is motivated by geometry, in particular the vector equation of a line in space, discussed in Section 5.1, and the idea of hopping on a line at a point (in the proof below represented by $\mathbf{x}_1$) and then moving backwards and forwards in the direction of the line (represented by scalar multiples of $\mathbf{x}_1 - \mathbf{x}_2$). This idea is precisely the backbone of the standard technique of solving a differential equation by finding a particular solution and adding it to the complementary function of the homogeneous part.

> **Theorem:** Every system of linear equations has either
>
> (A) no solutions,
>
> (B) exactly one solution, or
>
> (C) infinitely many solutions.

Proof using matrix arithmetic: A given system corresponds to a matrix equation

$$A\mathbf{x} \;=\; \mathbf{b}$$

where A is the matrix of coefficients, $\mathbf{x}$ is the column vector of variables and $\mathbf{b}$ is the column vector of constants on the right-hand sides. Suppose that neither case (A) nor case (B) holds. Hence, there are at least two distinct solutions, which correspond to column vectors $\mathbf{x}_1$, $\mathbf{x}_2$ respectively such that $\mathbf{x}_1 \neq \mathbf{x}_2$, yet $A\mathbf{x}_1 = \mathbf{b}$ and $A\mathbf{x}_2 = \mathbf{b}$. Hence,

$$A(\mathbf{x}_1 - \mathbf{x}_2) \;=\; A\mathbf{x}_1 - A\mathbf{x}_2 \;=\; \mathbf{b} - \mathbf{b} \;=\; \mathbf{0}\,,$$

where $\mathbf{0}$ denotes the zero column vector of the same length as $\mathbf{b}$. For a given scalar $\lambda \in \mathbb{R}$, put

$$\mathbf{s} \;=\; \mathbf{x}_1 + \lambda(\mathbf{x}_1 - \mathbf{x}_2)\,.$$

Observe that

$$A\mathbf{s} \;=\; A\left(\mathbf{x}_1 + \lambda(\mathbf{x}_1 - \mathbf{x}_2)\right) \;=\; A\mathbf{x}_1 + \lambda A(\mathbf{x}_1 - \mathbf{x}_2) \;=\; \mathbf{b} + \mathbf{0} \;=\; \mathbf{b}\,,$$

so that $\mathbf{s}$ corresponds to a solution of our original system. But as the scalar λ varies over the infinite set $\mathbb{R}$, the column vector $\mathbf{s}$ varies over infinitely many solutions (exercise). Thus, case (C) holds, and the theorem is proved.

Chapter 8: Important Ideas and Useful Facts

8.1 A *matrix* is an array of numbers, called *entries.* The plural of matrix is *matrices.* If a matrix M has m rows and n columns then we say that M is $m \times n$, and *square* if $m = n$. The entry lying in the ith row and jth column is called the (i, j)-*entry.* A matrix consisting of one row is called a *row vector.* A matrix consisting of one column is called a *column vector.*

8.2 To *add* or *subtract* using two matrices of the same size, simply add or subtract the corresponding entries. To form the *negative* of a matrix, simply take the negatives of its entries. To *multiply a matrix by a scalar*, simply multiply its entries by the scalar.

8.3 The *zero matrix* has all of its entries equal to 0 and is denoted by 0 or $0_{m\times n}$ if the size needs to be emphasised. The *identity matrix* is a square matrix with *diagonal* entries equal to 1 and all entries off the diagonal equal to 0. The identity matrix is denoted by I or I_n if it is $n \times n$ and the size needs to be emphasised.

8.4 If A is $m \times n$ and B is $n \times p$ then the *matrix product* AB is defined and is $m \times p$. The (i, k)-entry of AB is the "dot product" of the ith row of A with the kth column of B, which can be abbreviated using sigma notation:

$$\sum_{j=1}^{n} a_{ij}b_{jk} \;=\; a_{i1}b_{1k} + a_{i2}b_{2k} + \ldots + a_{in}b_{nk}$$

where a_{ij}, b_{jk} denote (i, j) and (j, k)-entries of A and B respectively.

8.5 If A, B, C are matrices of appropriate sizes for which the expressions make sense, and λ and μ are scalars, then the following properties hold:

$$A+B = B+A\,, \quad (A+B)+C = A+(B+C)\,, \quad A+0 = 0+A = A$$

$$-(-A) = A\,, \quad A+(-A) = A-A = 0\,, \quad \lambda(\mu A) = (\lambda\mu)A$$

$$\lambda(A+B) = \lambda A + \lambda B\,, \quad (\lambda+\mu)A = \lambda A + \mu A\,, \quad IA = AI = A$$

$$(AB)C = A(BC)\,, \quad A(B+C) = AB+AC\,, \quad (A+B)C = AC+BC$$

$$\lambda(BC) = (\lambda B)C = B(\lambda C)\,, \quad 0A = 0 = A0$$

8.6 Warning: Matrix multiplication is not in general commutative. Most of the time $AB \neq BA$. In (rare, but often important) instances when $AB = BA$, we say that A and B *commute.*

Exercise 8.1 Let $A = \begin{bmatrix} 2 & -1 \\ 1 & 0 \end{bmatrix}$, $B = \begin{bmatrix} 7 & 6 & 4 \end{bmatrix}$, $C = \begin{bmatrix} 2 & 1 \\ 5 & 0 \\ 0 & -1 \\ 1 & -6 \end{bmatrix}$.

Write down the sizes of A, B and C.

Exercise 8.2 For the matrix $M = \begin{bmatrix} 5 & -5 & -3 & -7 \\ 0 & 9 & 0 & -4 \\ -2 & -3 & 8 & 0 \end{bmatrix}$, locate the

(i) $(2,1)$-entry (ii) $(1,4)$-entry (iii) $(3,3)$-entry (iv) $(3,2)$-entry.

Exercise 8.3 Consider the following matrices:

$$A = \begin{bmatrix} 1 & 2 \\ 0 & 3 \end{bmatrix} \qquad B = \begin{bmatrix} -6 & 3 \\ 4 & 1 \end{bmatrix} \qquad C = \begin{bmatrix} 2 & 4 & 5 \end{bmatrix}$$

$$D = \begin{bmatrix} 1 \\ 0 \\ -1 \end{bmatrix} \qquad E = \begin{bmatrix} 1 & 2 & 0 & 4 \\ 6 & -1 & 2 & 3 \\ 0 & 0 & 1 & 2 \end{bmatrix} \qquad F = \begin{bmatrix} 0 \\ 1 \\ 2 \\ -3 \end{bmatrix}$$

Evaluate the following expressions:

(i) $2A$ (ii) $-B$ (iii) $A+B$ (iv) $A-B$ (v) $A^2 = AA$

(vi) AB (vii) BA (viii) CD (ix) $EF - 3D$ (x) CEF

Exercise 8.4 Evaluate and simplify

$$\begin{bmatrix} \cos\alpha & -\sin\alpha \\ \sin\alpha & \cos\alpha \end{bmatrix} \begin{bmatrix} \cos\beta & -\sin\beta \\ \sin\beta & \cos\beta \end{bmatrix}$$

where α and β are real numbers.

Exercise 8.5 Let X and Y be square matrices of the same size. Prove that X and Y commute if and only if

$$(X+Y)^2 = X^2 + 2XY + Y^2,$$

which in turn occurs if and only if

$$(X+Y)(X-Y) = X^2 - Y^2.$$

Since matrices rarely commute, these two identities usually fail.

Exercise 8.6 This is an exercise in interpreting notation. Suppose A is an $m \times n$ matrix where $m \neq n$. Explain why there is no paradox in the following assertions:

$$A0 = 0A = 0 \qquad \text{and} \qquad 0 \neq 0 \neq 0$$

Exercise 8.7 This exercise is an illustration of the correspondence between multiplication of matrices and composition of compatible systems of equations. Suppose

$$\begin{array}{rcrcl} 2x & + & 3y & = & u \\ x & - & 4y & = & v \end{array}$$

and further that

$$\begin{array}{rcrcl} 3u & - & 5v & = & c \\ 2u & + & 3v & = & d\,. \end{array}$$

Express c and d in terms of x and y by

(i) direct substitution (ii) matrix multiplication.

Exercise 8.8 Solve each of the following for x, y and z:

(i) $\begin{bmatrix} 1 & 2 & 3 \\ 4 & -1 & 2 \\ 0 & -6 & 1 \end{bmatrix} \begin{bmatrix} x \\ y \\ z \end{bmatrix} = \begin{bmatrix} 15 \\ 29 \\ 10 \end{bmatrix}$ (ii) $\begin{bmatrix} 2 & -3 & 3 \\ 4 & 9 & -4 \\ 2 & 0 & 1 \end{bmatrix} \begin{bmatrix} x \\ y \\ z \end{bmatrix} = \begin{bmatrix} 0 \\ 0 \\ 0 \end{bmatrix}$

Exercise 8.9 Explain briefly why the associative law for matrix multiplication implies that every square matrix commutes with its square.

Exercise 8.10 Which of the following do you know to be true or expect to be true for all square matrices A, B, C of the same size:

(i) $(AB)C = A(BC)$ (ii) $AB = BA$

(iii) $(AB)^2 = A^2B^2$ (iv) $A(B+C) = AB + AC$

(v) $(-A)(-B) = AB$ (vi) $A(B-C) = AB - AC$

(vii) $(A+I)^2 = A^2 + 2A + I$ (viii) $(A+I)(A-I) = A^2 - I$

Exercise 8.11 Find a 2×2 matrix M such that $M^2 = \begin{bmatrix} 0 & 0 \\ 0 & 0 \end{bmatrix}$ but every entry of M is nonzero.

Exercise 8.12 A square matrix is called *diagonal* if all entries away from the main diagonal are zero. Find a simple rule for multiplying diagonal matrices. More generally, describe in words as simply as you can what happens if you multiply any square matrix (of the same size) (i) on the left, or (ii) on the right, by a diagonal matrix.

Exercise 8.13* Consider the matrix $M = \begin{bmatrix} 3 & -1 \\ 4 & -1 \end{bmatrix}$.

(i) Verify that $M^2 = 2M - I$.

(ii) Deduce that $M^3 = 3M - 2I$ and guess a general formula for powers of M. (If you know the technique of proof by induction then you can verify that your guess is correct.)

(iii) Evaluate M^5, M^{10} and M^{100}.

Exercise 8.14* Find XY in each case, given that X is a row matrix and Y is a column matrix, both with the same number of entries:

(i) $YX = \begin{bmatrix} -2 & -3 \\ 2 & 3 \end{bmatrix}$ (ii) $YX = \begin{bmatrix} 3 & -3 & 6 \\ 4 & -4 & 8 \\ -2 & 2 & -4 \end{bmatrix}$

Exercise 8.15* Find necessary and sufficient conditions on a, b, c and d such that the matrix $B = \begin{bmatrix} a & b \\ c & d \end{bmatrix}$ commutes with the matrix A in each of the following cases:

(i) $A = \begin{bmatrix} 1 & 0 \\ 0 & -1 \end{bmatrix}$ (ii) $A = \begin{bmatrix} 0 & 7 \\ 7 & 0 \end{bmatrix}$ (iii) $A = \begin{bmatrix} 1 & 2 \\ 0 & 1 \end{bmatrix}$

Exercise 8.16* Let k be a constant and n a positive integer which varies. Guess what is meant by A^n where A is a square matrix. Find formulae for the following. Verify your formulae using mathematical induction.

(i) $\begin{bmatrix} 1 & k \\ 0 & 1 \end{bmatrix}^n$ (ii) $\begin{bmatrix} k & 1 \\ 0 & k \end{bmatrix}^n$ (iii) $\begin{bmatrix} k & 1 & 0 \\ 0 & k & 1 \\ 0 & 0 & k \end{bmatrix}^n$

Exercise 8.17* Verify that if $\mathbf{x}_1$ and $\mathbf{x}_2$ are different column vectors of the same length and λ is a scalar then there are infinitely many different column vectors of the following form:

$$\mathbf{s} \ = \ \mathbf{x}_1 + \lambda(\mathbf{x}_1 - \mathbf{x}_2)$$

Exercise 8.18* If $A = [a_{ij}]_{m \times n}$ is a matrix, define the *transpose* of A to be the matrix

$$A^T \ = \ [b_{ij}]_{n \times m}$$

where, for $i = 1, \ldots, n$ and $j = 1, \ldots, m$,

$$b_{ij} \ = \ a_{ji} \, .$$

For matrices X, Y and Z of dimensions for which the expressions make sense, verify the following:

$$(X^T)^T \ = \ X \, , \qquad (X + Y)^T \ = \ X^T + Y^T \, , \qquad (YZ)^T \ = \ Z^T Y^T$$

Exercise 8.19* Verify the distributive law for matrices:

$$A(B + C) \ = \ AB + AC$$

Use transpose and the equations of the previous exercise to deduce the other distributive law:

$$(A + B)C \ = \ AC + BC$$

Exercise 8.20* Verify the associative law for matrix multiplication.

Exercise 8.21* Evaluate and simplify the product

$$\begin{bmatrix} r\cos\alpha & -r\sin\alpha \\ r\sin\alpha & r\cos\alpha \end{bmatrix} \begin{bmatrix} s\cos\beta & -s\sin\beta \\ s\sin\beta & s\cos\beta \end{bmatrix}$$

where α, β are any real numbers and r, s are nonnegative real numbers. Relate your answer to multiplication of complex numbers (explained in Appendix 3). Is there any connection with addition of complex numbers?

Exercise 8.22** Prove that there are no square matrices A and B of the same size such that $AB - BA = I$.

9 Matrix Inverses

- **Motivating inverses** (Section 9.1)
  Time: 12.16
- **Inverting a three-by-three matrix** (Section 9.4)
  Time: 6.54

9 Matrix Inverses

In the last chapter, we introduced matrix arithmetic and explained the connection with systems of linear equations. Indeed, the problem of solving systems of equations is equivalent to algebraic manipulation of matrix equations. Of course, we always have access to Gaussian elimination, but matrix arithmetic provides a framework for sophisticated techniques, such as the theory of *eigenvalues* and *eigenvectors*, which we introduce in Chapter 11. Our aim is to simplify matrix arithmetic and apply it, if possible, using a process called *diagonalisation*, which is explained in Chapter 12.

To motivate the ideas in this chapter, we first remind ourselves of two important properties of the real number system $\mathbb{R}$ that we are in the habit of using automatically:

(1)	$ab = ac\ ,\ \ a \neq 0$	$\Longrightarrow$	$b = c$
(2)	$ab = 0$	$\Longrightarrow$	$a = 0$ or $b = 0$

Property (1) is called the *Cancellation Law*, because we think of "cancelling" the value a from both sides of the equation on the left of the implication. Property (2) says that there are no "proper divisors" of zero, in the sense that to "divide" zero into pieces, at least one of the pieces must remain zero.

Both properties (1) and (2) fail for matrix arithmetic!

Example: Consider the following matrices:

$$A = \begin{bmatrix} 0 & 1 \\ 0 & 2 \end{bmatrix}, \qquad B = \begin{bmatrix} 1 & 1 \\ 3 & 4 \end{bmatrix}, \qquad C = \begin{bmatrix} 2 & 5 \\ 3 & 4 \end{bmatrix}$$

Then

$$AB = \begin{bmatrix} 3 & 4 \\ 6 & 8 \end{bmatrix} = AC$$

yet $A \neq 0$ and $B \neq C$, so that the Cancellation Law fails. Further,

$$A(B - C) = AB - AC = 0$$

but $A \neq 0$ and $B - C \neq 0$, so we have proper divisors of zero!

Nevertheless, it is important to be able to recognise when cancellation may be used in matrix arithmetic. This is related to the *existence of inverses*, which we motivate briefly by considering the following manipulation of real numbers:

If $a, b, c \in \mathbb{R}$ and

$$ab = ac\,,$$

where $a \neq 0$, then multiplying by a^{-1} we get

$$a^{-1}ab \;=\; a^{-1}ac$$

so that

$$\boxed{1b \;=\; 1c}$$

so, finally,

$$b \;=\; c\,.$$

This is a verification of Property (1), the Cancellation Law. The argument utilised the real number 1, which is called the *multiplicative identity element* of $\mathbb{R}$, since it leaves things unchanged upon multiplication. It has a central role, highlighted above, in achieving the effect of cancellation. We also used the following element:

$$a^{-1} = \frac{1}{a}$$

known as the *multiplicative inverse* of a, both because it is upside-down in fractional notation (so is literally inverted), and because it has the property of "inverting" or "undoing" a, in the sense of getting back to 1:

$$(a^{-1})\,a \;=\; \frac{a}{a} \;=\; 1$$

9.1 Identity matrices and inverses

For matrices, we have an important analogue of the number 1, known as the *identity matrix*, which is square, say $n \times n$, with 1's on the diagonal and 0's everywhere else, and denoted by I or I_n if the size needs to be emphasised. Here are some examples, but identity matrices can be of any size:

$$I_2 = \begin{bmatrix} 1 & 0 \\ 0 & 1 \end{bmatrix}, \qquad I_3 = \begin{bmatrix} 1 & 0 & 0 \\ 0 & 1 & 0 \\ 0 & 0 & 1 \end{bmatrix}, \qquad I_4 = \begin{bmatrix} 1 & 0 & 0 & 0 \\ 0 & 1 & 0 & 0 \\ 0 & 0 & 1 & 0 \\ 0 & 0 & 0 & 1 \end{bmatrix}$$

The smallest example is

$$I_1 = [1] = 1 .$$

All of the nondiagonal zeros are needed, when $n > 1$, so that the following property holds:

> **Simple Fact:** If A is any $m \times n$ matrix then
>
> $$AI_n = A = I_m A .$$

(It is left as an exercise to verify that only identity matrices have this property.) For example, we have the following:

$$\begin{bmatrix} 1 & 0 \\ 0 & 1 \end{bmatrix} \begin{bmatrix} 2 & -3 & 4 \\ 1 & 2 & -6 \end{bmatrix} = \begin{bmatrix} 2 & -3 & 4 \\ 1 & 2 & -6 \end{bmatrix}$$

$$= \begin{bmatrix} 2 & -3 & 4 \\ 1 & 2 & -6 \end{bmatrix} \begin{bmatrix} 1 & 0 & 0 \\ 0 & 1 & 0 \\ 0 & 0 & 1 \end{bmatrix}$$

We now define *inverses of matrices*, by analogy with a property of inverses (reciprocals) of real numbers. But there is a sublety: our definition will be two-sided, because matrix multiplication is not in general commutative. (It is a theorem, quite difficult to prove, that we can make the following definition one-sided, but more about that later.)

The *inverse* of a matrix A is a matrix B such that, for some positive integer n,

$$AB = BA = I_n .$$

Comparing dimensions in the above equations forces both A and B to be $n \times n$. Thus, the concept of inverse is only applicable to square matrices:

Nonsquare matrices do not possess inverses.

We said the inverse in the above definition, because it implies there can only be one such matrix with the property of being an inverse of A. To see this is good practice in matrix arithmetic. Suppose B and C are two matrices such that

$$AB = BA = I \qquad \text{and} \qquad AC = CA = I .$$

Then, applying the

proof template: expand – apply something – contract

introduced in Section 1.5, we get

$$C = CI = C(AB) = (CA)B = IB = B ,$$

which proves all inverses of A are equal! The middle step applies associativity of matrix multiplication, the verification of which was left as a starred exercise in the previous chapter.

The following multiplicative inverse notation, inherited from $\mathbb{R}$, works like a charm, especially later when we define *powers of a matrix*:

Notation: The inverse of A, if it exists, is denoted by A^{-1}, so is characterised by the equations

$$AA^{-1} = A^{-1}A = I .$$

The existence of inverses is subtle. We will give a useful existence criterion in the next chapter, and fully examine the case of 2×2 matrices in the next section.

We finish this section by recording two extremely useful facts. The first is a theorem which says we only have to check one side of the definition of inverse for square matrices (the other side being implied). This fact is often taken for granted, but the proof is difficult and left as a starred exercise. The second fact is easy to check and left as an unstarred exercise.

Theorem: If A is a square matrix and $AB = I$ or $BA = I$ then

$$AB = BA = I$$

so that the inverse of A exists and equals B.

Simple Fact: If A and B are invertible matrices of the same size then AB is invertible and

$$(AB)^{-1} = B^{-1}A^{-1}.$$

Remember the matrix mantra: order is important. It is a common error for students to use $A^{-1}B^{-1}$ as the inverse of AB, which will be correct only when A and B commute, and that happens rarely.

9.2 Inverses of two-by-two matrices

Recall the inverse of a square matrix A, if it exists, is denoted by A^{-1} and satisfies the equations

$$AA^{-1} = A^{-1}A = I.$$

For 1×1 matrices this just becomes the usual inverse of a nonzero real number. In this section we describe fully the situation for 2×2 matrices. For square matrices of larger dimension, it takes considerable effort to find inverses, or even tell whether they exist.

Suppose throughout this section that

$$A = \begin{bmatrix} a & b \\ c & d \end{bmatrix}$$

and define $\det A$, the *determinant* of A, by the following formula:

$$\boxed{\det A = ad - bc}$$

It is also a useful convention to denote the determinant using vertical bars, replacing the square brackets of a matrix:

$$\begin{vmatrix} a & b \\ c & d \end{vmatrix} = ad - bc$$

The following fact tells the entire story for 2×2 matrices:

Useful Formula: The inverse A^{-1} exists if and only if

$$\det A \neq 0,$$

in which case

$$A^{-1} = \frac{1}{ad-bc} \begin{bmatrix} d & -b \\ -c & a \end{bmatrix}.$$

To remember this formula, just think of interchanging the diagonal elements, negating the off-diagonal elements, and dividing through by the determinant. Where does this formula come from, and how do we know it is true?

The formula can be derived fairly painlessly from scratch, as an exercise, using the method of Section 9.4. However, if someone provides a formula, regardless of where it came from, then, to know that it is true, it is simply a matter of checking that the equations for an inverse are satisfied. By the theorem above, it is enough to check one side:

$$\begin{bmatrix} a & b \\ c & d \end{bmatrix} \left(\frac{1}{ad-bc} \begin{bmatrix} d & -b \\ -c & a \end{bmatrix} \right) = \frac{1}{ad-bc} \begin{bmatrix} ad-bc & -ab+ba \\ cd-dc & -cb+da \end{bmatrix} \Big)$$

$$= \begin{bmatrix} 1 & 0 \\ 0 & 1 \end{bmatrix} \quad \checkmark$$

Example: Consider the following square matrices:

$$A = \begin{bmatrix} 2 & 1 \\ -3 & -4 \end{bmatrix}, \qquad B = \begin{bmatrix} 6 & 1 \\ 12 & 2 \end{bmatrix}$$

Then

$$A^{-1} = \frac{1}{2(-4) - 1(-3)} \begin{bmatrix} -4 & -1 \\ 3 & 2 \end{bmatrix} = \begin{bmatrix} 4/5 & 1/5 \\ -3/5 & -2/5 \end{bmatrix}$$

but

$$\det B = 6(2) - 1(12) = 0$$

so B^{-1} doesn't exist.

9.3 Powers of a matrix

The following notation is useful and leads to easy management of integer exponents in matrix expressions. Let A be a square matrix and make the following definitions:

$$\boxed{A^0 = I}$$

$$\boxed{A^n = \underbrace{A\,A\,\ldots\,A}_{n \text{ times}} \text{ if } n \text{ is positive}}$$

$$\boxed{A^{-n} = \underbrace{A^{-1}\,A^{-1}\,\ldots\,A^{-1}}_{n \text{ times}} \text{ if } n \text{ is positive}}$$

Each of the following is routine to check:

Simple Facts: If A is an invertible matrix then

(1) $A^m A^n = A^{m+n}$ for all integers m, n

(2) $(A^{-1})^{-1} = A$

(3) $(A^m)^n = A^{mn}$ for all integers m, n

(4) $(\lambda A)^{-1} = \frac{1}{\lambda} A^{-1}$ for all nonzero $\lambda \in \mathbb{R}$.

9.4 Using row reduction to find the inverse

The following recipe is guaranteed to find the inverse of a square matrix if it exists. We revive the earlier technique of augmenting a matrix, but this time not with just one column, but with an entire matrix of the same size as A. When you see the following method for the first time, it seems quite miraculous. The reason it works will be given in Section 9.6 after the introduction of *elementary matrices.*

Recipe for finding the inverse of a square matrix A:

(1) Augment A with the identity matrix I:

$$[\,A \mid I\,]$$

(2) Row reduce A to reduced row echelon form, **simultaneously** applying the same elementary row operations to I on the right-hand side.

(3) If A becomes I on the left-hand side then A is invertible; otherwise A is not invertible.

(4) If A becomes I on the left-hand side then I becomes A^{-1} on the right-hand side:

$$[\,I \mid A^{-1}\,]$$

Example: Use the recipe to invert $\begin{bmatrix} 1 & 2 \\ 1 & 3 \end{bmatrix}$.

Solution using recipe:

$$\left[\begin{array}{cc|cc} 1 & 2 & 1 & 0 \\ 1 & 3 & 0 & 1 \end{array}\right] \sim \left[\begin{array}{cc|cc} 1 & 2 & 1 & 0 \\ 0 & 1 & -1 & 1 \end{array}\right] \quad R_2 - R_1$$

$$\sim \left[\begin{array}{cc|cc} 1 & 0 & 3 & -2 \\ 0 & 1 & -1 & 1 \end{array}\right] \quad R_1 - 2R_2$$

This yields the following:

$$\begin{bmatrix} 1 & 2 \\ 1 & 3 \end{bmatrix}^{-1} = \begin{bmatrix} 3 & -2 \\ -1 & 1 \end{bmatrix}$$

Of course, this matches the formula from Section 9.2. $\checkmark$

Example: If possible, invert $A = \begin{bmatrix} 1 & 2 & 3 \\ 2 & 5 & 3 \\ 1 & 0 & 8 \end{bmatrix}$.

Solution using recipe:

$$\left[\begin{array}{ccc|ccc} 1 & 2 & 3 & 1 & 0 & 0 \\ 2 & 5 & 3 & 0 & 1 & 0 \\ 1 & 0 & 8 & 0 & 0 & 1 \end{array}\right] \sim \left[\begin{array}{ccc|ccc} 1 & 2 & 3 & 1 & 0 & 0 \\ 0 & 1 & -3 & -2 & 1 & 0 \\ 0 & -2 & 5 & -1 & 0 & 1 \end{array}\right] \begin{array}{l} \\ R_2 - 2R_1 \\ R_3 - R_1 \end{array}$$

$$\sim \left[\begin{array}{ccc|ccc} 1 & 0 & 9 & 5 & -2 & 0 \\ 0 & 1 & -3 & -2 & 1 & 0 \\ 0 & 0 & -1 & -5 & 2 & 1 \end{array}\right] \begin{array}{l} R_1 - 2R_2 \\ \\ R_3 + 2R_2 \end{array}$$

$$\sim \left[\begin{array}{ccc|ccc} 1 & 0 & 0 & -40 & 16 & 9 \\ 0 & 1 & 0 & 13 & -5 & -3 \\ 0 & 0 & 1 & 5 & -2 & -1 \end{array}\right] \begin{array}{l} R_1 + 9R_3 \\ R_2 - 3R_3 \\ -R_3 \end{array}$$

This yields the following:

$$A^{-1} = \begin{bmatrix} -40 & 16 & 9 \\ 13 & -5 & -3 \\ 5 & -2 & -1 \end{bmatrix}$$

Check:

$$\begin{bmatrix} 1 & 2 & 3 \\ 2 & 5 & 3 \\ 1 & 0 & 8 \end{bmatrix} \begin{bmatrix} -40 & 16 & 9 \\ 13 & -5 & -3 \\ 5 & -2 & -1 \end{bmatrix} = \begin{bmatrix} 1 & 0 & 0 \\ 0 & 1 & 0 \\ 0 & 0 & 1 \end{bmatrix} \qquad \checkmark$$

Example: If possible, invert $B = \begin{bmatrix} 1 & 6 & 4 \\ 2 & 4 & -1 \\ -1 & 2 & 5 \end{bmatrix}$.

Solution using recipe:

$$\left[\begin{array}{ccc|ccc} 1 & 6 & 4 & 1 & 0 & 0 \\ 2 & 4 & -1 & 0 & 1 & 0 \\ -1 & 2 & 5 & 0 & 0 & 1 \end{array}\right] \sim \left[\begin{array}{ccc|ccc} 1 & 6 & 4 & 1 & 0 & 0 \\ 0 & -8 & -9 & -2 & 1 & 0 \\ 0 & 8 & 9 & 1 & 0 & 1 \end{array}\right] \begin{array}{l} \\ R_2 - 2R_1 \\ R_3 + R_1 \end{array}$$

$$\sim \left[\begin{array}{ccc|ccc} 1 & 6 & 4 & 1 & 0 & 0 \\ 0 & -8 & -9 & -2 & 1 & 0 \\ 0 & 0 & 0 & -1 & 1 & 1 \end{array}\right] \begin{array}{l} \\ \\ R_3 + R_2 \end{array}$$

At this point we stop, because continuing all the way to reduced row echelon form will produce a matrix on the left-hand side which is not I (since it will have a row of zeros). We conclude that B is not invertible.

9.5 Using inverses to solve systems of equations

If A is an invertible matrix and $AX = B$ then

$$X = IX = A^{-1}AX = A^{-1}B\,.$$

This manipulation is precisely analogous to dividing through by A as if these were equations involving real numbers, except that we are careful about the order in which we multiply matrices. Thus, if information such as the inverse of A has been stored somewhere, we can use it to solve an associated system of equations:

> If $AX = B$ is the matrix equation corresponding to a system of equations, and A is invertible, then the (unique) solution is
>
> $$X = A^{-1}B\,,$$
>
> where X is the column vector of unknowns.

Example: Solve the following system:

$$\begin{array}{rcrcrcr} x_1 & + & 2x_2 & + & 3x_3 & = & 5 \\ 2x_1 & + & 5x_2 & + & 3x_3 & = & 3 \\ x_1 & & & + & 8x_3 & = & 17 \end{array}$$

Solution using the inverse of a matrix: The system can be expressed as a matrix equation

$$AX = B$$

where

$$A = \begin{bmatrix} 1 & 2 & 3 \\ 2 & 5 & 3 \\ 1 & 0 & 8 \end{bmatrix}, \qquad X = \begin{bmatrix} x_1 \\ x_2 \\ x_3 \end{bmatrix}, \qquad B = \begin{bmatrix} 5 \\ 3 \\ 17 \end{bmatrix}.$$

In an earlier example we found

$$A^{-1} = \begin{bmatrix} -40 & 16 & 9 \\ 13 & -5 & -3 \\ 5 & -2 & -1 \end{bmatrix}$$

so that

$$X = A^{-1}B = \begin{bmatrix} -40 & 16 & 9 \\ 13 & -5 & -3 \\ 5 & -2 & -1 \end{bmatrix} \begin{bmatrix} 5 \\ 3 \\ 17 \end{bmatrix} = \begin{bmatrix} 1 \\ -1 \\ 2 \end{bmatrix}.$$

This yields the following unique solution:

$$x_1 = 1, \quad x_2 = -1, \quad x_3 = 2$$

9.6 Elementary matrices

In this final section we introduce *elementary matrices*, and use them to explain the mystery of the recipe given earlier, for finding inverses of matrices. These are the simplest kinds of invertible matrices, and one can think of them as "atoms". At the end we will see that every invertible matrix is a product of elementary matrices (analogous to a molecule being built from atoms).

An $n \times n$ matrix is called *elementary* if it is the result of applying a single elementary row operation to the identity matrix I_n.

Below are some examples. Identity matrices appear on the left, and next to them are elementary matrices resulting from elementary row operations. The operation used in each case is indicated to the right in brackets.

As one can see, elementary matrices closely resemble identity matrices, just one step removed in terms of complexity:

$$\begin{bmatrix} 1 & 0 \\ 0 & 1 \end{bmatrix} \sim \begin{bmatrix} 1 & 0 \\ 0 & 3 \end{bmatrix} \qquad (\, R_2 \to 3R_2 \,)$$

$$\begin{bmatrix} 1 & 0 & 0 & 0 \\ 0 & 1 & 0 & 0 \\ 0 & 0 & 1 & 0 \\ 0 & 0 & 0 & 1 \end{bmatrix} \sim \begin{bmatrix} 1 & 0 & 0 & 0 \\ 0 & 0 & 1 & 0 \\ 0 & 1 & 0 & 0 \\ 0 & 0 & 0 & 1 \end{bmatrix} \qquad (\, R_2 \leftrightarrow R_3 \,)$$

$$\begin{bmatrix} 1 & 0 & 0 \\ 0 & 1 & 0 \\ 0 & 0 & 1 \end{bmatrix} \sim \begin{bmatrix} 1 & 0 & 3 \\ 0 & 1 & 0 \\ 0 & 0 & 1 \end{bmatrix} \qquad (\, R_1 \to R_1 + 3R_3 \,)$$

$$\begin{bmatrix} 1 & 0 & 0 \\ 0 & 1 & 0 \\ 0 & 0 & 1 \end{bmatrix} \sim \begin{bmatrix} 1 & 0 & 0 \\ 0 & 1 & 0 \\ -2 & 0 & 1 \end{bmatrix} \qquad (\, R_3 \to R_3 - 2R_1 \,)$$

The following fact establishes a crucial link between matrix multiplication and elementary row operations. It is not difficult to prove, but it is not obvious and needs checking. Some illustrations appear below, and it is left as an exercise to give a general proof.

> **Important Fact:** If E is the elementary matrix obtained by applying the elementary row operation ρ to I_n and A is any matrix with n rows then
>
> the matrix product EA
>
> is the matrix obtained by applying ρ to A.

Again, since we are dealing with matrix multiplication: order matters. In the matrix expression EA, the elementary matrix E appears on the left.

(If we placed E on the right of A, then it would have an effect on the *columns* of A, which is not what we want here. Opportunities to be ambidextrous, using both row and column operations, will arise in the next chapter on determinants.)

Examples: Each of the following equations is in the form $EA = B$ where E is elementary and B is obtained from A by applying the elementary row operation ρ that corresponds to E. The elementary matrices used here are precisely those exhibited earlier as illustrations of the definition.

$$\begin{bmatrix} 1 & 0 \\ 0 & 3 \end{bmatrix} \begin{bmatrix} a & b \\ c & d \end{bmatrix} = \begin{bmatrix} a & b \\ 3c & 3d \end{bmatrix} \quad 3R_2$$

$$\begin{bmatrix} 1 & 0 & 0 & 0 \\ 0 & 0 & 1 & 0 \\ 0 & 1 & 0 & 0 \\ 0 & 0 & 0 & 1 \end{bmatrix} \begin{bmatrix} 5 & 6 & 0 & 8 \\ 1 & 2 & 4 & -1 \\ 0 & 0 & 2 & 3 \\ 1 & 2 & -1 & 0 \end{bmatrix} = \begin{bmatrix} 5 & 6 & 0 & 8 \\ 0 & 0 & 2 & 3 \\ 1 & 2 & 4 & -1 \\ 1 & 2 & -1 & 0 \end{bmatrix} \quad R_2 \leftrightarrow R_3$$

$$\begin{bmatrix} 1 & 0 & 3 \\ 0 & 1 & 0 \\ 0 & 0 & 1 \end{bmatrix} \begin{bmatrix} 2 & 4 & 6 \\ 0 & 3 & 2 \\ 0 & 0 & -2 \end{bmatrix} = \begin{bmatrix} 2 & 4 & 0 \\ 0 & 3 & 2 \\ 0 & 0 & -2 \end{bmatrix} \quad R_1 + 3R_3$$

$$\begin{bmatrix} 1 & 0 & 0 \\ 0 & 1 & 0 \\ -2 & 0 & 1 \end{bmatrix} \begin{bmatrix} 1 & 2 & 3 \\ 4 & 5 & 6 \\ 2 & 4 & -1 \end{bmatrix} = \begin{bmatrix} 1 & 2 & 3 \\ 4 & 5 & 6 \\ 0 & 0 & -7 \end{bmatrix} \quad R_3 - 2R_1$$

Example: Now we illustrate the next level of complexity, that of using two elementary row operations, which corresponds to two multiplications by elementary matrices. In an earlier example we inverted

$$\begin{bmatrix} 1 & 2 \\ 1 & 3 \end{bmatrix}$$

by row reducing the augmented matrix

$$\left[\begin{array}{cc|cc} 1 & 2 & 1 & 0 \\ 1 & 3 & 0 & 1 \end{array}\right]$$

in just two steps.

The following matrix equation captures exactly the same process, but by two premultiplications by elementary matrices, in the correct order:

$$\underbrace{\begin{bmatrix} 1 & -2 \\ 0 & 1 \end{bmatrix}\begin{bmatrix} 1 & 0 \\ -1 & 1 \end{bmatrix}}\begin{bmatrix} 1 & 2 \\ 1 & 3 \end{bmatrix} = \begin{bmatrix} 1 & -2 \\ 0 & 1 \end{bmatrix}\begin{bmatrix} 1 & 2 \\ 0 & 1 \end{bmatrix} \quad R_2 - R_1$$

$$= \begin{bmatrix} 1 & 0 \\ 0 & 1 \end{bmatrix} \quad R_1 - 2R_1$$

But this is sufficient to tell us that the product of the two elementary matrices on the left, highlighted by the brace, is the inverse of our original matrix,

$$\begin{bmatrix} 1 & -2 \\ 0 & 1 \end{bmatrix}\begin{bmatrix} 1 & 0 \\ -1 & 1 \end{bmatrix} = \begin{bmatrix} 3 & -2 \\ -1 & 1 \end{bmatrix}$$

and indeed this is the case:

$$\begin{bmatrix} 1 & 2 \\ 1 & 3 \end{bmatrix}^{-1} = \begin{bmatrix} 3 & -2 \\ -1 & 1 \end{bmatrix}$$

The idea of the previous example can be turned into a general argument that explains why the row reduction method is guaranteed to find the inverse of a given matrix.

Proof that the recipe of Section 9.4 finds the inverse: Let A be a square matrix and suppose that A has been row reduced to I by elementary row operations

$$\rho_1\,,\ \rho_2\,,\ \ldots\,,\ \rho_m$$

in that order. Call the corresponding elementary matrices

$$E_1\,,\ E_2\,,\ \ldots\,,\ E_m\,.$$

By the Important Fact given above in this section, each of these produces the same effect as its corresponding elementary row operation, but by matrix multiplication on the left. Let B be the matrix obtained by applying $\rho_1\,,\ \rho_2\,,\ \ldots\,,\ \rho_m$ in the same order to the identity matrix I. This produces the following cascade of augmented matrices:

$$\begin{aligned}
\left[\,A\,\middle|\,I\,\right] &\sim \left[\,E_1A\,\middle|\,E_1\,\right] \quad \rho_1 \\
&\sim \left[\,E_2E_1A\,\middle|\,E_2E_1\,\right] \quad \rho_2 \\
&\sim \left[\,E_3E_2E_1A\,\middle|\,E_3E_2E_1\,\right] \quad \rho_3 \\
&\quad\ \vdots \qquad \vdots \qquad \vdots \\
&\sim \left[\,\underbrace{E_mE_{m-1}\ldots E_3E_2E_1A}_{I}\,\middle|\,\underbrace{E_mE_{m-1}\ldots E_3E_2E_1}_{B}\,\right] \quad \rho_m \\
&= \left[\,I\,\middle|\,B\,\right]
\end{aligned}$$

Thus, the following holds:

$$I \;=\; E_mE_{m-1}\ldots E_3E_2E_1A \;=\; \big(E_mE_{m-1}\ldots E_3E_2E_1\big)A \;=\; BA$$

By the Theorem from Section 9.1, this implies

$$I \;=\; BA \;=\; AB$$

and so

$$B \;=\; A^{-1}.$$

This completes the proof that the recipe finds the inverse of A when the process successfully row reduces the left-hand side of the augmented matrix to I.

We leave it as an exercise to show that if the process fails to row reduce the left-hand side of the augmented matrix to I then the inverse of A does not exist.

This proof also demonstrates the following (why?):

Every invertible matrix is a product of elementary matrices.

Chapter 9: Important Ideas and Useful Facts

9.1 The *inverse* of a matrix A is a matrix A^{-1} such that $AA^{-1} = A^{-1}A = I_n$ for some positive integer n. Only square matrices have inverses. When it exists, the inverse A^{-1} is unique.

9.2 Only half of the definition needs to be checked, in the following sense: if A is a square matrix and $AB = I$ or $BA = I$ then $AB = BA = I$ so that the inverse A^{-1} exists and equals B.

9.3 A matrix is *invertible* if its inverse exists. If A and B are invertible matrices of the same size then AB is invertible and $(AB)^{-1} = B^{-1}A^{-1}$.

9.4 Let $A = \begin{bmatrix} a & b \\ c & d \end{bmatrix}$. Define the *determinant* of A to be $\det A = ad - bc$. Then A is invertible if and only if $\det A \neq 0$, in which case

$$A^{-1} = \frac{1}{ad - bc}\begin{bmatrix} d & -b \\ -c & a \end{bmatrix}.$$

9.5 Let A be invertible. Then $(A^{-1})^{-1} = A$. If λ is a nonzero scalar then $(\lambda A)^{-1} = \frac{1}{\lambda}A^{-1}$. Define the following *powers* for any integer n:

$$A^n = \begin{cases} I & \text{if } n = 0 \\ \underbrace{AA\ldots A}_{n \text{ times}} & \text{if } n \text{ is positive} \\ \underbrace{A^{-1}A^{-1}\ldots A^{-1}}_{-n \text{ times}} & \text{if } n \text{ is negative} \end{cases}$$

If m and n are integers then $A^mA^n = A^{m+n}$ and $(A^m)^n = A^{mn}$.

9.6 A square matrix A is invertible if and only if the augmented matrix $[\,A \mid I\,]$ can be row reduced to $[\,I \mid B\,]$, in which case $A^{-1} = B$.

9.7 If a system of equations can be expressed in the form $A\mathbf{x} = \mathbf{b}$ where A is invertible, then $\mathbf{x} = A^{-1}\mathbf{b}$.

9.8 An $n \times n$ matrix is called *elementary* if it is the result of applying a single elementary row operation to the identity matrix I_n. If E is the elementary matrix obtained by applying the elementary row operation ρ to I_n, and A is any matrix with n rows, then the matrix product EA is the matrix obtained by applying ρ to A.

9.9 The inverse of an elementary matrix is elementary. Every invertible matrix is a product of elementary matrices.

Exercise 9.1 Explain why the matrix equations

$$AB = BA = I_n$$

imply that both A and B are square matrices of the same size.

Exercise 9.2 Find the inverse of each of the following matrices if it exists:

$$\text{(i)} \begin{bmatrix} 1 & 2 \\ 0 & 3 \end{bmatrix} \quad \text{(ii)} \begin{bmatrix} -6 & 3 \\ 4 & 1 \end{bmatrix} \quad \text{(iii)} \begin{bmatrix} 2 & 1 \\ -4 & -2 \end{bmatrix}$$

$$\text{(iv)} \begin{bmatrix} \cos\alpha & -\sin\alpha \\ \sin\alpha & \cos\alpha \end{bmatrix} \quad \text{(v)} \begin{bmatrix} 1 & 1 & 0 \\ 0 & -1 & 1 \\ 1 & 0 & 2 \end{bmatrix} \quad \text{(vi)} \begin{bmatrix} 0 & 1 & -3 \\ 1 & 3 & 0 \\ 2 & 2 & 11 \end{bmatrix}$$

$$\text{(vii)} \begin{bmatrix} 2 & -3 & 2 \\ 2 & 1 & -8 \\ 14 & -9 & -16 \end{bmatrix} \quad \text{(viii)} \begin{bmatrix} 0 & 1 & 1 \\ 1 & 0 & 1 \\ 1 & 1 & 0 \end{bmatrix} \quad \text{(ix)} \begin{bmatrix} 1 & 1 & 1 & 1 \\ 0 & 1 & 1 & 1 \\ 0 & 0 & 1 & 1 \\ 0 & 0 & 0 & 1 \end{bmatrix}$$

Exercise 9.3 Find the inverse of $\begin{bmatrix} 5 & -3 \\ 7 & -4 \end{bmatrix}$ and use it to solve for x, y, z and w where

$$\begin{bmatrix} 5 & -3 \\ 7 & -4 \end{bmatrix} \begin{bmatrix} x & y \\ z & w \end{bmatrix} = \begin{bmatrix} 11 & 4 \\ 15 & 5 \end{bmatrix}.$$

Exercise 9.4 Check that if A and B are invertible matrices of the same size then

$$(AB)^{-1} = B^{-1}A^{-1}.$$

Exercise 9.5 Suppose that A and D are invertible matrices. Explain briefly why the matrix equation $ABD = ACD$ implies $B = C$. Does it matter if A and D are of different sizes?

Exercise 9.6 Explain briefly why a square matrix with a row or column of zeros is not invertible.

Exercise 9.7 Use the method of row reduction to the identity matrix to derive (from scratch without looking at the formula given in Section 9.2) a criterion for invertibility and a general formula for the inverse of a 2×2 matrix.

Exercise 9.8 When is a *diagonal* matrix $\begin{bmatrix} d_1 & 0 & \cdots & 0 \\ 0 & d_2 & \cdots & 0 \\ \vdots & \vdots & \ddots & \vdots \\ 0 & 0 & \cdots & d_n \end{bmatrix}$ invertible, and what is its inverse?

Exercise 9.9 Let m and n be positive integers. Verify that if B and C are matrices such that

$$AB = A = CA$$

for all $m \times n$ matrices A, then $B = I_n$ and $C = I_m$.

Exercise 9.10* Which of the following are true for all invertible matrices A, B, C of the same size?

(i) $(ABC)^{-1} = A^{-1}B^{-1}C^{-1}$ (ii) $(ABA)^{-1} = A^{-1}B^{-1}A^{-1}$

(iii) $C^{-1}(ABC^{-1})^{-1}AB = I$ (iv) $(A+B)^{-1} = A^{-1} + B^{-1}$

(v) $A^{-1}(I+A)A = A + I$ (vi) $(A+I)(A^{-1} - I) = A^{-1} - A$

(vii) $A^2 - 2A + I = 0 \implies A^{-1} = 2I - A$

(viii) $A^2 - 2A + I = 0 \implies A = I$

Exercise 9.11* Let n be a positive integer and M_n the $n \times n$ matrix each of whose entries is 1. Verify that $I_n - M_n$ is invertible if and only if $n \geq 2$, in which case

$$(I_n - M_n)^{-1} = I_n - \frac{1}{n-1}M_n .$$

Exercise 9.12* Verify that if E is the elementary matrix obtained by applying the elementary row operation ρ to I_n, and A is any matrix, then the matrix product EA is the matrix obtained by applying ρ to A.

Exercise 9.13* Prove that the inverse of an elementary matrix is elementary, and deduce, from the proof at the end of the chapter, that every invertible matrix is a product of elementary matrices.

Exercise 9.14* Prove that if the recipe of Section 9.4 row reduces the augmented matrix $\left[\, A \mid I \,\right]$ so that a row of zeros appears on the left-hand side, then A is not invertible.

Exercise 9.15* Prove that if A is a square matrix and $AB = I$ then

$$AB \;=\; BA \;=\; I$$

so that the inverse of A exists and equals B.

Exercise 9.16* Suppose that A is a matrix such that

$$AB \;=\; I_n \qquad \text{and} \qquad BA \;=\; I_m$$

for some matrix B and positive integers m and n. Prove that $m = n$.

Exercise 9.17* A matrix A is called *symmetric* if $A = A^T$ and *skew-symmetric* if $A = -A^T$, where A^T is the transpose of A (see Exercise 8.18). Prove that inverses of symmetric and skew-symmetric matrices, when they exist, are symmetric and skew-symmetric, respectively.

Exercise 9.18* Prove that every square matrix is the sum of a symmetric matrix and a skew-symmetric matrix.

Exercise 9.19** Suppose that A and B are arrays of numbers where the rows and columns are indexed by positive integers, and each row and column contains finitely many nonzero entries. Then it is sensible to take the dot product of each row of A with each column of B (by ignoring summands involving infinitely many zeros) to form the product AB. Let I denote the infinite array with 1's down the diagonal and 0's elsewhere. Is it possible to find A and B such that $AB = I$ but $BA \neq I$?

10 Determinants

WATCH IT NOW on DVD VIDEO

- **Determinants** (Section 10.1)
  Time: 6.29
- **Determinant method for cross products** (Section 10.2)
  Time: 3.43
- **Orientation of a triangle** (Section 10.4)
  Time: 8.01
- **Numerical example of orientation** (Section 10.4)
  Time: 3.00

10 Determinants

In this chapter we associate with each square matrix, no matter how large, a single scalar (real number) called its *determinant*, in such a way that the following truly remarkable property holds:

Theorem: A square matrix is invertible if and only if its determinant is nonzero.

Despite the simplicity of the statement, this is one of the most difficult results in this book. An inductive definition and proof are outlined as a series of starred and double-starred exercises at the end of this chapter. It is well worth the effort for an advanced student to attempt these exercises or some part of them. It requires endurance and a very good grasp of the technique of proof by mathematical induction. This theorem will be an essential link in the chain for our discussion of the existence of eigenvalues in Chapter 11.

The determinant of a matrix will be an "alternating sum", which means that we alternate addition and subtraction a certain number of times. Alternating sums involve some kind of iterated "compensation" and are ubiquitous in mathematics. For example, the *Euler characteristic* in topology is an alternating sum. The *Inclusion/Exclusion Principle* is a counting technique in combinatorics that uses an alternating sum.

The history of the determinant is quite long and complicated. The mathematician Gauss, mentioned prominently in Chapter 7, was one of many key players in the evolution of the concept. Determinants historically preceded matrices and were related to systems of equations. Exercise 0.8, concerning the existence of solutions of two equations in two variables, is precisely how determinants emerged in embryonic form. Exercise 0.8 was starred, but the star can now be removed (see Exercise 10.11), in light of our characterisation of invertibility of 2×2 matrices in Section 9.2. Indeed, we have already seen the *determinant* of a 2×2 matrix:

$$\det \begin{bmatrix} a & b \\ c & d \end{bmatrix} = \begin{vmatrix} a & b \\ c & d \end{vmatrix} = ad - bc$$

and, when this is nonzero, the formula for the inverse of a 2×2 matrix:

$$\begin{bmatrix} a & b \\ c & d \end{bmatrix}^{-1} = \frac{1}{ad - bc} \begin{bmatrix} d & -b \\ -c & a \end{bmatrix}$$

This proves the "if" half of the above theorem in this small case. For 1×1 matrices we define the *determinant*

$$\det \begin{bmatrix} a \end{bmatrix} = a\,.$$

The theorem then is saying that a real number is invertible if and only if it is nonzero, which of course is true!

> **Important:** We never use $|a|$ to denote $\det \begin{bmatrix} a \end{bmatrix}$ because $|a|$ always denotes the magnitude of a when $a \in \mathbb{R}$.

For $n \times n$ matrices, where $n > 2$, the description of the determinant becomes more complicated, but complete familiarity with the 3×3 case will be sufficient for our purposes and will guide the reader towards the general case. It happens that there are many different ways of defining the determinant, all giving the same answer!

10.1 Determinant of a 3×3 matrix

Consider the following matrix:

$$A = \begin{bmatrix} a & b & c \\ d & e & f \\ g & h & k \end{bmatrix}$$

Define the *determinant* of A to be

$$\det A = |A| = a\begin{vmatrix} e & f \\ h & k \end{vmatrix} - b\begin{vmatrix} d & f \\ g & k \end{vmatrix} + c\begin{vmatrix} d & e \\ g & h \end{vmatrix}$$

called the *expansion along the first row*. There are several things to notice:

- each element of the first row of A is multiplied by the determinant of a 2×2 matrix
- there is an *adjustment factor* of a minus sign for the product involving the second element of the first row
- the 2×2 determinants, in each case, are obtained by deleting the row and column containing the entry that is being used as the other factor.

To emphasise the third and second points: if you cover up the first row and second column of A, then what remains forms a 2×2 matrix with determinant

$$\begin{vmatrix} d & f \\ g & k \end{vmatrix},$$

which is what b is being multiplied by, with an adjustment of -1. The adjustment factor is part of a general "chequerboard pattern" that starts with $+$ in the top left-hand corner:

$$\begin{bmatrix} + & - & + \\ - & + & - \\ + & - & + \end{bmatrix}$$

There are "hidden" $+1$ factors in the first and third summands in the above definition of the determinant.

When these features are understood, it is easy then to define

expansion along any row or column.

To do this, choose any particular row or column. Move along the row, or down the column, multiplying each entry by the determinant of the 2×2 matrix left over by deleting the row and column of that element. Add adjustment factors from the above pattern. Add up the three expressions for the given row or column, and the result is another definition of the determinant.

For example, here is the *expansion along the second row* of A:

$$\det A = |A| = -d\begin{vmatrix} b & c \\ h & k \end{vmatrix} + e\begin{vmatrix} a & c \\ g & k \end{vmatrix} - f\begin{vmatrix} a & b \\ g & h \end{vmatrix}$$

Here is the *expansion down the third column* of A:

$$\det A = |A| = c\begin{vmatrix} d & e \\ g & h \end{vmatrix} - f\begin{vmatrix} a & b \\ g & h \end{vmatrix} + k\begin{vmatrix} a & b \\ d & e \end{vmatrix}$$

There are three rows and three columns, so six of these expansions, so we won't go any further listing them.

> **Amazing Fact:** Expansion along any row or down any column always produces the same answer, and can be taken as the definition of the determinant.

Example: Consider the matrix $\begin{bmatrix} -2 & 3 & 2 \\ 4 & -1 & -1 \\ 1 & 2 & 0 \end{bmatrix}$.

Expanding along the first row gives

$$\begin{aligned} \begin{vmatrix} -2 & 3 & 2 \\ 4 & -1 & -1 \\ 1 & 2 & 0 \end{vmatrix} &= -2\begin{vmatrix} -1 & -1 \\ 2 & 0 \end{vmatrix} - 3\begin{vmatrix} 4 & -1 \\ 1 & 0 \end{vmatrix} + 2\begin{vmatrix} 4 & -1 \\ 1 & 2 \end{vmatrix} \\ &= -2\,(0+2) \;-\; 3\,(0+1) \;+\; 2\,(8+1) \\ &= -4-3+18 \;=\; 11\,. \end{aligned}$$

Expanding along the second row gives

$$\begin{aligned} \begin{vmatrix} -2 & 3 & 2 \\ 4 & -1 & -1 \\ 1 & 2 & 0 \end{vmatrix} &= -4\begin{vmatrix} 3 & 2 \\ 2 & 0 \end{vmatrix} - 1\begin{vmatrix} -2 & 2 \\ 1 & 0 \end{vmatrix} + 1\begin{vmatrix} -2 & 3 \\ 1 & 2 \end{vmatrix} \\ &= -4\,(0-4) \;-\; 1\,(0-2) \;+\; 1\,(-4-3) \\ &= 16+2-7 \;=\; 11\,. \end{aligned}$$

Expanding down the second column gives

$$\begin{aligned} \begin{vmatrix} -2 & 3 & 2 \\ 4 & -1 & -1 \\ 1 & 2 & 0 \end{vmatrix} &= -3\begin{vmatrix} 4 & -1 \\ 1 & 0 \end{vmatrix} - 1\begin{vmatrix} -2 & 2 \\ 1 & 0 \end{vmatrix} - 2\begin{vmatrix} -2 & 2 \\ 4 & -1 \end{vmatrix} \\ &= -3\,(0+1) \;-\; 1\,(0-2) \;-\; 2\,(2-8) \\ &= -3+2+12 \;=\; 11\,. \end{aligned}$$

Expanding down the third column gives

$$\begin{aligned} \begin{vmatrix} -2 & 3 & 2 \\ 4 & -1 & -1 \\ 1 & 2 & 0 \end{vmatrix} &= 2\begin{vmatrix} 4 & -1 \\ 1 & 2 \end{vmatrix} + 1\begin{vmatrix} -2 & 3 \\ 1 & 2 \end{vmatrix} + 0 \\ &= 2\,(8+1) \;+\; 1\,(-4-3) \;=\; 18-7 \;=\; 11\,. \end{aligned}$$

Notice that in this last case the zero in the third column abbreviated the calculation slightly.

> Usually, choose to expand along a row or down a column that contains a zero or zeros.

> The expansion method generalises to 4×4 matrices and beyond to give a well-defined definition of *determinant* for all square matrices!

For 4×4 matrices use the following "chequerboard":

$$\begin{bmatrix} + & - & + & - \\ - & + & - & + \\ + & - & + & - \\ - & + & - & + \end{bmatrix}$$

Example: Find the determinant of the matrix $\begin{bmatrix} 3 & 1 & 4 & 2 \\ -1 & 1 & 0 & 2 \\ 0 & 0 & -2 & 1 \\ 1 & 2 & 0 & 1 \end{bmatrix}$.

Solution: We choose to expand along the third row, because it contains two zeros, giving

$$\begin{vmatrix} 3 & 1 & 4 & 2 \\ -1 & 1 & 0 & 2 \\ 0 & 0 & -2 & 1 \\ 1 & 2 & 0 & 1 \end{vmatrix} = 0 - 0 - 2\begin{vmatrix} 3 & 1 & 2 \\ -1 & 1 & 2 \\ 1 & 2 & 1 \end{vmatrix} - 1\begin{vmatrix} 3 & 1 & 4 \\ -1 & 1 & 0 \\ 1 & 2 & 0 \end{vmatrix}$$

$$= -2\left(3\begin{vmatrix} 1 & 2 \\ 2 & 1 \end{vmatrix} - \begin{vmatrix} -1 & 2 \\ 1 & 1 \end{vmatrix} + 2\begin{vmatrix} -1 & 1 \\ 1 & 2 \end{vmatrix}\right)$$

$$- \left(4\begin{vmatrix} -1 & 1 \\ 1 & 2 \end{vmatrix} - 0 + 0\right)$$

$$= -2\Big(3(1-4) - (-1-2) + 2(-2-1)\Big) - 4(-2-1)$$

$$= -2(-9+3-6) + 12 = 24 + 12 = 36\,.$$

Working directly from an expansion is quite labour intensive, but we shall see a list of properties of determinants in Section 10.3 which greatly simplify the process.

10.2 Cross products revisited

The cross product of geometric vectors, the reader might recall, had a complicated algebraic definition. But on inspection, the reader can see that the formulae for the coefficients of $\mathbf{i}$, $\mathbf{j}$ and $\mathbf{k}$ look like 2×2 determinants! This resemblance is not an accident, and can be exploited to give yet another method of calculating cross products:

Determinant Method for Cross Products:

If

$$\mathbf{v} = a\,\mathbf{i} + b\,\mathbf{j} + c\,\mathbf{k}$$

and

$$\mathbf{w} = d\,\mathbf{i} + e\,\mathbf{j} + f\,\mathbf{k}$$

then

$$\mathbf{v} \times \mathbf{w} = \begin{vmatrix} \mathbf{i} & \mathbf{j} & \mathbf{k} \\ a & b & c \\ d & c & f \end{vmatrix}.$$

This is one instance where a matrix in this course has entries which are not all real numbers. When evaluating the 3×3 determinant above there are no products involving just $\mathbf{i}$, $\mathbf{j}$ and $\mathbf{k}$. The expression for the determinant can be interpreted sensibly and in fact evaluates to the usual cross product.

Example:

$$(\mathbf{i} + 2\,\mathbf{j} + 3\,\mathbf{k}) \times (4\,\mathbf{i} + 5\,\mathbf{j} + 6\,\mathbf{k}) = \begin{vmatrix} \mathbf{i} & \mathbf{j} & \mathbf{k} \\ 1 & 2 & 3 \\ 4 & 5 & 6 \end{vmatrix}$$

$$= \mathbf{i} \begin{vmatrix} 2 & 3 \\ 5 & 6 \end{vmatrix} - \mathbf{j} \begin{vmatrix} 1 & 3 \\ 4 & 6 \end{vmatrix} + \mathbf{k} \begin{vmatrix} 1 & 2 \\ 4 & 5 \end{vmatrix} = -3\,\mathbf{i} + 6\,\mathbf{j} - 3\,\mathbf{k}$$

10.3 Properties of determinants

The following properties are very difficult to prove, but you should have no hesitation in using them (more instances of the Plateau Principle!).

Properties of determinants: Let A and B be square matrices of the same size.

(1) The determinant is *multiplicative*, which means

$$\det(AB) = (\det A)(\det B)\,.$$

(2) If B is obtained from A by swapping two rows or swapping two columns then

$$\det B = -\det A\,.$$

(3) If B is obtained from A by multiplying a row or column by λ then

$$\det B = \lambda \det A\,.$$

(4) If B is obtained from A by adding a multiple of one row [column] to another row [column] then

$$\det B = \det A\,.$$

(5) If B is the *transpose* of A, that is, obtained from A by interchanging rows and columns, then

$$\det B = \det A\,.$$

As a consequence of property (1), we can relate the determinant of any invertible matrix to the determinant of its inverse:

Corollary: If A is invertible then $\det A \neq 0$ and

$$\det A^{-1} = \frac{1}{\det A}\,.$$

Proof: Suppose that A is invertible. Then $AA^{-1} = I$. It is easy to check that $\det I = 1$. By property (1),

$$1 = \det I = \det(AA^{-1}) = (\det A)(\det A^{-1}).$$

If $\det A = 0$ then

$$1 = 0(\det A^{-1}) = 0,$$

which is impossible. Hence, $\det A \neq 0$. By dividing through,

$$\det A^{-1} = \frac{1}{\det A}$$

and the corollary is proved.

In Chapter 7, we introduced elementary row operations on matrices, which were convenient for the application at the time, which was to manipulate augmented matrices associated with systems of linear equations. Working in general matrix arithmetic, it is very useful also to have access to *elementary column operations*, which are identical to their row counterparts, but just replacing the word "row" with "column". Caution needs to be exercised in using any new powerful tool, but in evaluating determinants, we can freely mix up both kinds of operations:

> Properties (2), (3) and (4) enable the simplification of evaluation of determinants by applying elementary row and column operations

Example: In the following calculation, the first step is to interchange the first and second columns, which multiplies the determinant by -1. All of the subsequent steps, except the last, employ row operations. The last step is simply to multiply diagonal entries together, using a general property about *triangular matrices* (below).

$$\begin{vmatrix} 3 & 1 & 1 \\ 1 & 2 & 3 \\ -2 & 1 & -4 \end{vmatrix} = -\begin{vmatrix} 1 & 3 & 1 \\ 2 & 1 & 3 \\ 1 & -2 & -4 \end{vmatrix} = -\begin{vmatrix} 1 & 3 & 1 \\ 0 & -5 & 1 \\ 0 & -5 & -5 \end{vmatrix}$$

$$= -\begin{vmatrix} 1 & 3 & 1 \\ 0 & -5 & 1 \\ 0 & 0 & -6 \end{vmatrix} = -(1)(-5)(-6) = -30$$

A matrix is called *triangular* if all entries above or below the diagonal are zero. The shading below represents elements that are possibly (but not necessarily) nonzero, and of course the large zero represents zero entries everywhere else.

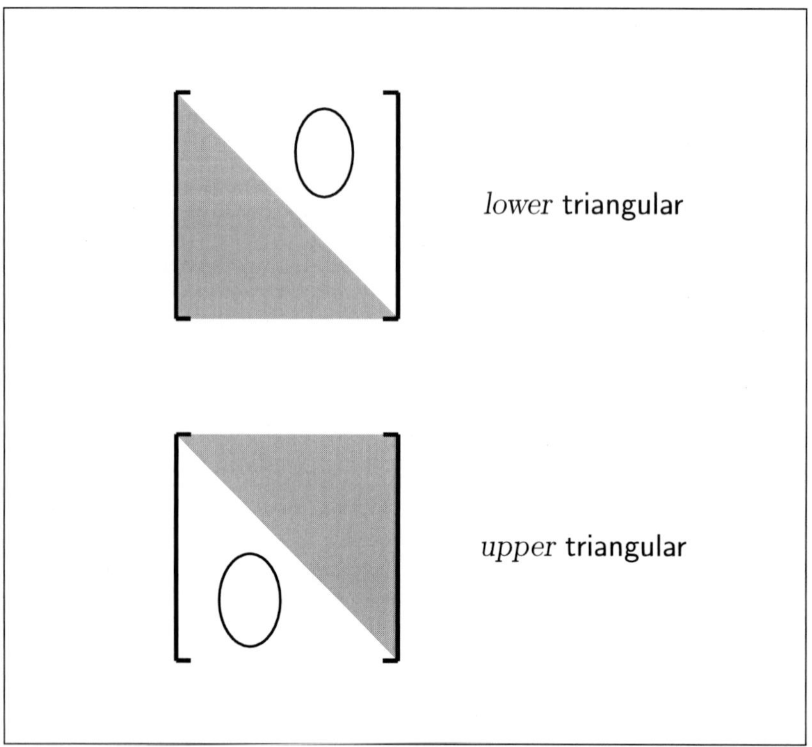

The last step in the previous example used the following property (which is easy to prove, for a change!):

> The determinant of a triangular matrix is the product of the diagonal elements.

In the previous example, the final matrix was upper triangular. Here is a 4×4 lower triangular example:

$$\begin{vmatrix} -1 & 0 & 0 & 0 \\ 3 & -1 & 0 & 0 \\ 2 & 0 & 4 & 0 \\ 0 & 6 & 5 & 8 \end{vmatrix} = (-1)(-1)(4)(8) = 32$$

Property (5) turns out again to be difficult to prove, but is very useful. It is so powerful that it can be used to explain (Exercise 10.18) why expansion down a column produces the same answer as expansion along a row. To transpose a matrix, the rows and columns are interchanged, which one can think of as a reflection in the diagonal:

$$\begin{bmatrix} a & b \\ c & d \end{bmatrix}^T = \begin{bmatrix} a & c \\ b & d \end{bmatrix}, \qquad \begin{bmatrix} a & b & c \\ d & e & f \\ g & h & i \end{bmatrix}^T = \begin{bmatrix} a & d & g \\ b & e & h \\ c & f & i \end{bmatrix}$$

$$\begin{vmatrix} a & b \\ c & d \end{vmatrix} = ad - bc = \begin{vmatrix} a & c \\ b & d \end{vmatrix}$$

$$\begin{vmatrix} 3 & 1 & 1 \\ 1 & 2 & 3 \\ -2 & 1 & -4 \end{vmatrix} = -30 = \begin{vmatrix} 3 & 1 & -2 \\ 1 & 2 & 1 \\ 1 & 3 & -4 \end{vmatrix}$$

10.4 Orientation of a triangle

We finish this chapter with an application to plane geometry. The following two figures are intended to represent the same triangle, except that the labels of the vertices B and C have been interchanged.

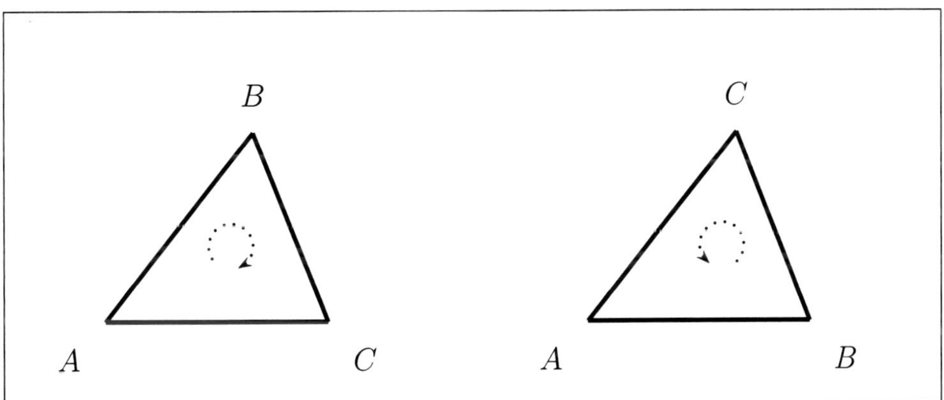

If we refer to the triangle as

$$\triangle ABC$$

then there is an implied *orientation*, as one follows the vertices in order. In the first case, one moves *clockwise* (direction of rotation of hands of a clock), and in the second case, one moves *anticlockwise*, the reverse direction. We derive a numerical test for determining orientation, given coordinates of the vertices, say

$$A = A(x_1, y_1), \qquad B = B(x_2, y_2), \qquad C = C(x_3, y_3).$$

Let δ be the determinant of the 3×3 matrix whose first row comprises 1's, the second row comprises the x-coordinates of our points, and the third row the y-coordinates:

$$\boxed{\delta = \delta_{ABC} = \begin{vmatrix} 1 & 1 & 1 \\ x_1 & x_2 & x_3 \\ y_1 & y_2 & y_3 \end{vmatrix}}$$

Armed with the techniques of the previous section, we can rewrite this determinant by the following sequence:

- applying two column operations, subtracting the first column from each of the second and third columns
- followed by expansion along the first row
- followed by transposing the matrix.

These do not change its value:

$$\begin{aligned} \delta &= \begin{vmatrix} 1 & 1 & 1 \\ x_1 & x_2 & x_3 \\ y_1 & y_2 & y_3 \end{vmatrix} \\ &= \begin{vmatrix} 1 & 0 & 0 \\ x_1 & x_2 - x_1 & x_3 - x_1 \\ y_1 & y_2 - y_1 & y_3 - y_1 \end{vmatrix} \\ &= \begin{vmatrix} x_2 - x_1 & x_3 - x_1 \\ y_2 - y_1 & y_3 - y_1 \end{vmatrix} = \begin{vmatrix} x_2 - x_1 & y_2 - y_1 \\ x_3 - x_1 & y_3 - y_1 \end{vmatrix} \end{aligned}$$

Now we are going to do something that looks strange (and probably looks unmotivated, but then mathematicians are fond of covering their tracks). We are going to partially "reverse" the above process, but from a different point of view.

Starting with a δ scalar multiple of the unit vector $\mathbf{k}$, we rewrite $\delta\,\mathbf{k}$ by

- reconstructing a 3×3 matrix, now with a first row consisting of vectors and a very simple third column
- and recognising the answer as a cross product of vectors in space:

$$\begin{aligned}\delta\,\mathbf{k} &= \begin{vmatrix} x_2 - x_1 & y_2 - y_1 \\ x_3 - x_1 & y_3 - y_1 \end{vmatrix}\mathbf{k} = \begin{vmatrix} \mathbf{i} & \mathbf{j} & \mathbf{k} \\ x_2 - x_1 & y_2 - y_1 & 0 \\ x_3 - x_1 & y_3 - y_1 & 0 \end{vmatrix} \\ &= \overrightarrow{AB} \times \overrightarrow{AC}\,,\end{aligned}$$

since

$$\overrightarrow{AB} = (x_2 - x_1)\,\mathbf{i} + (y_2 - y_1)\,\mathbf{j} + 0\,\mathbf{k}$$

and

$$\overrightarrow{AC} = (x_3 - x_1)\,\mathbf{i} + (y_3 - y_1)\,\mathbf{j} + 0\,\mathbf{k}\,,$$

where we think of our triangle $\triangle ABC$ in the plane embedded in space.

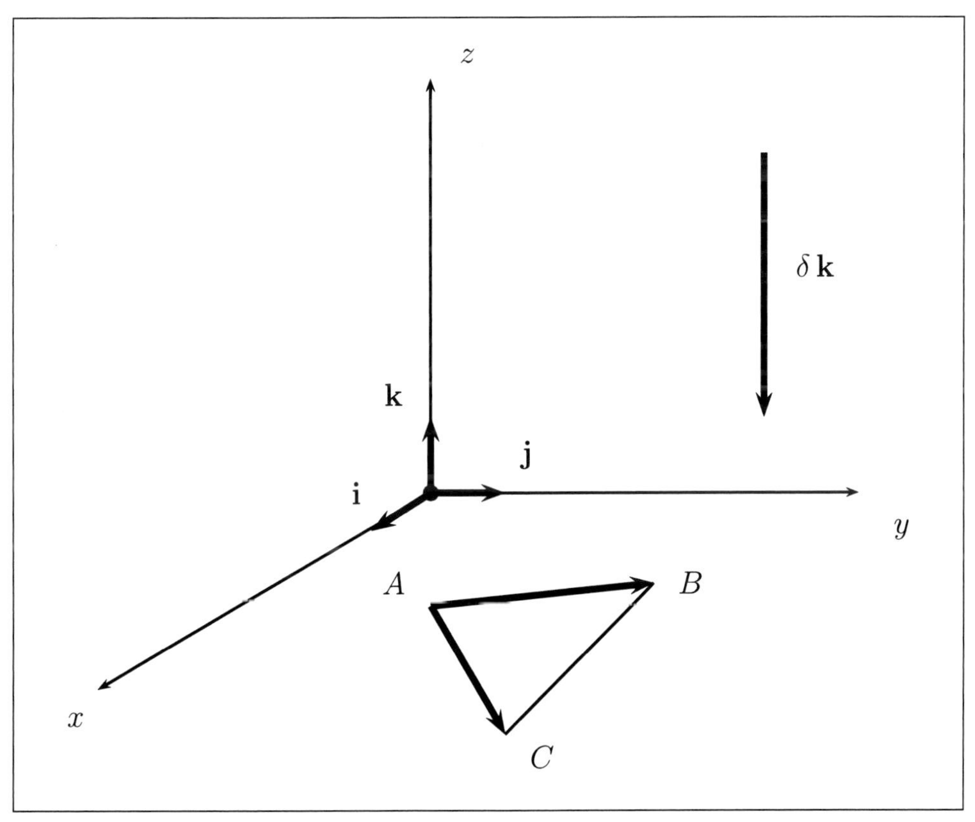

In the previous diagram, the triangle $\triangle ABC$ has been drawn oriented clockwise, if you are looking down "from above", that is, from the positive z-direction. Using the Right-Hand Rule, the cross product $\delta\,\mathbf{k} = \overrightarrow{AB} \times \overrightarrow{AC}$ points downwards, so that δ is negative.

On the other hand, if we drew the triangle oriented anticlockwise, as in the following diagram, then the cross product would point upwards, so that δ would be positive.

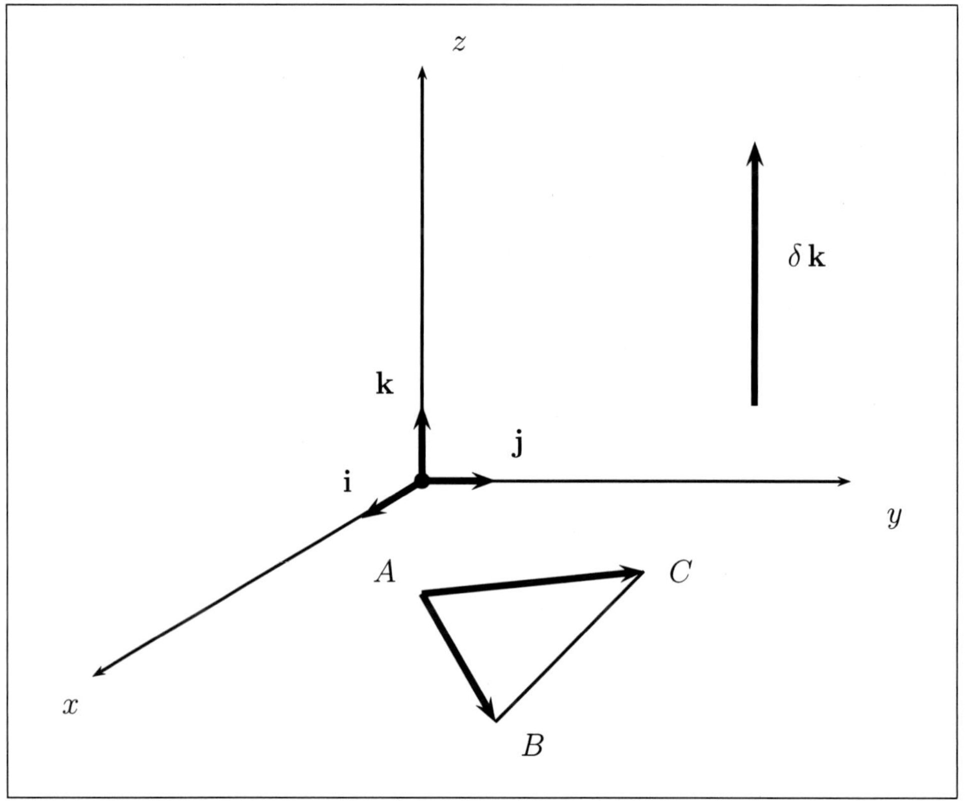

There is one other case, when A, B and C are collinear, so that the triangle becomes degenerate, and the cross product is the zero vector, so $\delta = 0$. We have thus derived the following:

Orientation criterion for a triangle:

$$\triangle ABC \text{ is } \begin{cases} \text{anticlockwise} & \text{if } \delta > 0 \\ \text{clockwise} & \text{if } \delta < 0 \\ \text{degenerate} & \text{if } \delta = 0 \end{cases}$$

This criterion is used extensively in computational geometry, when orientation of motion about the plane is important, for example in the development of efficient convex hull algorithms.

Example: Use the criterion to determine whether the triangle $\triangle PQR$ is oriented clockwise or anticlockwise, where

$$P = P(4,9)\,, \qquad Q = Q(11,20)\,, \qquad R = R(14,5)\,,$$

and whether the points $S(6,12)$ and $T(6,13)$ lie inside or outside this triangle.

Solution: We could of course plot the points on a graph and have a look. But imagine we are blind like a computer and only have access to the numerical information:

$$\delta_{PQR} = \begin{vmatrix} 1 & 1 & 1 \\ 4 & 11 & 14 \\ 9 & 20 & 5 \end{vmatrix} = \begin{vmatrix} 1 & 0 & 0 \\ 4 & 7 & 10 \\ 9 & 11 & -4 \end{vmatrix} = -28 - 110 = -138 < 0$$

$$\delta_{PQS} = \begin{vmatrix} 1 & 1 & 1 \\ 4 & 11 & 6 \\ 9 & 20 & 12 \end{vmatrix} = \begin{vmatrix} 1 & 0 & 0 \\ 4 & 7 & 2 \\ 9 & 11 & 3 \end{vmatrix} = 21 - 22 = -1 < 0$$

$$\delta_{PQT} = \begin{vmatrix} 1 & 1 & 1 \\ 4 & 11 & 6 \\ 9 & 20 & 13 \end{vmatrix} = \begin{vmatrix} 1 & 0 & 0 \\ 4 & 7 & 2 \\ 9 & 11 & 4 \end{vmatrix} = 28 - 22 = 6 > 0$$

From the first and third determinants, $\triangle PQR$ is oriented clockwise and $\triangle PQT$ anticlockwise, so certainly T is outside $\triangle PQR$. Since $\triangle PQR$ and $\triangle PQS$ have the same orientation (clockwise), we suspect that S lies (just) inside the triangle $\triangle PQR$, especially since S is close to T. However, to be sure, we calculate further:

$$\delta_{PSR} = \begin{vmatrix} 1 & 1 & 1 \\ 4 & 6 & 14 \\ 9 & 12 & 5 \end{vmatrix} = \begin{vmatrix} 1 & 0 & 0 \\ 4 & 2 & 10 \\ 9 & 3 & -4 \end{vmatrix} = -8 - 30 = -38 < 0$$

$$\delta_{SQR} = \begin{vmatrix} 1 & 1 & 1 \\ 6 & 11 & 14 \\ 12 & 20 & 5 \end{vmatrix} = \begin{vmatrix} 1 & 0 & 0 \\ 6 & 5 & 8 \\ 12 & 8 & -7 \end{vmatrix} = -35 - 64 = -99 < 0$$

The triangle $\triangle PQR$, and all of those obtained by replacing each of P, Q, R in turn by S, namely, $\triangle SQR$, $\triangle PSR$ and $\triangle PQS$, have the same orientation, so indeed S lies inside. The relative sizes of the determinants suggest that S is only just inside and close to the line joining P to Q.

Chapter 10: Important Ideas and Useful Facts

10.1 The *determinant* of a 1×1 matrix $[a]$ is simply the entry a.

10.2 The *determinant* of a 2×2 matrix $A = \begin{bmatrix} a & b \\ c & d \end{bmatrix}$ is

$$\det A = \begin{vmatrix} a & b \\ c & d \end{vmatrix} = ad - bc\,.$$

10.3 The *determinant* of a 3×3 matrix $A = \begin{bmatrix} a & b & c \\ d & e & f \\ g & h & k \end{bmatrix}$ is

$$\det A = |A| = a\begin{vmatrix} e & f \\ h & k \end{vmatrix} - b\begin{vmatrix} d & f \\ g & k \end{vmatrix} + c\begin{vmatrix} d & e \\ g & h \end{vmatrix}$$

called the *expansion along the first row*, where the smaller determinant arises by ignoring the row and column of the entry being used as a coefficient.

10.4 Expanding along any row or down any column of a square matrix A produces the same real number, called the *determinant* of A, denoted by $\det A$ or $|A|$, provided one uses adjustment factors given by the chequerboard patterns

$$\begin{bmatrix} + & - & + \\ - & + & - \\ + & - & + \end{bmatrix}, \quad \begin{bmatrix} + & - & + & - \\ - & + & - & + \\ + & - & + & - \\ - & + & - & + \end{bmatrix}$$

and so on to higher dimensions. Using sigma notation, if $A = [a_{ij}]$ is an $n \times n$ matrix and A_{ij} denotes the $(n-1)\times(n-1)$ matrix obtained by deleting the ith row and jth column of A, then expanding along the ith row (for fixed i) becomes

$$\det A = \sum_{j=1}^{n} (-1)^{i+j} a_{ij} \det A_{ij}\,,$$

and down the jth column (for fixed j) becomes

$$\det A = \sum_{i=1}^{n} (-1)^{i+j} a_{ij} \det A_{ij}\,.$$

10.5 Determinant method for cross products: If $\mathbf{v} = a\,\mathbf{i} + b\,\mathbf{j} + c\,\mathbf{k}$ and $\mathbf{w} = d\,\mathbf{i} + e\,\mathbf{j} + f\,\mathbf{k}$ then

$$\mathbf{v} \times \mathbf{w} = \begin{vmatrix} \mathbf{i} & \mathbf{j} & \mathbf{k} \\ a & b & c \\ d & e & f \end{vmatrix}.$$

10.6 Multiplicative property: $\det(AB) = (\det A)(\det B)$

10.7 Invertibility criterion: A square matrix is invertible if and only if its determinant is nonzero.

10.8 If B is obtained from A by swapping two rows or swapping two columns then

$$\det B = -\det A\,.$$

10.9 If B is obtained from A by multiplying a row or column by λ then

$$\det B = \lambda \det A\,.$$

10.10 If B is obtained from A by adding a multiple of one row [column] to another row [column] then

$$\det B = \det A\,.$$

10.11 If B is the *transpose* of A, that is, obtained by interchanging rows and columns, then

$$\det B = \det A\,.$$

10.12 If A is *triangular*, that is, all entries above or below the diagonal are zero, then $\det A$ is the product of the diagonal elements.

10.13 Orientation test for a triangle in the plane: If $A = A(x_1, y_1)$, $B = B(x_2, y_2)$ and $C = C(x_3, y_3)$ are points in the plane and

$$\delta = \delta_{ABC} = \begin{vmatrix} 1 & 1 & 1 \\ x_1 & x_2 & x_3 \\ y_1 & y_2 & y_3 \end{vmatrix}$$

then

$$\triangle ABC \text{ is } \begin{cases} \text{anticlockwise} & \text{if } \delta > 0 \\ \text{clockwise} & \text{if } \delta < 0 \\ \text{degenerate} & \text{if } \delta = 0\,. \end{cases}$$

Exercise 10.1 Find the following 2×2 determinants:

$$\begin{vmatrix} 1 & 0 \\ 0 & 2 \end{vmatrix}, \quad \begin{vmatrix} 5 & 4 \\ 3 & 3 \end{vmatrix}, \quad \begin{vmatrix} 1 & -2 \\ -1 & 3 \end{vmatrix}, \quad \begin{vmatrix} 1 & -1 \\ -2 & 3 \end{vmatrix}, \quad \begin{vmatrix} -1 & 1 \\ 3 & -2 \end{vmatrix}$$

Exercise 10.2 Find the 3×3 determinant $\begin{vmatrix} 2 & -3 & -2 \\ -1 & 3 & 4 \\ -7 & -2 & 8 \end{vmatrix}$ by expanding along the first row. Confirm your answer, also, by expanding (i) along the second row; (ii) along the third row; (iii) down the third column.

Exercise 10.3 Write down quickly the determinants of the following matrices:

(i) $\begin{bmatrix} 5 & 0 & 0 \\ 3 & -2 & 0 \\ 1 & -5 & -1 \end{bmatrix}$ (ii) $\begin{bmatrix} 3 & 3 & 8 \\ 0 & -6 & -7 \\ 0 & 0 & 2 \end{bmatrix}$ (iii) $\begin{bmatrix} -4 & -5 & 11 \\ 0 & 0 & 0 \\ 2 & -1 & 2 \end{bmatrix}$

(iv) $\begin{bmatrix} 0 & -1 & 0 \\ 0 & 0 & -2 \\ 1 & 0 & 0 \end{bmatrix}$ (v) $\begin{bmatrix} 0 & 0 & 5 \\ 6 & 0 & 0 \\ 0 & -3 & 0 \end{bmatrix}$ (vi) $\begin{bmatrix} 4 & 0 & 0 & 0 \\ 3 & -2 & 0 & 0 \\ 1 & -5 & 2 & 0 \\ -6 & -3 & -7 & -1 \end{bmatrix}$

Exercise 10.4 Use elementary row and column operations, or otherwise, to find the following:

(i) $\begin{vmatrix} 1 & 1 & 1 \\ -2 & 1 & 3 \\ 4 & 5 & 1 \end{vmatrix}$ (ii) $\begin{vmatrix} 1 & -1 & 1 \\ -1 & 1 & -1 \\ -1 & -1 & 1 \end{vmatrix}$ (iii) $\begin{vmatrix} 2 & 3 & 6 & 2 \\ 3 & 1 & 1 & -2 \\ 4 & 0 & 1 & 3 \\ 1 & 1 & 2 & -1 \end{vmatrix}$

Exercise 10.5 Evaluate the determinants of the following matrices:

(i) $\begin{bmatrix} 0 & 1 \\ 1 & 0 \end{bmatrix}$ (ii) $\begin{bmatrix} 0 & 1 & 0 \\ 0 & 0 & 1 \\ 1 & 0 & 0 \end{bmatrix}$ (iii) $\begin{bmatrix} 1 & 0 & 2 \\ 2 & -2 & 0 \\ 2 & 2 & 1 \end{bmatrix}$

(iv) $\begin{bmatrix} 2 & 0 & -3 \\ -4 & 0 & -1 \\ 6 & 9 & 3 \end{bmatrix}$ (v) $\begin{bmatrix} 7 & 2 & 4 \\ -1 & 6 & -2 \\ -1 & 1 & 1 \end{bmatrix}$ (vi) $\begin{bmatrix} a & b & c \\ d & 0 & d \\ a & b & d \end{bmatrix}$

(vii) $\begin{bmatrix} t+3 & -1 & 1 \\ 5 & t-3 & 1 \\ 6 & -6 & t+4 \end{bmatrix}$ (viii) $\begin{bmatrix} 2 & 5 & -3 & -2 \\ -2 & -3 & 2 & -5 \\ 1 & 3 & -2 & 2 \\ -1 & -6 & 4 & 3 \end{bmatrix}$

Exercise 10.6 Explain briefly why a square matrix with two identical rows or two identical columns has zero determinant.

Exercise 10.7 Decide whether the following equations are true for all 2×2 matrices A and B:

(i) $\det(AB) = (\det A)(\det B)$ (ii) $\det(-A) = \det A$

(iii) $\det(A+B) = (\det A) + (\det B)$ (iv) $\det(2A) = 2\det A$

Exercise 10.8 Find $\mathbf{v} \times \mathbf{w} = \begin{vmatrix} \mathbf{i} & \mathbf{j} & \mathbf{k} \\ v_1 & v_2 & v_3 \\ w_1 & w_2 & w_3 \end{vmatrix}$ in each of the following cases:

(i) $\mathbf{v} = \mathbf{i} - 2\mathbf{j} + 3\mathbf{k}$, $\mathbf{w} = 4\mathbf{i} + 5\mathbf{j} - 6\mathbf{k}$ (ii) $\mathbf{v} = \mathbf{i} + \mathbf{j}$, $\mathbf{w} = \mathbf{j} + \mathbf{k}$

(iii) $\mathbf{v} = 2\mathbf{i} - \mathbf{j} + 6\mathbf{k}$, $\mathbf{w} = -\mathbf{i} + \mathbf{j} - 3\mathbf{k}$ (iv) $\mathbf{v} = \mathbf{i} - \mathbf{j}$, $\mathbf{w} = \mathbf{j} - \mathbf{k}$

Exercise 10.9 Determine whether the triangle $\triangle PQR$ is oriented clockwise or anticlockwise in each case:

(i) $P(4,6)$, $Q(-7,0)$, $R(2,-5)$ (ii) $P(0,1)$, $Q(23,24)$, $R(-1,-3)$

Exercise 10.10 Let $P = (5,1)$, $Q = (7,9)$ and $R = (1,4)$. In each case, use determinants to deduce whether S lies inside or outside the triangle $\triangle PQR$:

(i) $S(3,3)$ (ii) $S(4,7)$ (iii) $S(6,5)$

Exercise 10.11 Prove that the lines $ax+by = k$ and $cx+dy = \ell$ intersect in a single point if and only if $ad - bc \neq 0$.

Exercise 10.12 Make sense of the expression $\mathbf{u} \times \mathbf{v} \cdot \mathbf{w}$ where $\mathbf{u}$, $\mathbf{v}$, $\mathbf{w}$ are geometric vectors. Calculate it using the formula

$$\mathbf{u} \times \mathbf{v} \cdot \mathbf{w} = \begin{vmatrix} u_1 & u_2 & u_3 \\ v_1 & v_2 & v_3 \\ w_1 & w_2 & w_3 \end{vmatrix}$$

in each of the following cases:

(i) $\mathbf{u} = \mathbf{i} - 3\mathbf{j} + \mathbf{k}$, $\mathbf{v} = 2\mathbf{i} + 3\mathbf{j} - 3\mathbf{k}$, $\mathbf{w} = -\mathbf{i} + 2\mathbf{j} - \mathbf{k}$

(ii) $\mathbf{u} = 2\mathbf{i} - \mathbf{j} - 2\mathbf{k}$, $\mathbf{v} = \mathbf{i} + 5\mathbf{j} + 6\mathbf{k}$, $\mathbf{w} = -\mathbf{i} - \mathbf{j} + \mathbf{k}$

Exercise 10.13* Explain how the formula $\mathbf{u} \times \mathbf{v} \cdot \mathbf{w} = \begin{vmatrix} u_1 & u_2 & u_3 \\ v_1 & v_2 & v_3 \\ w_1 & w_2 & w_3 \end{vmatrix}$ follows from the formula $\mathbf{u} \times \mathbf{v} = \begin{vmatrix} \mathbf{i} & \mathbf{j} & \mathbf{k} \\ u_1 & u_2 & u_3 \\ v_1 & v_2 & v_3 \end{vmatrix}$.

Exercise 10.14* In this and the following five exercises, we develop the general theory of the determinant from scratch. As usual we define the *determinant* of a 1×1 matrix:

$$\det[a] = a$$

for any scalar a. Suppose $A = [a_{ij}]$ is a square $n \times n$ matrix where $n > 1$. Define the *determinant* of A inductively:

$$\det A = \sum_{j=1}^{n} (-1)^{1+j} a_{1j} \det A_{1j}$$

where A_{ij} denotes the $(n-1) \times (n-1)$ matrix that results from deleting the ith row and jth column from A. This formula is precisely the meaning of *expansion along the first row*. Verify the following:

(i) If A has a row or column of zeros then $\det A = 0$.

(ii) If A is upper or lower triangular, then $\det A$ is the product of the diagonal entries.

(iii) If A is a *diagonal sum* of square matrices B and C, that is, has the shape

$$A = \begin{bmatrix} B & 0 \\ 0 & C \end{bmatrix},$$

then $\det A = \det B \det C$.

Exercise 10.15* Prove that $\det A = (-1)^{n-1}$ where A is the matrix

$$\begin{bmatrix} 0 & 1 & & & & \\ & 0 & 1 & & & \\ & & \cdot & \cdot & & \\ & & & \cdot & \cdot & \\ & & & & \cdot & \cdot \\ & & & & 0 & 1 \\ 1 & & & & & 0 \end{bmatrix}$$

and it is understood that any blank entry is zero.

Exercise 10.16* Prove that

$$\det E = \begin{cases} \lambda \\ 1 \\ -1 \end{cases}$$

if E is the elementary matrix corresponding to the elementary row operation ρ in the same order:

$$\rho : \begin{cases} R_i \to \lambda R_i \\ R_i \to R_i + \lambda R_j \\ R_i \leftrightarrow R_j \end{cases}$$

Exercise 10.17** Prove that if E is elementary and A is any square matrix then

$$\det(EA) \;=\; (\det E)(\det A)\,.$$

Exercise 10.18** Use the previous exercise to deduce the following:

(i) the determinant is multiplicative, that is,

$$\det(AB) \;=\; (\det A)(\det B)$$

for any square matrices A and B of the same size

(ii) the determinant of a matrix equals the determinant of its transpose

(iii) the determinant can be found by expanding along any row or down any column, that is, for $A = [a_{ij}]$,

$$\begin{aligned} \det A \;&=\; \sum_{j=1}^{n} (-1)^{i+j} a_{ij} \det A_{ij} \qquad (i \text{ fixed}) \\ &=\; \sum_{i=1}^{n} (-1)^{i+j} a_{ij} \det A_{ij} \qquad (j \text{ fixed}) \end{aligned}$$

Exercise 10.19* Use the multiplicative property of the determinant and elementary matrices to prove that if the determinant of a matrix is nonzero then the matrix is invertible. (This completes the proof of the theorem that was stated at the beginning of the chapter.)

Exercise 10.20** Consider the following $n \times n$ matrix:

$$V = \begin{bmatrix} 1 & x_1 & x_1^2 & \dots & x_1^{n-1} \\ 1 & x_2 & x_2^2 & \dots & x_2^{n-1} \\ 1 & x_3 & x_3^2 & \dots & x_3^{n-1} \\ \vdots & \vdots & \vdots & \ddots & \vdots \\ 1 & x_n & x_n^2 & \dots & x_n^{n-1} \end{bmatrix}$$

Verify the formula

$$\det V = \prod_{1 \le i < j \le n} x_j - x_i$$

known as *Vandermonde's determinant*. Use it to deduce that a polynomial

$$p(x) = a_0 + a_1 x + a_2 x^2 + \ldots + a_{n-1} x^{n-1}$$

of degree $n-1$ (where $a_{n-1} \neq 0$) has at most $n-1$ roots.

Exercise 10.21** Let X be a finite set with n elements. If $f : X \to X$ is a function, then call f a *permutation* if f is one-one and onto, and a *transposition* if f is a permutation that leaves all but two elements of X fixed and interchanges those two elements. It is straightforward to show that any permutation of X is a composite of transpositions (and you can take this fact as granted). We call a permutation *even* if it is a composite of an even number of transpositions, and *odd* if it is a composite of an odd number. Prove that no permutation can be both even and odd. [Hint: relate this to elementary row operations on I_n, and apply determinants.]

11 Eigenvalues and Eigenvectors

- **Eigenvalues and eigcnvcctors** (Section 11.1)
  Time: 6.55
- **Finding eigenvectors** (Section 11.2)
  Time: 5.34

11 Eigenvalues and Eigenvectors

We are getting close to the end of this book, but still we haven't fathomed the real reason why mathematicians need linear algebra. Certainly, lines are the prime motivation, because of the simplicity of the arithmetic involved and the useful way the equation for a line can be generalised to any number of variables or quantities. That led us to systems of equations and the theory of matrices.

But, from a physical perspective, precisely what is it about lines that makes them important in our lives? A common definition of a line is the shortest or most direct path between two points in space. Indeed, if we want to travel from A to B in a hurry, we choose a straight line. But what if we meet an obstacle, such as an abyss or a mountain range? Naturally we avoid obstacles by altering our position, shifting out of the way, sufficiently far that the abyss or mountain range no longer prevents us from travelling straight ahead. But then we have to compensate for having shifted out of our way, and come back to the original pathway from A to B.

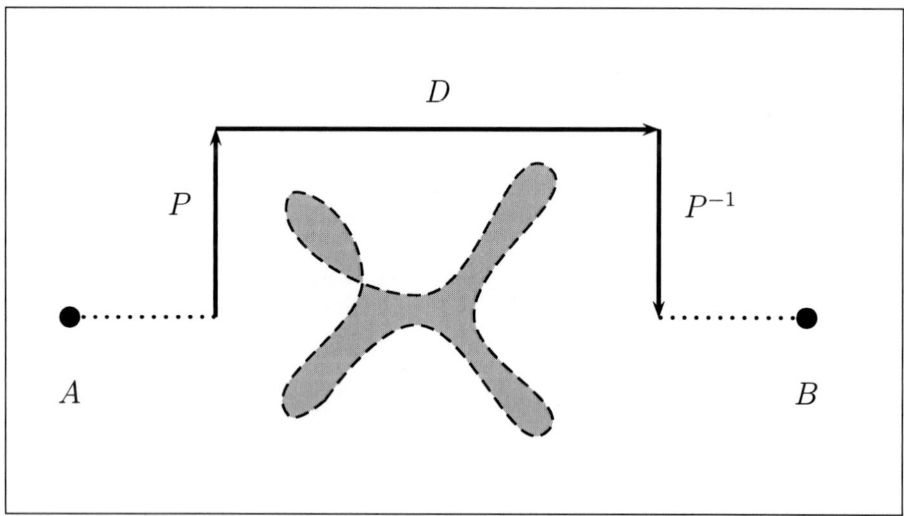

A general principle is lurking there somewhere. Call the overall act of movement past this obstacle, which lies between A and B, the transformation M. Unfortunately for us, M is not straightforward. Call the act of passing sideways to avoid the abyss or mountain range P, and the act of travelling directly ahead, uninhibited by obstacles, D.

We have to return to the pathway by undoing the sideways action, that is, by performing P^{-1}. In totality then, we have achieved M as a composite of actions:

$$M = PDP^{-1}$$

This is a famous equation, and can be captured in words:

The Conjugation Principle: To do something difficult, change position so that things get easier, and then return.

"Getting easier" in this context means "going straight ahead", in other words, following the path of a line, which leads us to

The Linear Algebra Principle: Arrange matters to move linearly.

Now a system of linear equations is summed up succinctly by the matrix equation

$$M\mathbf{x} = \mathbf{b}$$

where M is a matrix. We can think of M as "transforming" $\mathbf{x}$ into $\mathbf{b}$ by matrix multiplication on the left. There is a shift or movement from the vector $\mathbf{x}$ to the vector $\mathbf{b}$ and this movement may be complicated. Matrices can be very large with a variety of different entries. Indeed, most matrices one encounters in natural settings are messy. The matrix product $M\mathbf{x}$ can be a vector that has little resemblance to $\mathbf{x}$. It is "easiest" if the movement, the act of matrix multiplication, happens to be "in the path of a line". If we think of $\mathbf{x}$ as a position vector anchored at the origin, then it is precisely scalar multiples of $\mathbf{x}$ that result in the tips of the vectors following a line through the origin. Thus, the easiest type of "matrix movement" should be of the form

$$M\mathbf{x} = \lambda\mathbf{x}$$

for some scalar λ. Now this is also a famous equation, which leads to one of the most important definitions in mathematics:

When the following equation holds, and the vector $\mathbf{x}$ is nonzero,

$$M\mathbf{x} = \lambda\mathbf{x}$$

we call the scalar λ an *eigenvalue* of M and the vector $\mathbf{x}$ an *eigenvector* of M associated with the eigenvalue λ.

The prefix "eigen" is German for "special", "nice", "characteristic", even "ticklish", and in this context was coined by David Hilbert (1862–1943), a German mathematician, who set the stage and led twentieth century mathematics with his famous list of problems announced at the International Congress of Mathematicians in Paris, 1900.

This is a long introduction to a chapter, but we haven't finished yet, and want to gel together the Conjugation and Linear Algebra Principles. Suppose our messy matrix M is introspective, and instead of moving vectors around, is happy to move itself around by matrix multiplication. Consider the following 2×2 matrix, and start forming powers:

$$M = \begin{bmatrix} 4 & -1 \\ 2 & 1 \end{bmatrix}$$

$$M^2 = \begin{bmatrix} 4 & -1 \\ 2 & 1 \end{bmatrix}\begin{bmatrix} 4 & -1 \\ 2 & 1 \end{bmatrix} = \begin{bmatrix} 14 & -5 \\ 10 & -1 \end{bmatrix}$$

$$M^3 = M^2M = \begin{bmatrix} 14 & -5 \\ 10 & -1 \end{bmatrix}\begin{bmatrix} 4 & -1 \\ 2 & 1 \end{bmatrix} = \begin{bmatrix} 46 & -19 \\ 38 & -11 \end{bmatrix}$$

$$M^4 = M^3M = \begin{bmatrix} 46 & -19 \\ 38 & -11 \end{bmatrix}\begin{bmatrix} 4 & -1 \\ 2 & 1 \end{bmatrix} = \begin{bmatrix} 146 & -65 \\ 130 & -49 \end{bmatrix}$$

There is no obvious pattern to the numbers, except perhaps that the first column comprises larger and larger positive even integers, and the second column larger and larger negative odd integers. But, there appears to be no way of writing down quickly the precise entries of say M^{10} or say M^{50}? Now this is a 2×2 example. Imagine the seemingly hopeless complexity of powers of a matrix M with thousands of entries, say used to model economics or population dynamics to predict what will happen in 10 or 50 years! (This oversimplification assumes for purposes of illustration of a hypothetical model that one iteration of matrix multiplication by M corresponds to one time interval, such as a season or year.)

Returning to our 2×2 example, the precise pattern in fact is the following, which no-one would have any hope of guessing:

Claim:

$$M^n = \begin{bmatrix} -2^n + 2(3^n) & 2^n - 3^n \\ -2^{n+1} + 2(3^n) & 2^{n+1} - 3^n \end{bmatrix}$$

Where does the Claim come from? The messy nature of M creates an obstacle to finding the pattern. We can in fact "step or pass sideways" and then move in a straight line, by using the following instance of the Conjugation Principle:

SubClaim:

$$M = P \begin{bmatrix} 2 & 0 \\ 0 & 3 \end{bmatrix} P^{-1}$$

where

$$P = \begin{bmatrix} 1 & 1 \\ 2 & 1 \end{bmatrix} \quad \text{and} \quad P^{-1} = \begin{bmatrix} -1 & 1 \\ 2 & -1 \end{bmatrix}$$

Where does the SubClaim come from? It comes from the theory of eigenvalues and eigenvectors. In this example, the numbers 2 and 3 that appear down the diagonal of the middle matrix on the right-hand side of the equation are the eigenvalues of M. Corresponding eigenvectors turn out to be the two column vectors that make up the matrix P. The derivation of eigenvalues and eigenvectors will be explained in Section 11.1. The mystery of the form of P will be explained in Chapter 12. The inverse of P we can write down by the formula for a 2×2 matrix.

Notice that the middle matrix

$$D = \begin{bmatrix} 2 & 0 \\ 0 & 3 \end{bmatrix}$$

is *diagonal*, that is, there are zeros everywhere off the main diagonal. It is straightforward to find powers of a diagonal matrix. In this case

$$D^n = \begin{bmatrix} 2^n & 0 \\ 0 & 3^n \end{bmatrix}$$

and so, by the SubClaim,

$$\begin{aligned}
M^n &= \underbrace{(PDP^{-1})(PDP^{-1})(PDP^{-1})\cdots(PDP^{-1})}_{n \text{ times}} \\
&= P\,D\,(P^{-1}P)\,D\,(P^{-1}P)\,D\,(P^{-1}P)\,\cdots\,(P^{-1}P)\,D\,P^{-1} \\
&= P\,D\,I\,D\,I\,D\,I\cdots I\,D\,P^{-1} \\
&= P\,\underbrace{DDD\cdots D}_{n \text{ times}}\,P^{-1} \\
&= PD^nP^{-1} \\
&= \begin{bmatrix} 1 & 1 \\ 2 & 1 \end{bmatrix}\begin{bmatrix} 2^n & 0 \\ 0 & 3^n \end{bmatrix}\begin{bmatrix} -1 & 1 \\ 2 & -1 \end{bmatrix} \\
&= \begin{bmatrix} 2^n & 3^n \\ 2^{n+1} & 3^n \end{bmatrix}\begin{bmatrix} -1 & 1 \\ 2 & -1 \end{bmatrix} \\
&= \begin{bmatrix} -2^n + 2(3^n) & 2^n - 3^n \\ -2^{n+1} + 2(3^n) & 2^{n+1} - 3^n \end{bmatrix}.
\end{aligned}$$

The Claim is not only proved, but **derived**. (It is an easy exercise to **check** it using *mathematical induction*, if you are familiar with that technique.)

The key to the derivation of the above Claim was the ease with which we took powers of the diagonal matrix D. In general, a *diagonal matrix* has the following form:

$$\begin{bmatrix} \lambda_1 & & & \\ & \lambda_2 & & \\ & & \ddots & \\ & & & \lambda_n \end{bmatrix}$$

Multiplication of diagonal matrices is particularly easy: just multiply corresponding entries on the diagonal. For example,

$$\begin{bmatrix} 2 & 0 & 0 \\ 0 & 4 & 0 \\ 0 & 0 & 3 \end{bmatrix} \begin{bmatrix} -1 & 0 & 0 \\ 0 & 8 & 0 \\ 0 & 0 & 7 \end{bmatrix} = \begin{bmatrix} -2 & 0 & 0 \\ 0 & 32 & 0 \\ 0 & 0 & 21 \end{bmatrix},$$

$$\begin{bmatrix} 2 & 0 & 0 \\ 0 & 4 & 0 \\ 0 & 0 & 3 \end{bmatrix}^2 = \begin{bmatrix} 4 & 0 & 0 \\ 0 & 16 & 0 \\ 0 & 0 & 9 \end{bmatrix}, \quad \begin{bmatrix} 2 & 0 & 0 \\ 0 & 4 & 0 \\ 0 & 0 & 3 \end{bmatrix}^k = \begin{bmatrix} 2^k & 0 & 0 \\ 0 & 4^k & 0 \\ 0 & 0 & 3^k \end{bmatrix},$$

$$\begin{bmatrix} \lambda_1 & & & \\ & \lambda_2 & & \\ & & \ddots & \\ & & & \lambda_n \end{bmatrix}^k = \begin{bmatrix} \lambda_1^k & & & \\ & \lambda_2^k & & \\ & & \ddots & \\ & & & \lambda_n^k \end{bmatrix}.$$

Now consider the simplest possible effects of "pushing" vectors around by multiplying on the left by a diagonal matrix. Observe, for example,

$$\begin{bmatrix} 2 & 0 & 0 \\ 0 & 4 & 0 \\ 0 & 0 & 3 \end{bmatrix} \begin{bmatrix} 1 \\ 0 \\ 0 \end{bmatrix} = \begin{bmatrix} 2 \\ 0 \\ 0 \end{bmatrix} = 2 \begin{bmatrix} 1 \\ 0 \\ 0 \end{bmatrix},$$

$$\begin{bmatrix} 2 & 0 & 0 \\ 0 & 4 & 0 \\ 0 & 0 & 3 \end{bmatrix} \begin{bmatrix} 0 \\ 1 \\ 0 \end{bmatrix} = \begin{bmatrix} 0 \\ 4 \\ 0 \end{bmatrix} = 4 \begin{bmatrix} 0 \\ 1 \\ 0 \end{bmatrix},$$

$$\begin{bmatrix} 2 & 0 & 0 \\ 0 & 4 & 0 \\ 0 & 0 & 3 \end{bmatrix} \begin{bmatrix} 0 \\ 0 \\ 1 \end{bmatrix} = \begin{bmatrix} 0 \\ 0 \\ 3 \end{bmatrix} = 3 \begin{bmatrix} 0 \\ 0 \\ 1 \end{bmatrix}.$$

Each of the vectors

$$\begin{bmatrix} 1 \\ 0 \\ 0 \end{bmatrix}, \quad \begin{bmatrix} 0 \\ 1 \\ 0 \end{bmatrix}, \quad \begin{bmatrix} 1 \\ 0 \\ 0 \end{bmatrix},$$

which correspond in some sense to the usual geometric unit vectors

$$\mathbf{i}, \quad \mathbf{j}, \quad \mathbf{k},$$

are moved by the matrix parallel to themselves. The scaling factor is precisely the corresponding entry along the diagonal. It is as though this matrix "moves things", in space, in straight lines along the x-, y- and z-axes, corresponding to the directions of $\mathbf{i}$, $\mathbf{j}$ and $\mathbf{k}$.

In general, if

$$D = \begin{bmatrix} \lambda_1 & & & \\ & \lambda_2 & & \\ & & \ddots & \\ & & & \lambda_n \end{bmatrix}$$

is diagonal, then

$$\boxed{D\,\mathbf{v}_i = \lambda_i\,\mathbf{v}_i}$$

where

$$\mathbf{v}_i = \begin{bmatrix} 0 \\ \vdots \\ 0 \\ 1 \\ 0 \\ \vdots \\ 0 \end{bmatrix} \longleftarrow \text{ith place}$$

for $i = 1, \ldots, n$. This says precisely that λ_i is an eigenvalue for D with corresponding eigenvector $\mathbf{v}_i$.

In the substantive examples in this book, n will be 2 or 3, but n can be any positive integer. If one thinks of $\mathbf{v}_1, \ldots, \mathbf{v}_n$ as pointing in directions of "coordinate axes" then D is a transformation that moves things in straight lines along these axes. The whole point of the theory is to take an arbitrary $n \times n$ matrix M and, by the device of moving backwards and forwards using a "passing sideways" matrix P, reduce the arithmetic to that of a diagonal matrix D. This process is called *diagonalisation*, formalised in Chapter 12. Complete diagonalisation is not always possible, but something very close can be obtained, called the *Jordan Normal Form* (a topic in an advanced course on linear algebra and alluded to in the final chapter).

11.1 Existence of eigenvalues

The previous introduction was very long, and the reader may have skipped it, so here is the definition again:

> If M is a square matrix, $\mathbf{x}$ a nonzero column vector and λ a scalar such that
>
> $$M\mathbf{x} = \lambda\mathbf{x}$$
>
> then λ is called an *eigenvalue* of M and the vector $\mathbf{x}$ an *eigenvector* of M associated with the eigenvalue λ.

We give two examples now, labelled (A) and (B), which will be referred to several times for purposes of illustration in this chapter and the next.

Example (A): Observe that

$$\begin{bmatrix} 1 & 6 \\ 5 & 2 \end{bmatrix}\begin{bmatrix} 1 \\ 1 \end{bmatrix} = \begin{bmatrix} 7 \\ 7 \end{bmatrix} = 7\begin{bmatrix} 1 \\ 1 \end{bmatrix},$$

$$\begin{bmatrix} 1 & 6 \\ 5 & 2 \end{bmatrix}\begin{bmatrix} 6 \\ -5 \end{bmatrix} = \begin{bmatrix} -24 \\ 20 \end{bmatrix} = -4\begin{bmatrix} 6 \\ -5 \end{bmatrix},$$

so

> $\begin{bmatrix} 1 \\ 1 \end{bmatrix}$ and $\begin{bmatrix} 6 \\ -5 \end{bmatrix}$ are eigenvectors of $\begin{bmatrix} 1 & 6 \\ 5 & 2 \end{bmatrix}$
>
> corresponding to eigenvalues 7 and -4 respectively.

It is still a mystery where these vectors and scalars come from (to be explained shortly). Indeed, randomly picked vectors are extremely unlikely to be eigenvectors. Here is a nonexample:

$$\begin{bmatrix} 1 & 6 \\ 5 & 2 \end{bmatrix}\begin{bmatrix} 1 \\ 0 \end{bmatrix} = \begin{bmatrix} 1 \\ 5 \end{bmatrix} \neq \lambda\begin{bmatrix} 1 \\ 0 \end{bmatrix} = \begin{bmatrix} \lambda \\ 0 \end{bmatrix}$$

for all $\lambda \in \mathbb{R}$, so

> $\begin{bmatrix} 1 \\ 0 \end{bmatrix}$ is not an eigenvector of $\begin{bmatrix} 1 & 6 \\ 5 & 2 \end{bmatrix}$

Example (B): Observe that

$$\begin{bmatrix} 3 & 3 & 2 \\ 2 & 4 & 2 \\ -1 & -3 & 0 \end{bmatrix} \begin{bmatrix} -1 \\ 0 \\ 1 \end{bmatrix} = \begin{bmatrix} -1 \\ 0 \\ 1 \end{bmatrix} = 1 \begin{bmatrix} -1 \\ 0 \\ 1 \end{bmatrix},$$

$$\begin{bmatrix} 3 & 3 & 2 \\ 2 & 4 & 2 \\ -1 & -3 & 0 \end{bmatrix} \begin{bmatrix} 1 \\ 1 \\ -2 \end{bmatrix} = \begin{bmatrix} 2 \\ 2 \\ -4 \end{bmatrix} = 2 \begin{bmatrix} 1 \\ 1 \\ -2 \end{bmatrix},$$

$$\begin{bmatrix} 3 & 3 & 2 \\ 2 & 4 & 2 \\ -1 & -3 & 0 \end{bmatrix} \begin{bmatrix} -1 \\ -1 \\ 1 \end{bmatrix} = \begin{bmatrix} -4 \\ -4 \\ 4 \end{bmatrix} = 4 \begin{bmatrix} -1 \\ -1 \\ 1 \end{bmatrix},$$

so

> $\begin{bmatrix} -1 \\ 0 \\ 1 \end{bmatrix}$, $\begin{bmatrix} 1 \\ 1 \\ -2 \end{bmatrix}$ and $\begin{bmatrix} -1 \\ -1 \\ 1 \end{bmatrix}$ are eigenvectors of
>
> $$\begin{bmatrix} 3 & 3 & 2 \\ 2 & 4 & 2 \\ -1 & -3 & 0 \end{bmatrix}$$
>
> corresponding to eigenvalues 1, 2 and 4 respectively.

We now explain where all of the information in these examples comes from. The following double implications hold by matrix arithmetic and serve to rearrange the equation used in the eigenvalue definition:

$$\begin{aligned} & M\mathbf{v} = \lambda\mathbf{v} \\ \Longleftrightarrow \quad & M\mathbf{v} - \lambda\mathbf{v} = \mathbf{0} \\ \Longleftrightarrow \quad & M\mathbf{v} - \lambda I\mathbf{v} = \mathbf{0} \\ \Longleftrightarrow \quad & \big(M - \lambda I\big)\mathbf{v} = \mathbf{0} \end{aligned}$$

This last matrix equation corresponds to a *homogeneous* system of linear equations (where the right-hand constants are all zero) with coefficient matrix $M - \lambda I$.

The collection of all column vectors which satisfy this equation has a special name:

We call

$$\left\{ \mathbf{v} \,\middle|\, M\mathbf{v} = \lambda\mathbf{v} \right\} = \left\{ \mathbf{v} \,\middle|\, (M - \lambda I)\,\mathbf{v} = \mathbf{0} \right\}$$

the

eigenspace of M corresponding to λ.

Note that the eigenspace always contains the zero column vector of length n if M is $n \times n$, but

by definition, the zero vector is **never** an eigenvector.

This last point is really only for convenience. Building into the definition of an eigenvector that it be nonzero produces cleaner statements of theorems.

In principle then,

finding eigenspaces (and hence eigenvectors) reduces to solving homogeneous systems of equations.

But we can't get started until we know the values of the eigenvalues λ. This is where our friend the determinant comes in. The following is the main result we need, and we will sketch a proof later, after a couple of examples:

Theorem: A scalar λ is an eigenvalue of a square matrix M if and only if

$$\det(M - \lambda I) = 0\,.$$

Example (A) revisited: Consider the matrix

$$M = \begin{bmatrix} 1 & 6 \\ 5 & 2 \end{bmatrix}.$$

Then

$$M - \lambda I = \begin{bmatrix} 1 & 6 \\ 5 & 2 \end{bmatrix} - \begin{bmatrix} \lambda & 0 \\ 0 & \lambda \end{bmatrix} = \begin{bmatrix} 1-\lambda & 6 \\ 5 & 2-\lambda \end{bmatrix}$$

so that

$$\begin{aligned} \det(M - \lambda I) &= (1-\lambda)(2-\lambda) - 6(5) \\ &= 2 - 3\lambda + \lambda^2 - 30 = \lambda^2 - 3\lambda - 28 \\ &= (\lambda - 7)(\lambda + 4). \end{aligned}$$

Hence,

$$\det(M - \lambda I) = 0 \quad \Longleftrightarrow \quad \lambda = 7 \ \text{ or } \ \lambda = -4$$

so that

the eigenvalues of M are 7 and -4.

Example (B) revisited: Consider the matrix

$$M = \begin{bmatrix} 3 & 3 & 2 \\ 2 & 4 & 2 \\ -1 & -3 & 0 \end{bmatrix}.$$

Then

$$M - \lambda I = \begin{bmatrix} 3 & 3 & 2 \\ 2 & 4 & 2 \\ -1 & -3 & 0 \end{bmatrix} - \begin{bmatrix} \lambda & 0 & 0 \\ 0 & \lambda & 0 \\ 0 & 0 & \lambda \end{bmatrix} = \begin{bmatrix} 3-\lambda & 3 & 2 \\ 2 & 4-\lambda & 2 \\ -1 & -3 & -\lambda \end{bmatrix}$$

so that, using a mixture of elementary row and column operations,

$$\det(M-\lambda I) = \begin{vmatrix} 3-\lambda & 3 & 2 \\ 2 & 4-\lambda & 2 \\ -1 & -3 & -\lambda \end{vmatrix}$$

$$= \begin{vmatrix} 2-\lambda & 0 & 2-\lambda \\ 0 & -2-\lambda & 2-2\lambda \\ -1 & -3 & -\lambda \end{vmatrix} \quad \begin{matrix} R_1 \to R_1 + R_3 \\ R_2 \to R_2 + 2R_3 \end{matrix}$$

$$= \begin{vmatrix} 2-\lambda & 0 & 0 \\ 0 & -2-\lambda & 2-2\lambda \\ -1 & -3 & 1-\lambda \end{vmatrix} \quad C_3 \to C_3 - C_1$$

$$= (2-\lambda)\begin{vmatrix} -2-\lambda & 2-2\lambda \\ -3 & 1-\lambda \end{vmatrix}$$

$$= (2-\lambda)\begin{vmatrix} 4-\lambda & 0 \\ -3 & 1-\lambda \end{vmatrix} \quad R_1 \to R_1 - 2R_2$$

$$= (2-\lambda)(4-\lambda)(1-\lambda)\,.$$

Hence,

$$\det(M-\lambda I) = 0 \quad \Longleftrightarrow \quad \lambda = 2\,,\ 4 \text{ or } 1$$

so that

> the eigenvalues of M are 1, 2 and 4.

Notice, in this example, that a deliberate attempt was made to simplify the determinant $\det(M-\lambda I)$, so that it factorised at the end. This made it easy to read off the eigenvalues.

Sketch of the proof of the theorem: The link between existence of eigenvalues λ of M and the vanishing of $\det(M-\lambda I)$ is related to our characterisation, in the last chapter, of invertibility of square matrices. Recall from there that a matrix is invertible if and only if its determinant is nonzero. That result provides the last step in the following sequence of implications.

$$\begin{aligned} & \lambda \text{ is an eigenvalue of } M \\ \iff \quad & M\mathbf{v} = \lambda\mathbf{v} \quad \text{for some } \textbf{nonzero} \text{ vector } \mathbf{v} \\ \iff \quad & (M - \lambda I)\,\mathbf{v} = \mathbf{0} \quad \text{for some } \textbf{nonzero} \text{ vector } \mathbf{v} \\ \implies \quad & (M - \lambda I) \quad \text{is } \textbf{not} \text{ invertible} \\ \iff \quad & \det(M - \lambda I) = 0 \end{aligned}$$

Now, the first two double implications are just restatements of the definition of eigenvalue. The third implication needs to be double, to get the full theorem, but this is only a sketch, and the reverse implication is left as a starred exercise. To see the third implication, suppose that

$$(M - \lambda I)\,\mathbf{v} = \mathbf{0}$$

for some nonzero vector $\mathbf{v}$. If $M - \lambda I$ is invertible, then $(M - \lambda I)^{-1}$ exists and so, multiplying through both sides of the previous equation, we get

$$\mathbf{0} = (M - \lambda I)^{-1}\mathbf{0} = (M - \lambda I)^{-1}(M - \lambda I)\,\mathbf{v} = I\mathbf{v} = \mathbf{v}\,,$$

which contradicts that $\mathbf{v}$ is nonzero. Hence, $M - \lambda I$ is not invertible, and we have proved the third implication above. Our sketch is complete.

The following fact is very easy to verify for 2×2 and 3×3 matrices (and follows by induction for $n \times n$ matrices):

Easy Fact: $\det(M - \lambda I)$ is always a polynomial in λ, and is called the *characteristic polynomial* of M.

For 2×2 matrices the characteristic polynomial is a quadratic, which can always be factorised (over the complex numbers) using the quadratic formula. For 3×3 matrices it becomes a cubic, which also can always be factorised, and formulae exist, but they are complicated. The theory of existence of roots of polynomials, and formulae for them, is a fascinating subject and leads quickly into advanced mathematics. For example, the Fundamental Theorem of Algebra says that every polynomial with complex numbers as coefficients *always* has roots (which are complex numbers) and therefore can be factorised completely. However, this is an existential result and there are no general formulae, in terms of usual arithmetic operations and taking kth roots (k a positive integer), for polynomials of degree 5 or more! These remarkable facts are usually explained in a course on Galois Theory, which employs Group Theory. The Fundamental Theorem of Algebra, a proof of which is sketched in Appendix 3, implies that the characteristic polynomial of a matrix with real entries has at least complex number roots (which may or may not be real). We will see an example in the last section where the roots are not real. This implies the following important result:

Theorem: Every matrix with complex (and hence also with real) entries has eigenvalues, which are complex numbers, and these are the roots of its characteristic polynomial.

11.2 Finding eigenvectors

Once an eigenvalue λ has been found or exhibited for a given matrix M, it is then a straightforward process to find the eigenvectors. One is solving an associated homogeneous system of linear equations corresponding to the matrix equation

$$(M - \lambda I)\,\mathbf{v} \;=\; \mathbf{0}\,.$$

Point of technique: To find particular eigenvectors of a matrix it is usual to find the entire eigenspace (solution of the homogeneous system) and then evaluate at particular values of the parameters.

Example (A) revisited: Earlier we found that 7 and -4 are the eigenvalues of the matrix

$$M \;=\; \begin{bmatrix} 1 & 6 \\ 5 & 2 \end{bmatrix}.$$

We perform separate calculations to find the eigenspaces and particular eigenvectors. We spell out most of the details, though in practice one writes only a small fraction of this, and takes shortcuts wherever possible.

Finding the eigenspace for $\lambda = 7$: We are solving the system associated with the matrix equation

$$(M - 7I)\mathbf{v} \;=\; \mathbf{0}\,,$$

so that, writing

$$\mathbf{v} \;=\; \begin{bmatrix} x \\ y \end{bmatrix}$$

this becomes

$$\begin{bmatrix} -6 & 6 \\ 5 & -5 \end{bmatrix}\begin{bmatrix} x \\ y \end{bmatrix} \;=\; \begin{bmatrix} 0 \\ 0 \end{bmatrix}.$$

Row reducing the augmented matrix produces

$$\left[\begin{array}{cc|c} -6 & 6 & 0 \\ 5 & -5 & 0 \end{array}\right] \sim \left[\begin{array}{cc|c} 1 & -1 & 0 \\ 1 & -1 & 0 \end{array}\right] \sim \left[\begin{array}{cc|c} 1 & -1 & 0 \\ 0 & 0 & 0 \end{array}\right],$$

which corresponds to the system with one nontrivial equation:

$$x - y \;=\; 0$$

with solution

$$x = t\,, \;\; y = t \qquad (t \in \mathbb{R})\,.$$

Thus,

> the eigenspace for $\lambda = 7$ is $\left\{ \begin{bmatrix} t \\ t \end{bmatrix} \;\middle|\; t \in \mathbb{R} \right\}$.

In particular, taking $t = 1$,

> $\begin{bmatrix} 1 \\ 1 \end{bmatrix}$ is an eigenvector of M corresponding to the eigenvalue 7.

Finding the eigenspace for $\lambda = -4$: We are solving the system associated with the matrix equation

$$(M + 4I)\mathbf{v} = \mathbf{0},$$

so that, writing

$$\mathbf{v} = \begin{bmatrix} x \\ y \end{bmatrix}$$

this becomes

$$\begin{bmatrix} 5 & 6 \\ 5 & 6 \end{bmatrix} \begin{bmatrix} x \\ y \end{bmatrix} = \begin{bmatrix} 0 \\ 0 \end{bmatrix}.$$

Row reducing the augmented matrix produces

$$\left[\begin{array}{cc|c} 5 & 6 & 0 \\ 5 & 6 & 0 \end{array}\right] \sim \left[\begin{array}{cc|c} 5 & 6 & 0 \\ 0 & 0 & 0 \end{array}\right],$$

which corresponds to the system with one nontrivial equation:

$$5x + 6y = 0$$

with solution

$$x = -\tfrac{6}{5}t, \quad y = t \qquad (t \in \mathbb{R}).$$

Thus,

> the eigenspace for $\lambda = -4$ is $\left\{ \begin{bmatrix} -\frac{6}{5}t \\ t \end{bmatrix} \,\middle|\, t \in \mathbb{R} \right\}$.

In particular, taking $t = -5$ to avoid a fraction,

> $\begin{bmatrix} 6 \\ -5 \end{bmatrix}$ is an eigenvector of M corresponding to the eigenvalue -4.

Example (B) revisited: Earlier we found that 1, 2 and 4 are the eigenvalues of the matrix

$$M = \begin{bmatrix} 3 & 3 & 2 \\ 2 & 4 & 2 \\ -1 & -3 & 0 \end{bmatrix}.$$

We will gradually abbreviate the amount written down, dispensing with the augmented column of zeros in the second case (which is the usual practice with homogeneous systems), and just give the answer in the third case (to be checked by the reader).

Finding the eigenspace for $\lambda = 1$: We are solving

$$(M - I)\mathbf{v} = \mathbf{0},$$

that is,

$$\begin{bmatrix} 2 & 3 & 2 \\ 2 & 3 & 2 \\ -1 & -3 & -1 \end{bmatrix} \begin{bmatrix} x \\ y \\ z \end{bmatrix} = \begin{bmatrix} 0 \\ 0 \\ 0 \end{bmatrix}.$$

Row reducing the augmented matrix produces

$$\left[\begin{array}{ccc|c} 2 & 3 & 2 & 0 \\ 2 & 3 & 2 & 0 \\ -1 & -3 & -1 & 0 \end{array}\right] \sim \left[\begin{array}{ccc|c} 2 & 3 & 2 & 0 \\ 0 & 0 & 0 & 0 \\ 1 & 3 & 1 & 0 \end{array}\right] \sim \left[\begin{array}{ccc|c} 1 & 3 & 1 & 0 \\ 2 & 3 & 2 & 0 \\ 0 & 0 & 0 & 0 \end{array}\right]$$
$$\sim \left[\begin{array}{ccc|c} 1 & 3 & 1 & 0 \\ 0 & -3 & 0 & 0 \\ 0 & 0 & 0 & 0 \end{array}\right] \sim \left[\begin{array}{ccc|c} 1 & 0 & 1 & 0 \\ 0 & 1 & 0 & 0 \\ 0 & 0 & 0 & 0 \end{array}\right],$$

which corresponds to the system

$$\begin{array}{ccccc} x & & + z & = & 0 \\ & y & & = & 0 \end{array}$$

with solution

$$x = -t, \quad y = 0, \quad z = t \qquad (t \in \mathbb{R}).$$

Thus,

> the eigenspace for $\lambda = 1$ is $\left\{ \begin{bmatrix} -t \\ 0 \\ t \end{bmatrix} \,\middle|\, t \in \mathbb{R} \right\}$.

In particular, taking $t = 1$,

> $\begin{bmatrix} -1 \\ 0 \\ 1 \end{bmatrix}$ is an eigenvector of M corresponding to the eigenvalue 1.

Finding the eigenspace for $\lambda = 2$: We are solving the system corresponding to

$$(M - 2I)\mathbf{v} = \mathbf{0}$$

with coefficient matrix

$$\begin{bmatrix} 1 & 3 & 2 \\ 2 & 2 & 2 \\ -1 & -3 & -2 \end{bmatrix} \sim \begin{bmatrix} 1 & 3 & 2 \\ 0 & -4 & -2 \\ 0 & 0 & 0 \end{bmatrix} \sim \begin{bmatrix} 1 & -1 & 0 \\ 0 & 2 & 1 \\ 0 & 0 & 0 \end{bmatrix},$$

which corresponds to the system

$$\begin{array}{rcrcrcl} x & - & y & & & = & 0 \\ & & 2y & + & z & = & 0 \end{array}$$

with solution

$$x = -t/2\,, \quad y = -t/2\,, \quad z = t \qquad (t \in \mathbb{R})\,.$$

Thus,

> the eigenspace for $\lambda = 2$ is $\left\{ \begin{bmatrix} -t/2 \\ -t/2 \\ t \end{bmatrix} \middle| \, t \in \mathbb{R} \right\}$.

In particular, taking $t = -2$,

> $\begin{bmatrix} 1 \\ 1 \\ -2 \end{bmatrix}$ is an eigenvector of M corresponding to the eigenvalue 2.

Finding the eigenspace for $\lambda = 4$: By the same method,

> the eigenspace for $\lambda = 4$ is $\left\{ \begin{bmatrix} -t \\ -t \\ t \end{bmatrix} \middle| \, t \in \mathbb{R} \right\}$

so

> $\begin{bmatrix} -1 \\ -1 \\ 1 \end{bmatrix}$ is an eigenvector of M corresponding to the eigenvalue 4.

11.3 Reflections and rotations in the plane

In this section, we will explore the idea of a 2×2 matrix

$$M = \begin{bmatrix} a & b \\ c & d \end{bmatrix}$$

"pushing" or "moving" a vector $\mathbf{v}$ in the plane around by multiplication on the left, that is,

$$\mathbf{v} \mapsto M\,\mathbf{v}$$

where the symbol $\mapsto$ means "goes to" or "maps to" (in the language of *functions*).

A geometric vector $\mathbf{v}$ in the plane is a directed line segment, which is not a matrix, so how can we even consider multiplication with M? Well, a geometric vector has a Cartesian form (Section 2.3), say

$$\mathbf{v} = x\,\mathbf{i} + y\,\mathbf{j}$$

which comes from regarding $\mathbf{v}$ as the position vector of the point (x, y). We can identify both $\mathbf{v}$ and the point (x, y) with the column vector having entries x and y:

$$\mathbf{v} \equiv (x, y) \equiv \begin{bmatrix} x \\ y \end{bmatrix}$$

The symbol $\equiv$ is being used in a very strong, artificial form of "equals", in which all three objects are forcibly made the same.

This form of equality is not the result of a line of reasoning or manipulations of equations, but simply decreed! Now, of course there is a danger in decreeing things to be equal, because then anything is possible, and you can get absurd identifications like $0 \equiv 1$ where contradictions result, and the mathematics ends up modelling nothing of any use or interest. But in cases like the above, mathematicians are very careful only to identify objects that have the same algebraic properties, for the purpose of the mathematical model that they are using. In our case, geometric vectors in the plane and column vectors with two entries use operations that match perfectly under the above identification; in both cases addition and scalar multiplication are "coordinate-wise". (In more advanced linear algebra, this identification is formally described as an *isomorphism between vector spaces.*)

However you wish to view it, under these identifications, the matrix equation

$$M\begin{bmatrix} x \\ y \end{bmatrix} = \begin{bmatrix} a & b \\ c & d \end{bmatrix}\begin{bmatrix} x \\ y \end{bmatrix} = \begin{bmatrix} x' \\ y' \end{bmatrix}$$

where

$$x' = ax + by \ , \quad y' = cx + dy$$

becomes a "transformation" of geometric vectors or points in the plane. (In advanced linear algebra, M corresponds to a *linear transformation of vector spaces.*) Observe that this transformation fixes the origin, since

$$M\begin{bmatrix} 0 \\ 0 \end{bmatrix} = \begin{bmatrix} 0 \\ 0 \end{bmatrix}.$$

In the following diagram, the position vector $\begin{bmatrix} x \\ y \end{bmatrix}$ is in the first quadrant and its image $\begin{bmatrix} x' \\ y' \end{bmatrix}$ under the transformation is in the second quadrant, but any mixture of quadrants is possible (as we shall see in a moment).

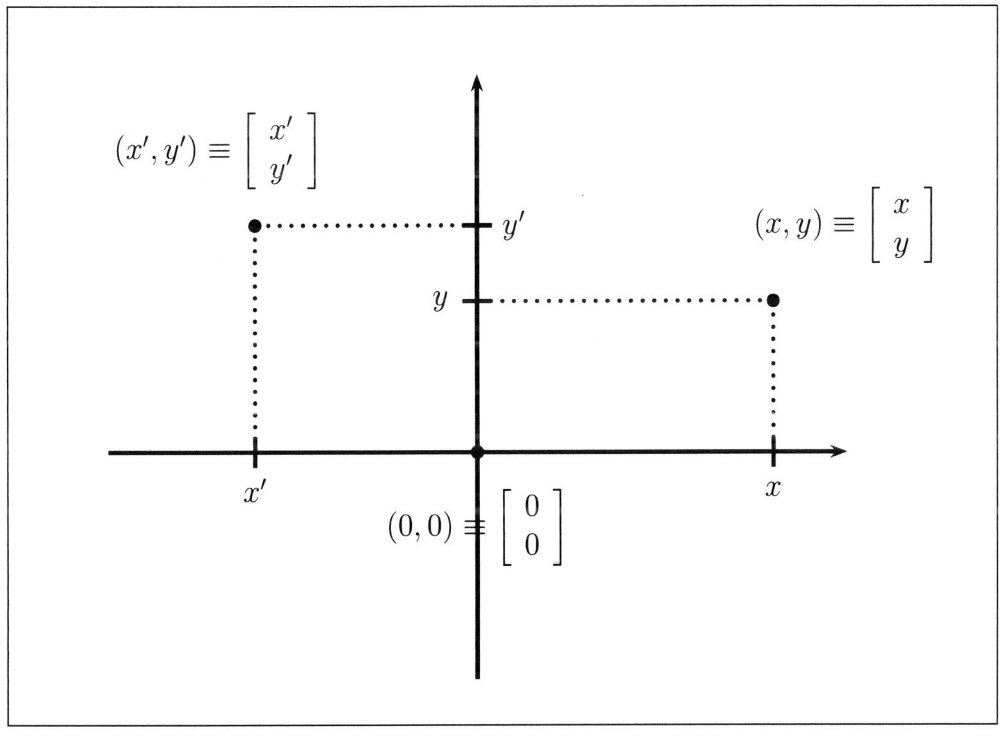

In the special "eigen" case, when

$$\begin{bmatrix} x' \\ y' \end{bmatrix} = M \begin{bmatrix} x \\ y \end{bmatrix} = \lambda \begin{bmatrix} x \\ y \end{bmatrix},$$

then $\begin{bmatrix} x \\ y \end{bmatrix}$ is an eigenvector of M with eigenvalue λ, provided of course that the vector is nonzero (away from the origin). In this case, M pushes or moves the tip of the vector along the line through the origin in the direction of the vector.

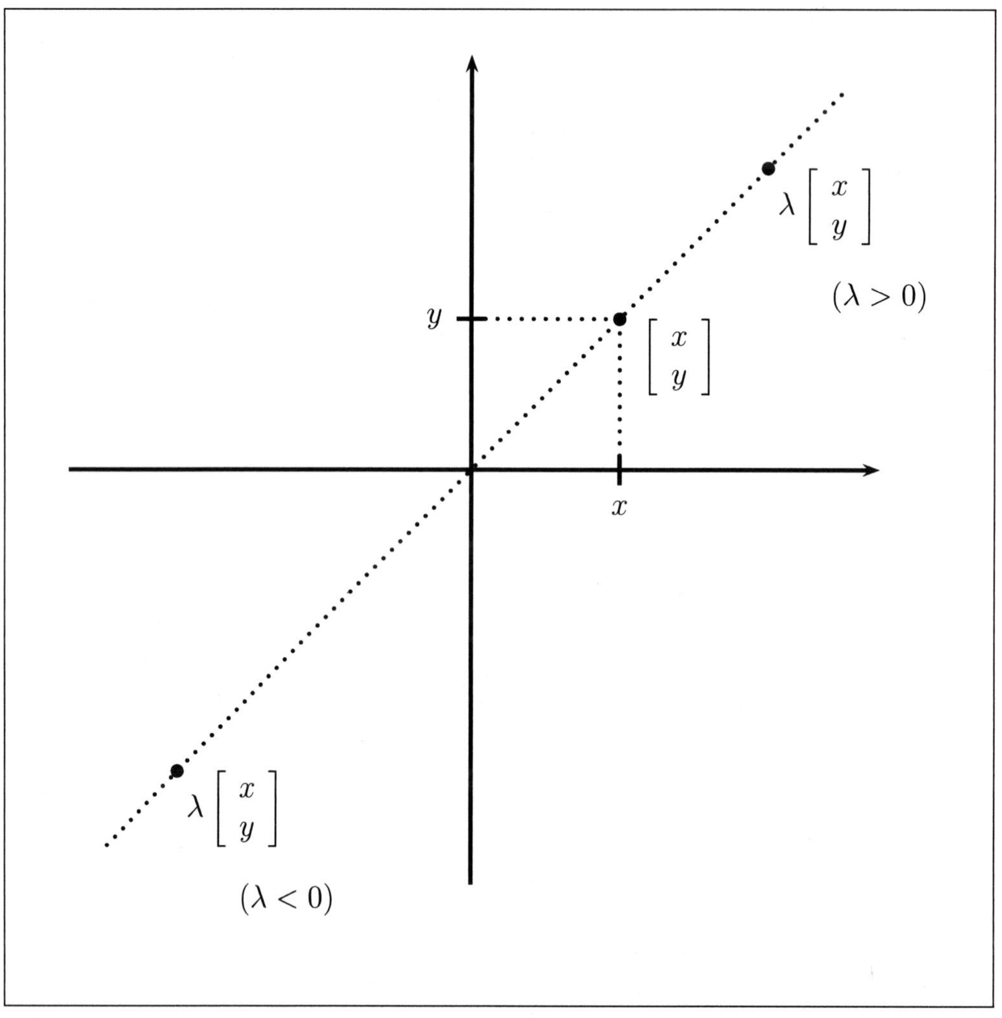

We will consider two special cases, where M yields a transformation that is a *reflection* in a line through the origin, or a *rotation* about the origin. There is another special type of transformation in the plane, known as a *shear*, which we will mention in Chapter 12 (associated with the simplest example of a matrix whose *Jordan form* is not diagonal).

We call M a *reflection* matrix if it has the following form for some real number (angle) θ:

Reflection matrix:

$$M = \begin{bmatrix} \cos 2\theta & \sin 2\theta \\ \sin 2\theta & -\cos 2\theta \end{bmatrix}$$

It is a nice exercise in trigonometry to show that the transformation associated with M is a reflection in the line $\mathcal{L}$ making an angle θ with the x-axis. (When one learns about *linear transformations* in a more advanced course, there is a slick, short verification by considering the effect of M on *standard basis vectors*.)

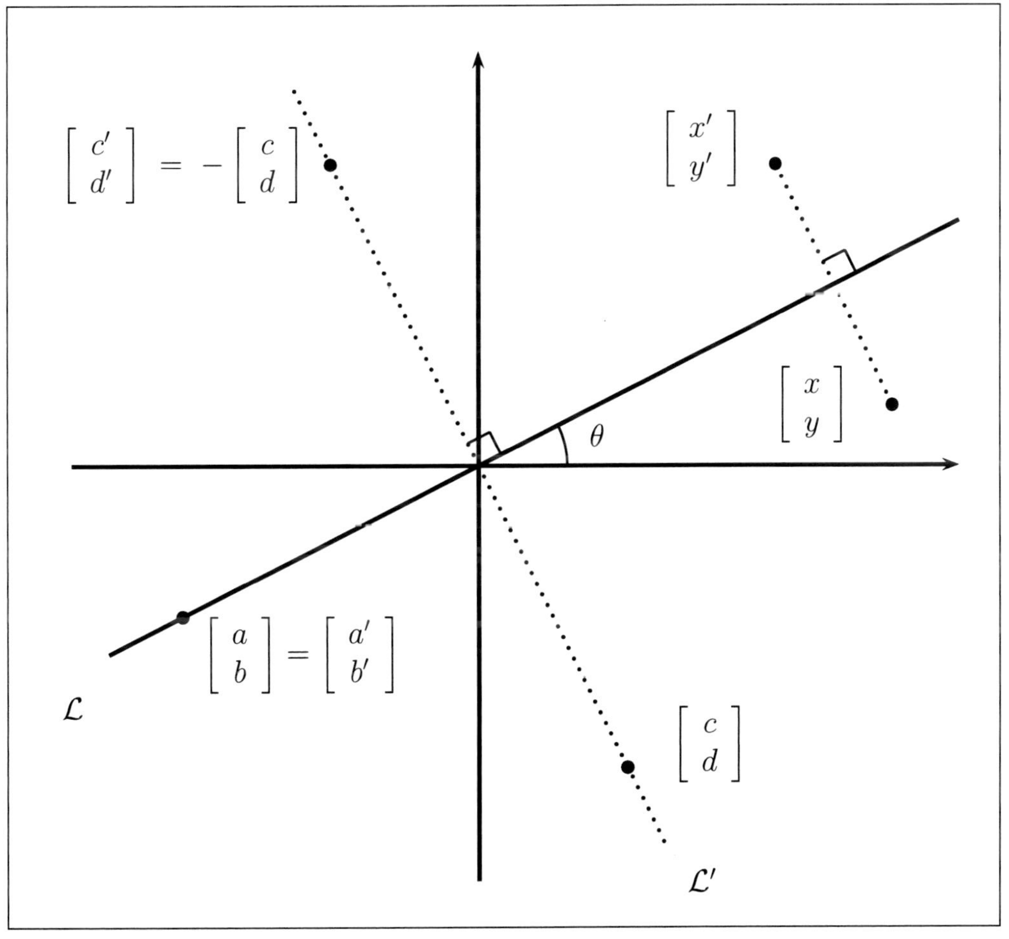

Which points in the plane are "moved" along lines through the origin by a reflection through $\mathcal{L}$? The easiest case to consider are points on $\mathcal{L}$ itself. These are not moved at all by reflection in $\mathcal{L}$, so they correspond to eigenvectors of M with eigenvalue 1.

What about points not lying on $\mathcal{L}$? Clearly, anything not on $\mathcal{L}$ is moved to the other side of $\mathcal{L}$. Thus the only case, when the reflected point lands on the same line $\mathcal{L}'$ through the origin as the original point, will be in the extreme case, when that line $\mathcal{L}'$ is perpendicular to $\mathcal{L}$. Such points then correspond to eigenvectors with eigenvalue -1. This is because multiplying by the scalar -1 reverses direction, which is precisely the effect of reflecting $\mathcal{L}'$ in $\mathcal{L}$.

> The eigenvalues of a reflection matrix are 1 and -1.
>
> Eigenvectors corresponding to 1 are nonzero position vectors of points on the line of reflection.
>
> Eigenvectors corresponding to -1 are nonzero position vectors of the line through the origin which is perpendicular to the line of reflection.

These considerations have been purely geometric, and they should be able to be confirmed using our algebraic machinery:

$$\begin{aligned}
\det(M-\lambda I) &= \begin{vmatrix} \cos 2\theta - \lambda & \sin 2\theta \\ \sin 2\theta & -\cos 2\theta - \lambda \end{vmatrix} \\
&= -(\cos 2\theta - \lambda)(\cos 2\theta + \lambda) - \sin^2 2\theta \\
&= \lambda^2 - (\cos^2 2\theta + \sin^2 2\theta) \\
&= \lambda^2 - 1 \;=\; (\lambda - 1)(\lambda + 1)\,,
\end{aligned}$$

so indeed the eigenvalues are 1 and -1. We leave it to the reader to go further and check algebraically that the position vectors for $\mathcal{L}$ comprise the eigenspace for 1, and the position vectors for $\mathcal{L}'$ comprise the eigenspace for -1.

Which reflection matrices are diagonal? The line of reflection must use an angle θ such that

$$\sin 2\theta \;=\; 2\sin\theta\cos\theta \;=\; 0\,,$$

so $\sin\theta = 0$ or $\cos\theta = 0$. We can assume $0 \leq \theta < \pi$, so the solutions are $\theta = 0$ and $\theta = \pi/2$. We get the diagonal reflection matrices

$$\begin{bmatrix} 1 & 0 \\ 0 & -1 \end{bmatrix} \quad \text{and} \quad \begin{bmatrix} -1 & 0 \\ 0 & 1 \end{bmatrix},$$

which correspond to reflections in the coordinate axes. From the introduction to this chapter, and the discussion of the Conjugation Principle, we would expect a close relationship between a general reflection matrix and either of these diagonal matrices, say the first, of the form

$$\begin{bmatrix} \cos 2\theta & \sin 2\theta \\ \sin 2\theta & -\cos 2\theta \end{bmatrix} \;=\; P \begin{bmatrix} 1 & 0 \\ 0 & -1 \end{bmatrix} P^{-1}$$

for some invertible matrix P. This equation "straightens out" the reflection matrix. (Of course, we don't need this equation to find powers of a reflection matrix M, because M^n equals I if n is even and M if n is odd.) If we put

$$P \;=\; \begin{bmatrix} \cos\theta & \sin\theta \\ \sin\theta & -\cos\theta \end{bmatrix}$$

then it is easy to check directly (using trigonometric identities) that the above equation holds. Notice that $(\cos\theta, \sin\theta)$ lies on $\mathcal{L}$ and $(\sin\theta, -\cos\theta)$ lies on $\mathcal{L}'$, so in fact the first column of P is an eigenvector corresponding to 1 and the second column an eigenvector corresponding to -1. None of this is accidental, but, rather, part of a general phenomenon explained in the next chapter.

We come to our final example in this chapter, which establishes a connection with the arithmetic of complex numbers. We call M a *rotation* matrix if it has the following form for some real number (angle) θ:

Rotation matrix:

$$M \;=\; \begin{bmatrix} \cos\theta & -\sin\theta \\ \sin\theta & \cos\theta \end{bmatrix}$$

Now it becomes an exercise in trigonometry to show that the transformation associated with M is an anticlockwise rotation θ about the origin. (Again, this follows quickly using facts about *linear transformations* and *standard basis vectors*, from more advanced theory.)

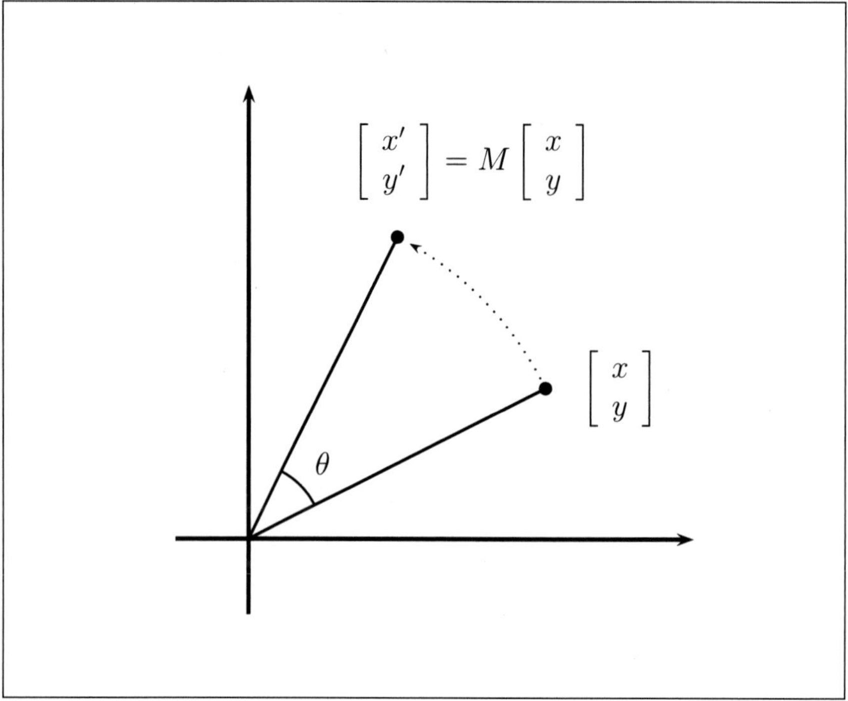

Thinking now geometrically, what can the eigenvalues and eigenvectors possibly be? How can M rotate a vector parallel to itself? Surely it must change direction! There are two extreme cases. If the rotation is $\theta = 0$ then everything is fixed, so we have an eigenvalue of 1 and all nonzero vectors are eigenvectors. If the rotation is $\theta = \pi$, then every vector is reversed, so we have an eigenvalue of -1, and again all nonzero vectors are eigenvectors. For any other value of θ, for example the one depicted in the above diagram, there is no eigenvalue. Or is there?!

Let's see what happens algebraically. In this case,

$$\begin{aligned}
\det(M - \lambda I) &= \begin{vmatrix} \cos\theta - \lambda & -\sin\theta \\ \sin\theta & \cos\theta - \lambda \end{vmatrix} \\
&= (\cos\theta - \lambda)^2 + \sin^2\theta \\
&= \cos^2\theta - 2\lambda\cos\theta + \lambda^2 + \sin^2\theta \\
&= \lambda^2 - 2\lambda\cos\theta + 1\,.
\end{aligned}$$

By the quadratic formula, $\det(M - \lambda I) = 0$ if and only if

$$\begin{aligned}\lambda &= \frac{2\cos\theta \pm \sqrt{4\cos^2\theta - 4}}{2} \\ &= \cos\theta \pm \sqrt{\cos^2\theta - 1} \\ &= \cos\theta \pm \sqrt{-\sin^2\theta} \\ &= \cos\theta \pm \sin\theta\sqrt{-1} \\ &= \cos\theta \pm i\sin\theta \\ &= e^{\pm i\theta}\end{aligned}$$

where $i = \sqrt{-1}$. These last steps, and the exponential notation, presupposes that you know a little bit about the arithmetic of $\mathbb{C}$, the complex numbers. (An introduction to complex number arithmetic is provided in Appendix 3.)

So, in fact, there are always two distinct eigenvalues when θ is not 0 or π, contrary to our earlier geometrical intuition which told us there could be no eigenvalues! Of course, there is no paradox, because these eigenvalues are complex numbers, which are not real. So, if you are prepared to expand your arithmetic to the complex numbers, then a rotation really does move in straight lines after all! (Exercise: find the straight lines.)

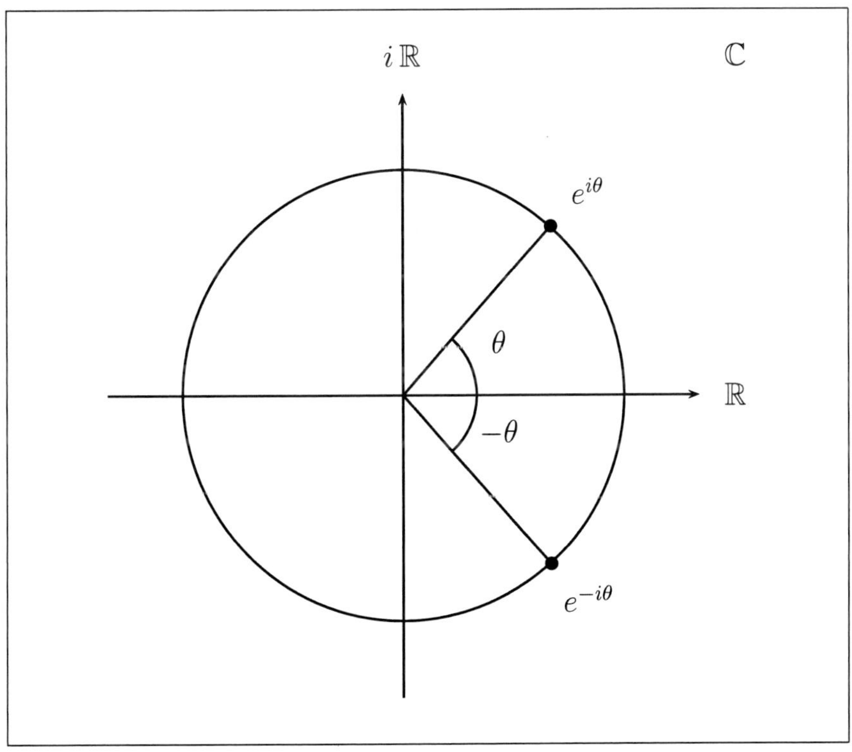

Chapter 11: Important Ideas and Useful Facts

11.1 Let M be a square matrix, $\mathbf{x}$ a nonzero column vector and λ a scalar such that

$$M\mathbf{x} = \lambda\mathbf{x}.$$

Then λ is called an *eigenvalue* of M and $\mathbf{x}$ is called an *eigenvector* of M associated with the eigenvalue λ.

11.2 The *eigenspace* of M associated with an eigenvalue λ is the collection

$$\left\{ \mathbf{v} \,\middle|\, M\mathbf{v} = \lambda\mathbf{v} \right\} = \left\{ \mathbf{v} \,\middle|\, (M - \lambda I)\mathbf{v} = \mathbf{0} \right\}$$

comprising all the eigenvectors of M associated with λ and the zero vector (which is never an eigenvector).

11.3 A scalar λ is an eigenvalue of a square matrix M if and only if

$$\det(M - \lambda I) = 0.$$

11.4 The expression $\det(M - \lambda I)$ is always a polynomial in λ and is called the *characteristic polynomial* of M. Thus, the eigenvalues of a matrix are precisely the roots of its characteristic polynomial.

11.5 Finding the eigenspace corresponding to the eigenvalue λ of a matrix M is equivalent to solving the homogeneous system with coefficient matrix $M - \lambda I$. After the eigenspace has been found, substituting particular values of the parameters yields particular eigenvectors.

11.6 The eigenvalues of a triangular matrix are simply the diagonal entries.

11.7 A *reflection matrix* has the form

$$M = \begin{bmatrix} \cos 2\theta & \sin 2\theta \\ \sin 2\theta & -\cos 2\theta \end{bmatrix}$$

and corresponds to reflection in the plane through a line through the the origin making an angle θ with the positive x-axis.

11.8 The eigenvalues of a reflection matrix are 1 and -1. Eigenvectors corresponding to the eigenvalue 1 are nonzero position vectors of points on the line of reflection. Eigenvectors corresponding to the eigenvalue -1 are nonzero position vectors of points on the line through the origin which is perpendicular to the line of reflection.

11.9 A *rotation matrix* has the form

$$M = \begin{bmatrix} \cos\theta & -\sin\theta \\ \sin\theta & \cos\theta \end{bmatrix}$$

and correponds to rotation in the plane anticlockwise through an angle θ about the origin.

11.10 The eigenvalues of a rotation matrix through θ about the origin are the complex numbers

$$e^{\pm i\theta} = \cos\theta \pm i\sin\theta$$

where $i = \sqrt{-1}$. The eigenvalues are real if and only if θ is an integer multiple of π. When $\theta = 0$ and $\theta = \pi$, the eigenvalues are both equal to 1 and -1 respectively, and all nonzero vectors are eigenvectors.

11.11 The Fundamental Theorem of Algebra: Every nonconstant polynomial with complex number coefficients has a root in the arithmetic of complex numbers. This implies that every square matrix with complex (and hence real) entries has eigenvalues.

11.12 The Cayley-Hamilton Theorem: Every square matrix is a root of its own characteristic polynomial. This means that if A is a square matrix with characteristic polynomial $p(\lambda) = \det(A - \lambda I)$, then $p(M) = 0$, the zero matrix, where M is substituted for the indeterminate λ, the identity matrix I is substituted for the real number 1, and the resulting expression $p(M)$ is evaluated using matrix arithmetic.

Exercise 11.1 Find $A\mathbf{v}$ and $A\mathbf{w}$ where

$$A = \begin{bmatrix} 1 & 2 \\ 2 & 1 \end{bmatrix}, \quad \mathbf{v} = \begin{bmatrix} 1 \\ 1 \end{bmatrix}, \quad \mathbf{w} = \begin{bmatrix} -1 \\ 1 \end{bmatrix}.$$

By inspection, write down the two eigenvalues of A.

Exercise 11.2 Factorise the determinant $\begin{vmatrix} 1-\lambda & 2 \\ 2 & 1-\lambda \end{vmatrix}$, which is a quadratic in λ, find its roots and relate them to the matrix A from the previous exercise.

Exercise 11.3 Find the characteristic polynomial $\det(M - \lambda I)$ and its roots in each case:

$$\text{(i)} \quad M = \begin{bmatrix} 1 & 0 \\ 0 & 2 \end{bmatrix} \qquad \text{(ii)} \quad M = \begin{bmatrix} 0 & -1 \\ -1 & 0 \end{bmatrix} \qquad \text{(iii)} \quad M = \begin{bmatrix} -1 & 2 \\ 3 & 0 \end{bmatrix}$$

Exercise 11.4 Using answers from the previous exercise, write down the eigenvalues of M in each case, and then find the corresponding eigenspaces:

$$\text{(i)} \quad M = \begin{bmatrix} 1 & 0 \\ 0 & 2 \end{bmatrix} \qquad \text{(ii)} \quad M = \begin{bmatrix} 0 & -1 \\ -1 & 0 \end{bmatrix} \qquad \text{(iii)} \quad M = \begin{bmatrix} -1 & 2 \\ 3 & 0 \end{bmatrix}$$

Exercise 11.5 Find $B\mathbf{v}_1$, $B\mathbf{v}_2$ and $B\mathbf{v}_3$ where

$$B = \begin{bmatrix} 0 & 1 & 1 \\ 1 & 2 & 1 \\ -1 & 1 & 2 \end{bmatrix}, \quad \mathbf{v}_1 = \begin{bmatrix} 1 \\ -1 \\ 1 \end{bmatrix}, \quad \mathbf{v}_2 = \begin{bmatrix} 0 \\ -1 \\ 1 \end{bmatrix}, \quad \mathbf{v}_3 = \begin{bmatrix} 1 \\ 2 \\ 1 \end{bmatrix}.$$

By inspection, write down the three eigenvalues of B.

Exercise 11.6 Factorise the determinant $\begin{vmatrix} -\lambda & 1 & 1 \\ 1 & 2-\lambda & 1 \\ -1 & 1 & 2-\lambda \end{vmatrix}$, which is a cubic in λ, find its roots, and relate them to the matrix B from the previous exercise.

Exercise 11.7 Write down the eigenvalues immediately for the following triangular matrices, and then find all of the corresponding eigenspaces.

$$\text{(i)} \ M = \begin{bmatrix} 1 & 1 \\ 0 & 1 \end{bmatrix} \qquad \text{(ii)} \ M = \begin{bmatrix} 2 & 0 \\ -1 & -1 \end{bmatrix} \qquad \text{(iii)} \ M = \begin{bmatrix} 1 & 1 & 1 \\ 0 & 7 & 1 \\ 0 & 0 & 7 \end{bmatrix}$$

Exercise 11.8 Find the eigenvalues and eigenspaces for the matrices

(i) $\begin{bmatrix} 2 & 0 \\ 1 & 1 \end{bmatrix}$ (ii) $\begin{bmatrix} 1 & 1 \\ -2 & 4 \end{bmatrix}$ (iii) $\begin{bmatrix} 1 & 1 & 1 \\ 0 & 2 & -1 \\ 0 & 0 & 3 \end{bmatrix}$

(iv) $\begin{bmatrix} 1 & 1 \\ -1 & 3 \end{bmatrix}$ (v) $\begin{bmatrix} -7 & -2 & 6 \\ -2 & 1 & 2 \\ -10 & -2 & 9 \end{bmatrix}$ (vi) $\begin{bmatrix} 0 & 1 & 0 \\ -2 & 3 & 0 \\ 0 & 0 & 2 \end{bmatrix}$

Exercise 11.9 Let λ be an eigenvalue of M. Show that λ^k is an eigenvalue of M^k for all positive integers k.

Exercise 11.10 Let λ be an eigenvalue of M and suppose that M is invertible. Show that λ is nonzero and λ^{-1} is an eigenvalue of M^{-1}.

Exercise 11.11 In the introduction of this chapter we derived the formula

$$\begin{bmatrix} 4 & -1 \\ 2 & 1 \end{bmatrix}^n = \begin{bmatrix} -2^n + 2(3^n) & 2^n - 3^n \\ -2^{n+1} + 2(3^n) & 2^{n+1} - 3^n \end{bmatrix}$$

where n is a positive integer. What happens to the formula when $n \leq 0$?

Exercise 11.12* We know that the determinant of a square matrix is equal to the determinant of its transpose. Deduce that a square matrix and its transpose have the same eigenvalues.

Exercise 11.13* Use the multiplicative property of the determinant to verify that if A and B are square matrices of the same size, and B is invertible, then A and $B^{-1}AB$ have the same eigenvalues.

Exercise 11.14* Explain how the proof template

expand – apply something – contract

relates to the Conjugation Principle. Explain how the Conjugation Principle applies to using a train to get between Bondi Junction and Redfern railway stations. In making a cake, one mixes ingredients on the bench and places the raw ingredients in the oven. When the cake is cooked one takes it out of the oven, but it is hot, so one has to avoid burning one's fingers. Explain what happens symbolically using a nested application of the Conjugation Principle. Identify any operation which is not invertible.

Exercise 11.15* Let A be a square matrix with eigenvalue λ. Prove the following implications:

(i) $A^2 = 0 \quad \Longrightarrow \quad \lambda = 0$

(ii) $A^2 = A \quad \Longrightarrow \quad \lambda = 0 \text{ or } \lambda = 1$

(iii) $A^2 = I \quad \Longrightarrow \quad \lambda = 1 \text{ or } \lambda = -1$

Exercise 11.16* Suppose $A = \begin{bmatrix} a & b \\ c & d \end{bmatrix}$. Verify that the characteristic polynomial of A is

$$\lambda^2 - (a+d)\lambda + ad - bc\,.$$

Now also verify that

$$A^2 - (a+d)A + (ad - bc)I = 0\,.$$

This result says that, in matrix arithmetic, A is a root of its own characteristic polynomial, a special instance of the celebrated *Cayley-Hamilton Theorem.*

Exercise 11.17* Verify by induction that the determinant $\det(M - \lambda I)$ is always a polynomial in λ for any square matrix M.

Exercise 11.18* Prove that if A is any square matrix that is not invertible then there exists a nonzero column vector $\mathbf{v}$ such that $A\mathbf{v} = \mathbf{0}$.

Exercise 11.19* Let M be the rotation matrix

$$M = \begin{bmatrix} \cos\theta & -\sin\theta \\ \sin\theta & \cos\theta \end{bmatrix}$$

where θ is a rational number. Prove that if $M^n = I$ for some positive integer n then π would be a rational number. It is a theorem that π is irrational, so we deduce that no power of M is the identity matrix.

Exercise 11.20* Find the (complex) eigenspaces of a rotation matrix.

Exercise 11.21* Find the characteristic polynomial of the matrix

$$M = \begin{bmatrix} -7 & -2 & 6 \\ -2 & 1 & 2 \\ -10 & -2 & 9 \end{bmatrix},$$

and use the Cayley-Hamilton Theorem, and manipulate a matrix equation, to find M^{-1}.

Exercise 11.22* Criticise the following "proof" of the Cayley-Hamilton Theorem, where $p(\lambda) = \det(M - \lambda I)$ is the characteristic polynomial of a square matrix M:

$$p(M) = \det(M - MI) = \det(M - M) = \det(0) = 0 .$$

Are there any matrices M for which this reasoning is valid?

Exercise 11.23* Let A be a square matrix. Denote by A_{ij} the square matrix that results by deleting the ith row and jth column of A. Define the *adjugate matrix* $\operatorname{adj} A$ to be the square matrix, of the same size as A, whose (j, i)-entry is $(-1)^{i+j} \det A_{ij}$. Quoting results about determinants, verify that

$$A(\operatorname{adj} A) = (\operatorname{adj} A)A = (\det A)I .$$

Deduce that A is invertible if and only if $\det A \neq 0$, in which case

$$A^{-1} = \frac{1}{\det A} \operatorname{adj} A .$$

Exercise 11.24* Modify the following proof, for 3×3 matrices, of the Cayley-Hamilton Theorem, so that it works for $n \times n$ matrices.

> Write the characteristic polynomial of M as $p(\lambda) = b_0 + b_1\lambda + b_2\lambda^2 + \lambda^3$ and put $B = \operatorname{adj}(\lambda I - M) = B_0 + \lambda B_1 + \lambda^2 B_2$, the adjugate matrix, where B_0, B_1, B_2 are matrices whose entries do not contain any expressions involving λ. By properties of the adjugate, we have
>
> $$(\lambda I - M)(B_0 + \lambda B_1 + \lambda^2 B_2) = (b_0 + b_1\lambda + b_2\lambda^2 + \lambda^3)I .$$
>
> Equating coefficients yields
>
> $$-MB_0 = b_0 I , \quad B_0 - MB_1 = b_1 I , \quad B_1 - MB_2 = b_2 I , \quad B_2 = I .$$
>
> But $p(M) = b_0 I + b_1 M + b_2 M^2 + M^3$, so that, from the above equations, $p(M) = -MB_0 + M(B_0 - MB_1) + M^2(B_1 - MB_2) + M^3 B_2 = 0$, the zero matrix, and the proof of the Cayley-Hamilton Theorem is complete.

Exercise 11.25* Use the adjugate matrix to verify the following fact, known as *Cramer's Rule* (in honour of the eighteenth century Swiss mathematician Gabriel Cramer): if M is an invertible $n \times n$ matrix and $\mathbf{c}$ a column vector, then the equation $M\mathbf{x} = \mathbf{c}$ has a unique solution $\mathbf{x}$ whose ith entry is

$$x_i = \frac{\det M_i}{\det M}$$

where M_i is the matrix obtained by replacing the ith column of M by $\mathbf{c}$. Use Cramer's Rule to solve the following system of equations:

$$\begin{array}{rcrcrcr} 2x & + & 3y & + & 4z & = & -4 \\ 5x & + & 5y & + & 6z & = & -3 \\ 3x & + & y & + & 2z & = & -1 \end{array}$$

12 Diagonalising a Matrix

- **Diagonalising a matrix**
  Time: 4.00
- **Finding powers of a matrix**
  Time: 5.03
- **An example of a Markov process** (Section 12.2)
  Time: 8.18

12 Diagonalising a Matrix

In the introduction to Chapter 11, we examined the rough-looking 2×2 matrix

$$M = \begin{bmatrix} 4 & -1 \\ 2 & 1 \end{bmatrix}$$

and explained how we could find a formula for powers M^n by exploiting the following relationship with a diagonal matrix D:

$$M = PDP^{-1}$$

where

$$D = \begin{bmatrix} 2 & 0 \\ 0 & 3 \end{bmatrix} \qquad \text{and} \qquad P = \begin{bmatrix} 1 & 1 \\ 2 & 1 \end{bmatrix}$$

We mentioned that the diagonal elements of D turn out to be the eigenvalues of M and the columns of P are corresponding eigenvectors. Of course, one can easily check that this matrix equation holds, but that doesn't explain where it comes from. It is time now to unravel the mystery and explain why these particular ingredients do the job of "putting M in line".

The first thing to notice is the following double implication:

$$M = PDP^{-1} \qquad \Longleftrightarrow \qquad MP = PD$$

The equation on the right is telling us that M and P do not commute. If they did commute, then by cancellation (since P is invertible), M and D would be equal, and certainly they are not: M is in a mess and D is organised and diagonal. The noncommutativity of matrix multiplication is what gives matrix arithmetic its spice.

Let's think about what the matrix product MP really "means". We will not just think of the specific example of M above, but more generally. We will engage in a thought experiment and see where it leads.

Let M be any square $n \times n$ matrix for which we have found n eigenvalues $\lambda_1, \lambda_2, \ldots, \lambda_n$ and n corresponding eigenvectors $\mathbf{v}_1, \mathbf{v}_2, \ldots, \mathbf{v}_n$. These are column vectors and we can line them up one by one to create a new $n \times n$ matrix:

$$P = \begin{bmatrix} \mathbf{v}_1 & \mathbf{v}_2 & \cdots & \mathbf{v}_n \end{bmatrix}$$

It is understood in this notation that the square brackets surrounding the individual column vectors have been thrown away so that just the elements remain, in order.

But by definition of eigenvalues and eigenvectors,

$$M\,\mathbf{v}_i \;=\; \lambda_i\,\mathbf{v}_i$$

for $i = 1, \ldots, n$. Putting this together means that half of our hypothetical equation $MP = PD$ is accounted for:

$$\begin{aligned} MP \;&=\; M\left[\begin{array}{cccc} \mathbf{v}_1 & \mathbf{v}_2 & \cdots & \mathbf{v}_n \end{array}\right] \\ &=\; \left[\begin{array}{cccc} M\,\mathbf{v}_1 & M\,\mathbf{v}_2 & \cdots & M\,\mathbf{v}_n \end{array}\right] \\ &=\; \left[\begin{array}{cccc} \lambda_1\,\mathbf{v}_1 & \lambda_2\,\mathbf{v}_2 & \cdots & \lambda_n\,\mathbf{v}_n \end{array}\right] \end{aligned}$$

What about the other half of the equation $MP = PD$? There is an important principle used in combinatorics, especially, but it applies to all of mathematics:

> **The Counting Principle:** It is useful to count the same things in two completely different ways.

First form the diagonal matrix

$$D \;=\; \begin{bmatrix} \lambda_1 & & & \\ & \lambda_2 & & \\ & & \ddots & \\ & & & \lambda_n \end{bmatrix}$$

and observe that it is a product of elementary matrices

$$D \;=\; E_1 E_2 \ldots E_n$$

where E_i corresponds to the elementary row operation $R_i \to \lambda_i R_i$, which multiplies the ith row through by λ_i. (In fact, $E_1, \ldots, E_n$ commute because it doesn't matter which order these elementary row operations are undertaken; the overall result on rows is the same.)

But, in the matrix expression PD, the matrix D is on the **right** (not the left), so the corresponding effects of the elementary operations are on **columns** (not rows). Thus, PD is the matrix resulting from multiplying the respective columns of P by the scalars $\lambda_1, \ldots, \lambda_n$ in turn. Thus,

$$PD = \begin{bmatrix} \lambda_1 \mathbf{v}_1 & \lambda_2 \mathbf{v}_2 & \cdots & \lambda_n \mathbf{v}_n \end{bmatrix}.$$

This is exactly the same outcome, using a different route, as the matrix product MP above. We have derived the following result:

Fact: If M is a square $n \times n$ matrix with eigenvalues $\lambda_1, \ldots, \lambda_n$ and corresponding eigenvectors $\mathbf{v}_1, \ldots, \mathbf{v}_n$ then

$$\boxed{MP = PD}$$

where D is the diagonal matrix with eigenvalues down the diagonal, and P the matrix with corresponding eigenvectors as columns.

In Example (A) of Section 11.1, this equation amounts to

$$\begin{bmatrix} 1 & 6 \\ 5 & 2 \end{bmatrix}\begin{bmatrix} 1 & 6 \\ 1 & -5 \end{bmatrix} = \begin{bmatrix} 7 & -24 \\ 7 & 20 \end{bmatrix} = \begin{bmatrix} 1 & 6 \\ 1 & -5 \end{bmatrix}\begin{bmatrix} 7 & 0 \\ 0 & -4 \end{bmatrix}$$

and, in Example (B),

$$\begin{bmatrix} 3 & 3 & 2 \\ 2 & 4 & 2 \\ -1 & -3 & 0 \end{bmatrix}\begin{bmatrix} -1 & 1 & -1 \\ 0 & 1 & -1 \\ 1 & -2 & 1 \end{bmatrix} = \begin{bmatrix} -1 & 2 & -4 \\ 0 & 2 & -4 \\ 1 & -4 & 4 \end{bmatrix}$$

$$= \begin{bmatrix} -1 & 1 & -1 \\ 0 & 1 & -1 \\ 1 & -2 & 1 \end{bmatrix}\begin{bmatrix} 1 & 0 & 0 \\ 0 & 2 & 0 \\ 0 & 0 & 4 \end{bmatrix}.$$

There is one potential snag, however. In the statement of the above fact, there is no requirement even that the eigenvalues and eigenvectors be distinct. It is easy to show (using determinants), for example, that a matrix P with two identical columns is not invertible. If P^{-1} happens to exist then the above equation becomes

$$M = PDP^{-1}$$

leading to the following natural terminology:

When P is invertible, and D is diagonal, we say that

$$M = PDP^{-1}$$

has been *diagonalised.*

This is an important idea, and it focuses attention on knowing when P is invertible. Indeed, linear algebra textbooks have an obsession, well justified, for listing a huge variety of tests for matrix invertibility. We will not pursue this general issue here, but settle for a *sufficient* condition for invertibility of P, the proof of which is left as a starred exercise:

Theorem: If the eigenvalues $\lambda_1, \ldots, \lambda_n$ of M are all different, then any matrix P, whose columns are corresponding eigenvectors, is invertible, so M is diagonalisable.

We will give an example shortly of a matrix where the eigenvalues are not distinct, and it is impossible to find such a matrix P which is invertible.

In both Examples (A) and (B) of Section 11.1, the eigenvalues are distinct, so the matrices are diagonalisable. To give the reader practice, we will work through the technique, that this implies, for finding arbitrary powers of these matrices.

Example (A) revisited: Diagonalise

$$M = \begin{bmatrix} 1 & 6 \\ 5 & 2 \end{bmatrix}$$

and find M^k for any k.

Solution: Put

$$P = \begin{bmatrix} 1 & 6 \\ 1 & -5 \end{bmatrix} \quad \text{and} \quad D = \begin{bmatrix} 7 & 0 \\ 0 & -4 \end{bmatrix}$$

so

$$M = PDP^{-1}$$

where

$$P^{-1} = -\frac{1}{11}\begin{bmatrix} -5 & -6 \\ -1 & 1 \end{bmatrix} = \frac{1}{11}\begin{bmatrix} 5 & 6 \\ 1 & -1 \end{bmatrix}.$$

Hence

$$M^k = PD^kP^{-1} = \begin{bmatrix} 1 & 6 \\ 1 & -5 \end{bmatrix} \begin{bmatrix} 7^k & 0 \\ 0 & (-4)^k \end{bmatrix} \frac{1}{11} \begin{bmatrix} 5 & 6 \\ 1 & -1 \end{bmatrix}$$

$$= \frac{1}{11} \begin{bmatrix} 7^k & 6(-4)^k \\ 7^k & -5(-4)^k \end{bmatrix} \begin{bmatrix} 5 & 6 \\ 1 & -1 \end{bmatrix}$$

$$= \frac{1}{11} \begin{bmatrix} 5(7^k) + 6(-4)^k & 6(7^k) - 6(-4)^k \\ 5(7^k) - 5(-4)^k & 6(7^k) + 5(-4)^k \end{bmatrix}.$$

[It is easy to make an algebraic slip, so we quickly check, say, for $k = 1$,

$$\frac{1}{11} \begin{bmatrix} 35-24 & 42+24 \\ 35+20 & 42-20 \end{bmatrix} = \frac{1}{11} \begin{bmatrix} 11 & 66 \\ 55 & 22 \end{bmatrix} = \begin{bmatrix} 1 & 6 \\ 5 & 2 \end{bmatrix} = M \quad \checkmark$$]

Example (B) revisited: Diagonalise

$$M = \begin{bmatrix} 3 & 3 & 2 \\ 2 & 4 & 2 \\ -1 & -3 & 0 \end{bmatrix}$$

and find M^k for any k.

Solution: Put

$$P = \begin{bmatrix} -1 & 1 & -1 \\ 0 & 1 & -1 \\ 1 & -2 & 1 \end{bmatrix} \quad \text{and} \quad D = \begin{bmatrix} 1 & 0 & 0 \\ 0 & 2 & 0 \\ 0 & 0 & 4 \end{bmatrix}$$

so

$$M = PDP^{-1}.$$

By the theorem, we know P^{-1} exists, but there is still some effort to find it, using the method of row reducing an augmented matrix:

$$\left[\begin{array}{ccc|ccc} -1 & 1 & -1 & 1 & 0 & 0 \\ 0 & 1 & -1 & 0 & 1 & 0 \\ 1 & -2 & 1 & 0 & 0 & 1 \end{array}\right] \sim \left[\begin{array}{ccc|ccc} 1 & -1 & 1 & -1 & 0 & 0 \\ 0 & 1 & -1 & 0 & 1 & 0 \\ 0 & -1 & 0 & 1 & 0 & 1 \end{array}\right]$$

$$\sim \left[\begin{array}{ccc|ccc} 1 & 0 & 0 & -1 & 1 & 0 \\ 0 & 1 & -1 & 0 & 1 & 0 \\ 0 & 0 & -1 & 1 & 1 & 1 \end{array}\right] \sim \left[\begin{array}{ccc|ccc} 1 & 0 & 0 & -1 & 1 & 0 \\ 0 & 1 & 0 & -1 & 0 & -1 \\ 0 & 0 & 1 & -1 & -1 & -1 \end{array}\right]$$

This gives

$$P^{-1} = \begin{bmatrix} -1 & 1 & 0 \\ -1 & 0 & -1 \\ -1 & -1 & -1 \end{bmatrix}.$$

[At this stage, one should do a quick numerical check: $PP^{-1} = I$ $\surd$]

Hence

$$M^k = PD^kP^{-1} = \begin{bmatrix} -1 & 1 & -1 \\ 0 & 1 & -1 \\ 1 & -2 & 1 \end{bmatrix} \begin{bmatrix} 1 & 0 & 0 \\ 0 & 2^k & 0 \\ 0 & 0 & 4^k \end{bmatrix} \begin{bmatrix} -1 & 1 & 0 \\ -1 & 0 & -1 \\ -1 & -1 & -1 \end{bmatrix}$$

$$= \begin{bmatrix} -1 & 2^k & -4^k \\ 0 & 2^k & -4^k \\ 1 & -2^{k+1} & 4^k \end{bmatrix} \begin{bmatrix} -1 & 1 & 0 \\ -1 & 0 & -1 \\ -1 & -1 & -1 \end{bmatrix}$$

$$= \begin{bmatrix} 1-2^k+4^k & -1+4^k & -2^k+4^k \\ -2^k+4^k & 4^k & -2^k+4^k \\ -1+2^{k+1}-4^k & 1-4^k & 2^{k+1}-4^k \end{bmatrix}.$$

[Again a quick check to help guard against an algebraic slip, say for $k = 1$:

$$\begin{bmatrix} 1-2+4 & -1+4 & -2+4 \\ -2+4 & 4 & -2+4 \\ -1+4-4 & 1-4 & 4-4 \end{bmatrix} = \begin{bmatrix} 3 & 3 & 2 \\ 2 & 4 & 2 \\ -1 & -3 & 0 \end{bmatrix} = M \quad \surd$$]

12.1 An example which cannot be diagonalised

If all matrices could be diagonalised then the method just described would be the end of the story. In fact, this is just the beginning. In this section we discuss in detail the simplest example of a matrix for which the diagonalisation process fails completely. (It is an example of a *Jordan form* which is not diagonal. Jordan forms are introduced in Section 12.3.)

Example: The matrix $M = \begin{bmatrix} 1 & 1 \\ 0 & 1 \end{bmatrix}$ cannot be diagonalised.

Exploration leading to a proof: We will look for an *ad hoc* proof by contradiction. We suppose

$$M = PDP^{-1}$$

for some invertible matrix P and diagonal matrix D and look for nonsense.

(The proof is not planned: we just interpret all of the information implied by this equation, calculate and compare what is left. The virtue of proofs like this is that we are told in advance that the statement in the box is true. This proof will almost certainly just fall out, as a consequence of technique. In research mathematics, by contrast, one rarely knows what is true or what to take as a starting point, with no guarantee of success.)

We know the entries of M and can give names to the entries of P, say

$$P = \begin{bmatrix} a & b \\ c & d \end{bmatrix}.$$

Since P is invertible, $ad - bc \neq 0$ and

$$P^{-1} = \frac{1}{ad-bc}\begin{bmatrix} d & -b \\ -c & a \end{bmatrix}.$$

From our assumption, we can rewrite D in terms of M and P:

$$\begin{aligned} D = P^{-1}MP &= \frac{1}{ad-bc}\begin{bmatrix} d & -b \\ -c & a \end{bmatrix}\begin{bmatrix} 1 & 1 \\ 0 & 1 \end{bmatrix}\begin{bmatrix} a & b \\ c & d \end{bmatrix} \\ &= \frac{1}{ad-bc}\begin{bmatrix} d & d-b \\ -c & -c+a \end{bmatrix}\begin{bmatrix} a & b \\ c & d \end{bmatrix} \\ &= \frac{1}{ad-bc}\begin{bmatrix} * & d^2 \\ -c^2 & * \end{bmatrix} \end{aligned}$$

where the asterisks $*$ denote some complicated elements we don't care about. This matrix is diagonal, so the off-diagonal elements must be zero:

$$d^2 = -c^2 = 0,$$

which implies $c = d = 0$, and so

$$\det P = ad - bc = 0.$$

But this is nonsense, because P is invertible and has nonzero determinant. This completes the proof that it is impossible to diagonalise M.

This proof is indisputable, and one can easily follow each step. But that in itself does not alleviate the mystery! (Many proofs in mathematics are like that: we follow them at a *syntactic* level, but do not necessarily reach a *semantic* level.)

The real issue is finding where the method of diagonalisation breaks down. Let's examine the eigenvalues of M:

$$\det(M - \lambda I) \;=\; \begin{vmatrix} 1-\lambda & 1 \\ 0 & 1-\lambda \end{vmatrix} \;=\; (1-\lambda)^2$$

yielding only one eigenvalue $\lambda = 1$.

(This accords with a general fact we already know: if there were distinct eigenvalues then the matrix would be diagonalisable.)

Let's now calculate the associated eigenspace:

$$M - \lambda I \;=\; \begin{bmatrix} 1 & 1 \\ 0 & 1 \end{bmatrix} - \begin{bmatrix} 1 & 0 \\ 0 & 1 \end{bmatrix} \;=\; \begin{bmatrix} 0 & 1 \\ 0 & 0 \end{bmatrix},$$

which is already row reduced. The associated homogeneous system just says

$$y \;=\; 0$$

and so the eigenspace for $\lambda = 1$ is

$$\left\{ \begin{bmatrix} t \\ 0 \end{bmatrix} \;\middle|\; t \in \mathbb{R} \right\}.$$

Hence the only eigenvectors of M are

$$\begin{bmatrix} t \\ 0 \end{bmatrix} \quad \text{for} \quad t \neq 0\,.$$

If we use two of these as columns for a matrix P we get

$$P \;=\; \begin{bmatrix} t_1 & t_2 \\ 0 & 0 \end{bmatrix}.$$

Indeed, as we expect,

$$MP = \begin{bmatrix} 1 & 1 \\ 0 & 1 \end{bmatrix} \begin{bmatrix} t_1 & t_2 \\ 0 & 0 \end{bmatrix} = \begin{bmatrix} t_1 & t_2 \\ 0 & 0 \end{bmatrix}$$

$$= \begin{bmatrix} t_1 & t_2 \\ 0 & 0 \end{bmatrix} \begin{bmatrix} 1 & 0 \\ 0 & 1 \end{bmatrix} = PD$$

where

$$D = \begin{bmatrix} 1 & 0 \\ 0 & 1 \end{bmatrix} = I$$

is the diagonal matrix with eigenvalues down the diagonal. Now here's the rub. We can't proceed any further towards diagonalising M because

$$\det P = 0 - 0 = 0$$

so that P is not invertible. No matter how hard we try to construct P in this example, the expression PDP^{-1} can never make sense.

There is something else peculiar about this example. The process of diagonalisation gets its impetus from the fact that matrix multiplication is not commutative. In this example, the eigenvalues of M are both equal to 1, so placing them on the diagonal gives $D = I$, the identity matrix, which commutes with everything. So even if PIP^{-1} made sense for some choice of P, this expression would evaluate just to I again, and one would make no progress towards reconstructing M (or anything else not equal to I).

12.2 An example of a Markov process

The connection between linear algebra and calculus was alluded to in the Introduction, but has only occasionally been exploited in this book. In this section we give an example that mixes matrix arithmetic with the process of taking limits. We use without comment the usual limit laws for numbers, which apply equally to matrices. (The reason this works is that matrix arithmetic uses real arithmetic, and the operations of addition and multiplication are *continuous*. This fact was used in Chapter 4 when we explained the universality of the Right-Hand Rule.)

A.A. Markov (1856–1922) was a Russian mathematician who studied stochastic processes and introduced in the early twentieth century a particular kind that bears his name, known as *Markov chains* or *Markov processes*. We won't try to define them in general. Rather, as an introduction to the topic, we will explore just one example, making some remarks along the way.

The example is a toy model for the behaviour of two companies A and B who between them control the market for a certain product. For example, A and B could be the only publishing companies on an island like Australia, insulated from the rest of the world, both wanting to control the market for university mathematics textbooks.

The model is governed by two rules:

> **Rule 1:** Each year, Company A retains $\frac{7}{10}$ of its custom, and loses $\frac{3}{10}$ of its custom to Company B.

> **Rule 2:** Each year, Company B retains $\frac{6}{10}$ of its custom, and loses $\frac{4}{10}$ of its custom to Company A.

Assuming the model is valid, the question is, in two parts:

> What happens to the market
>
> (i) after n years, and
>
> (ii) after a very long time (as $n \to \infty$)?

It turns out that we can answer (ii) almost immediately, given a certain theorem about Markov processes, which we will explain shortly. We will then reconstruct the answer using the theory of eigenvalues and eigenvectors. The diagonalisation method we employ can be used to answer (i) in a way very similar to the earlier examples in this chapter, so we will leave those details as an exercise.

To turn this problem into matrix arithmetic, let x denote the fraction of the market held by A, and y the fraction held by B at any given time. Then

$$x + y = 1$$

since the companies have the market all to themselves. The fractions given in the two rules of the model can be placed in a *transition* matrix:

$$M = \begin{bmatrix} \frac{7}{10} & \frac{4}{10} \\ \frac{3}{10} & \frac{6}{10} \end{bmatrix}$$

This yields the following matrix equation:

$$M \begin{bmatrix} x \\ y \end{bmatrix} = \begin{bmatrix} \frac{7}{10}x + \frac{4}{10}y \\ \frac{3}{10}x + \frac{6}{10}y \end{bmatrix}$$

$M \begin{bmatrix} x \\ y \end{bmatrix}$ tells us the respective shares of the market, held by Companies A and B, **one year later**.

Repeating this process yields the share of the market after two years:

$$M\left(M \begin{bmatrix} x \\ y \end{bmatrix}\right) = M^2 \begin{bmatrix} x \\ y \end{bmatrix}$$

after three years:

$$M\left(M\left(M \begin{bmatrix} x \\ y \end{bmatrix}\right)\right) = M^3 \begin{bmatrix} x \\ y \end{bmatrix}$$

after n years:

$$M\left(\cdots\left(M\left(M\left(M \begin{bmatrix} x \\ y \end{bmatrix}\right)\right)\right)\cdots\right) = M^n \begin{bmatrix} x \\ y \end{bmatrix}$$

Thus, to answer Part (A) of our question,

we need information about the matrix M^n.

If we can diagonalise M then we should be able to find a formula for M^n. Taking limits as $n \to \infty$ should lead us to the answer to (ii).

We pause first to get the answer directly using an important result about Markov processes. Observe that our matrix M above has the form

$$M = \begin{bmatrix} p & q \\ 1-p & 1-q \end{bmatrix}$$

where $0 < p, q < 1$. This is an instance of a *regular stochastic matrix*. A *probability vector* is a column vector with two nonnegative entries adding up to 1. Thus the two columns of M are probability vectors with positive entries. Call a probability vector $\mathbf{v}$ a *steady state vector* for M if it is fixed by multiplication by M, that is,

$$M\mathbf{v} = \mathbf{v}.$$

Notice that this $\mathbf{v}$ is an eigenvector of M corresponding to the eigenvalue 1.

> **Theorem:** If M is a regular stochastic matrix then there exists a unique steady state vector $\mathbf{v}$ for M, and, for any probability vector $\mathbf{x}$,
>
> $$\lim_{n\to\infty} M^n \mathbf{x} = \mathbf{v}.$$

The proof of this theorem for 2×2 matrices is left as a starred exercise.

In our question about the companies, the steady state vector $\mathbf{v}$ will be the answer to part (ii). We find it now, by calculating the eigenspace of M corresponding to 1 and using the constraint that the entries of $\mathbf{v}$ add up to 1. In our company example,

$$M - I = \begin{bmatrix} -\frac{3}{10} & \frac{4}{10} \\ \frac{3}{10} & -\frac{4}{10} \end{bmatrix} \sim \begin{bmatrix} 3 & -4 \\ 0 & 0 \end{bmatrix}$$

from which it follows quickly that the eigenspace corresponding to the eigenvalue 1 is

$$\left\{ \begin{bmatrix} 4t/3 \\ t \end{bmatrix} \,\middle|\, t \in \mathbb{R} \right\}.$$

We want $\mathbf{v}$ such that

$$\mathbf{v} = \begin{bmatrix} 4t/3 \\ t \end{bmatrix} \qquad \text{and} \qquad 1 = \frac{4t}{3} + t = \frac{4t + 3t}{3} = \frac{7t}{3}$$

so that $t = 3/7$. Hence our steady state vector is

$$\mathbf{v} = \begin{bmatrix} 4/7 \\ 3/7 \end{bmatrix}$$

yielding our solution to (ii):

In the long run, Company A will capture $4/7$ and Company B will capture $3/7$ of the market.

To practise our technique, we will see if we can get this result from first principles. We first calculate eigenvalues (knowing in advance that one of them will be 1):

$$\begin{aligned} \det(M - \lambda I) &= \begin{vmatrix} \frac{7}{10} - \lambda & \frac{4}{10} \\ \frac{3}{10} & \frac{6}{10} - \lambda \end{vmatrix} = \frac{1}{100} \begin{vmatrix} 7 - 10\lambda & 4 \\ 3 & 6 - 10\lambda \end{vmatrix} \\ &= \frac{1}{100}\Big((7 - 10\lambda)(6 - 10\lambda) - 12 \Big) \\ &= \lambda^2 - \frac{13}{10}\lambda + \frac{3}{10} = (\lambda - 1)(\lambda - 3/10) \end{aligned}$$

This yields the following eigenvalues:

$$\lambda = 1 \quad \text{and} \quad \frac{3}{10}$$

These are different, so we know for sure that we can diagonalise M.

The eigenspace for $\lambda = 1$ we have already calculated. For $\lambda = 3/10$:

$$M - \lambda I = \begin{bmatrix} \frac{4}{10} & \frac{4}{10} \\ \frac{3}{10} & \frac{3}{10} \end{bmatrix} \sim \begin{bmatrix} 1 & 1 \\ 0 & 0 \end{bmatrix}$$

from which it follows quickly that the eigenspace is

$$\left\{ \begin{bmatrix} -t \\ t \end{bmatrix} \,\middle|\, t \in \mathbb{R} \right\}.$$

We choose, say, eigenvectors $\begin{bmatrix} 4 \\ 3 \end{bmatrix}$ for $\lambda = 1$ and $\begin{bmatrix} -1 \\ 1 \end{bmatrix}$ for $\lambda = 3/10$, and use them to form

$$P = \begin{bmatrix} 4 & -1 \\ 3 & 1 \end{bmatrix}.$$

Hence

$$M = PDP^{-1}$$

where

$$D = \begin{bmatrix} 1 & 0 \\ 0 & 3/10 \end{bmatrix} \qquad \text{and} \qquad P^{-1} = \frac{1}{7}\begin{bmatrix} 1 & 1 \\ -3 & 4 \end{bmatrix}.$$

Thus

$$M^n = PD^nP^{-1} = P\begin{bmatrix} 1 & 0 \\ 0 & (3/10)^n \end{bmatrix}P^{-1}$$

and, taking limits,

$$\begin{aligned}
\lim_{n\to\infty} M^n &= \lim_{n\to\infty}\left(P\begin{bmatrix} 1 & 0 \\ 0 & (3/10)^n \end{bmatrix}P^{-1}\right) \\
&= P\left(\lim_{n\to\infty}\begin{bmatrix} 1 & 0 \\ 0 & (3/10)^n \end{bmatrix}\right)P^{-1} \\
&= P\begin{bmatrix} 1 & 0 \\ 0 & \lim_{n\to\infty}(3/10)^n \end{bmatrix}P^{-1} \\
&= P\begin{bmatrix} 1 & 0 \\ 0 & 0 \end{bmatrix}P^{-1} \\
&= \begin{bmatrix} 4 & -1 \\ 3 & 1 \end{bmatrix}\begin{bmatrix} 1 & 0 \\ 0 & 0 \end{bmatrix}\frac{1}{7}\begin{bmatrix} 1 & 1 \\ -3 & 4 \end{bmatrix} \\
&= \frac{1}{7}\begin{bmatrix} 4 & 0 \\ 3 & 0 \end{bmatrix}\begin{bmatrix} 1 & 1 \\ -3 & 4 \end{bmatrix} \\
&= \begin{bmatrix} \frac{4}{7} & \frac{4}{7} \\ \frac{3}{7} & \frac{3}{7} \end{bmatrix}.
\end{aligned}$$

Hence, if x and y are the initial market shares of Companies A and B respectively, then

$$x + y = 1$$

and the long-term outlook is expressed by the vector

$$\lim_{n\to\infty} M^n \begin{bmatrix} x \\ y \end{bmatrix} = \begin{bmatrix} \frac{4}{7} & \frac{4}{7} \\ \frac{3}{7} & \frac{3}{7} \end{bmatrix} \begin{bmatrix} x \\ y \end{bmatrix}$$

$$= \begin{bmatrix} \frac{4}{7}(x+y) \\ \frac{3}{7}(x+y) \end{bmatrix}$$

$$= \begin{bmatrix} 4/7 \\ 3/7 \end{bmatrix}$$

as we predicted before, using the theorem. Notice that the long-term outcome is independent of the starting values x and y.

> In the long run, **regardless of the starting conditions**, Company A will capture $4/7$ and Company B will capture $3/7$ of the market.

The phenomenon in this example, of a steady state vector, independent of the starting conditions, predicted by the theorem listed earlier, occurs for a *stochastic matrix* (nonnegative entries, and columns sum to 1) of any size which is *regular* (some positive power of the matrix has all of its entries positive). This regularity condition is very strong. In practice, one may want to include large numbers, perhaps thousands (or billions in the case of web search engines) of variables, which may only be occasionally related. The large coefficient matrix of the linear system will be *sparse*, which means that most of the entries are zero. It is rare for some positive power of a sparse matrix to have all of its entries nonzero. Every stochastic matrix has a steady state vector (probability eigenvector corresponding to eigenvalue 1), but it need not be unique and, even if unique, need not have the convergence property of the example discussed in this section.

12.3 The Jordan form of a matrix

The Linear Algebra Principle tells us to arrange things to move as much as possible in straight lines. But, in Section 12.1, we examined in detail the matrix

$$\begin{bmatrix} 1 & 1 \\ 0 & 1 \end{bmatrix}$$

and explained in several ways why there is an obstruction to it being diagonalised. Yes, the 1 above the diagonal turns out to be the spanner in the works, but only because the elements on the diagonal are equal. The same arguments in fact apply to

$$\begin{bmatrix} \lambda & 1 \\ 0 & \lambda \end{bmatrix}$$

for any $\lambda \in \mathbb{C}$, and this is as bad as it gets for 2×2 matrices, as we make precise in a moment.

Call matrices A and B *similar* if

$$A = PBP^{-1}$$

for some invertible matrix P. Thus, to say that a matrix is diagonalisable is to say that it is similar to a diagonal matrix. The following theorem tells the whole story for 2×2 matrices where we allow entries to be complex numbers. The proof is left as a starred exercise.

Theorem (2×2 complex matrices): Every 2×2 matrix with complex number entries is similar either to a diagonal matrix, or to a matrix of the form

$$\begin{bmatrix} \lambda & 1 \\ 0 & \lambda \end{bmatrix}$$

for some $\lambda \in \mathbb{C}$.

Recall from Section 11.3 that rotation matrices

$$\begin{bmatrix} \cos\theta & -\sin\theta \\ \sin\theta & \cos\theta \end{bmatrix}$$

have complex eigenvalues which are not real. The same is clearly true for scalar multiples of rotation matrices

$$\begin{bmatrix} \lambda\cos\theta & -\lambda\sin\theta \\ \lambda\sin\theta & \lambda\cos\theta \end{bmatrix}$$

where $\lambda \in \mathbb{R}$, and again that is as bad as it gets. The following can be deduced from the previous theorem (Exercise 12.15) using the fact that real quadratics have solutions which come in complex conjugate pairs.

Theorem (2×2 real matrices): Every 2×2 matrix with real number entries is similar either to a diagonal matrix, to a matrix of the form

$$\begin{bmatrix} \lambda & 1 \\ 0 & \lambda \end{bmatrix}$$

for some $\lambda \in \mathbb{R}$, or to a matrix of the form

$$\begin{bmatrix} \lambda\cos\theta & -\lambda\sin\theta \\ \lambda\sin\theta & \lambda\cos\theta \end{bmatrix}$$

for some $\lambda, \theta \in \mathbb{R}$.

(The reader may notice a similarity to the standard recipe for solving homogeneous second order differential equations. This is not a coincidence.)

What happens for larger square matrices? We need the notion of a *Jordan block* J, which has the form

$$J = [\lambda]$$

in the trivial 1×1 case, and for nontrivial cases,

$$J = \begin{bmatrix} \lambda & 1 & & & & \\ & \lambda & 1 & & & \\ & & \lambda & 1 & & \\ & & & \ddots & \ddots & \\ & & & & \lambda & 1 \\ & & & & & \lambda \end{bmatrix},$$

which has the same number λ down the diagonal, and 1's sitting just above the diagonal, as if they are handcuffing the λ's together in a chain. These 1's above the diagonal create an "obstruction" that prevents J from getting any closer to being purely diagonal. A *diagonal sum* of Jordan blocks $J_1, \ldots, J_k$ has the form

$$M = \begin{bmatrix} J_1 & & & \\ & J_2 & & \\ & & \ddots & \\ & & & J_k \end{bmatrix}.$$

One can think of M as a "molecule" and the Jordan blocks as "atoms" (in a completely different sense to the discussion about elementary matrices). If 1×1 Jordan blocks are "hydrogen" atoms then here are examples of "helium", "lithium", "beryllium" and so on:

$$\begin{bmatrix} \lambda & 1 \\ 0 & \lambda \end{bmatrix} \quad \begin{bmatrix} \lambda & 1 & 0 \\ 0 & \lambda & 1 \\ 0 & 0 & \lambda \end{bmatrix} \quad \begin{bmatrix} \lambda & 1 & 0 & 0 \\ 0 & \lambda & 1 & 0 \\ 0 & 0 & \lambda & 1 \\ 0 & 0 & 0 & \lambda \end{bmatrix} \quad \cdots$$

In this language, our earlier theorem about complex 2×2 matrices is a special case of the following theorem, which is one of the pinnacles of linear algebra. It says that matrix arithmetic is essentially "molecular":

Jordan Normal Form Theorem: Every square matrix with complex entries is similar to a diagonal sum of Jordan blocks.

The simplicity of the statement belies the technical and conceptual difficulty of the proof. To prove the 3×3 case is sufficiently difficult to capture the main ideas of the general case and is usually sketched in advanced courses in linear algebra. The theorem is not true as it stands if "complex" is replaced by "real", though we get a more complicated Real Jordan Form Theorem, which uses modified Jordan blocks involving rotation matrices.

We return to plane geometry for one last time and our matrix

$$M = \begin{bmatrix} 1 & 1 \\ 0 & 1 \end{bmatrix},$$

which corresponds to a transformation of the plane

$$\begin{bmatrix} x \\ y \end{bmatrix} \mapsto \begin{bmatrix} x' \\ y' \end{bmatrix} = M \begin{bmatrix} x \\ y \end{bmatrix} = \begin{bmatrix} x+y \\ y \end{bmatrix}$$

called a *standard shear*. This fixes points on the x-axis and moves points off the x-axis parallel to the y-axis an amount equal to their y-coordinate. If one imagines the wind blowing a building over sideways, the part of the building fixed to the ground remains stable whilst the rest of the building is shifted over proportional to the height off the ground. In the standard shear the constant of proportionality is one.

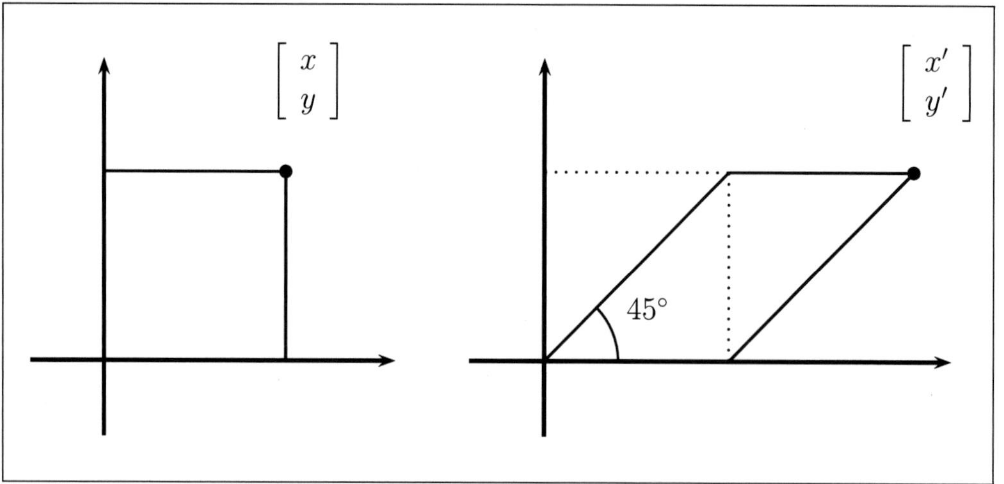

A general *shear* is a transformation where the matrix has the form

$$S = \begin{bmatrix} 1 & k \\ 0 & 1 \end{bmatrix} : \begin{bmatrix} x \\ y \end{bmatrix} \mapsto \begin{bmatrix} x' \\ y' \end{bmatrix} = \begin{bmatrix} x + ky \\ y \end{bmatrix},$$

but now the constant of proportionality for the sideways shift is $k \neq 0$. In the following diagram k is positive, but it could be negative. The larger k is, the steeper the incline after the shear has been applied.

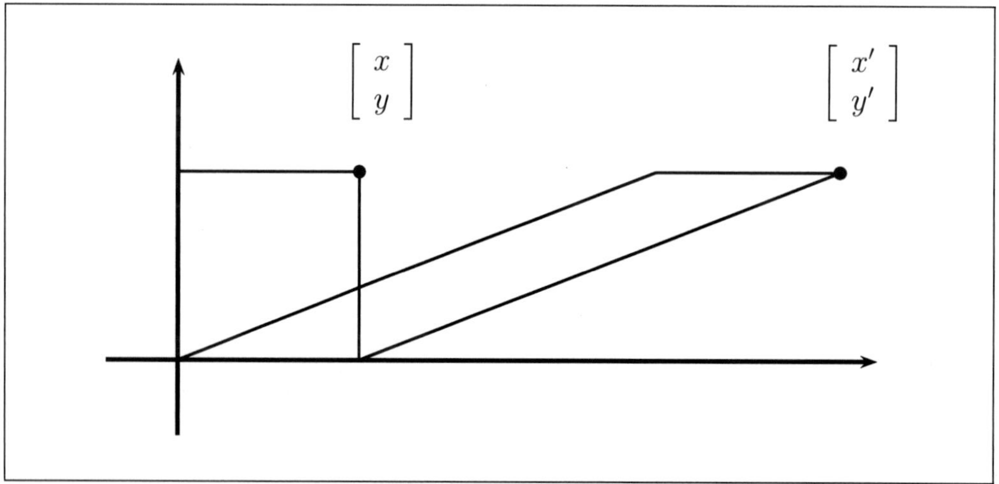

One can have general shears fixed along any line, but we will stick to shears along the x-axis for the purposes of this illustration. Standard shears are the "easiest" from the point of view of angle of inclination, because vertical walls simply end up inclining at 45°. Suppose you have built a shearing machine that only knows how to incline walls to 45°. How would you use your machine to achieve a general shear?

Observe, first, that S is not a Jordan block, since the "handcuff" is k, not 1. Nevertheless, the sole eigenvalue of S is 1, so it follows easily from the Jordan Normal Form Theorem that S is similar to M, where the "handcuff" is 1, so

$$S \;=\; PMP^{-1}$$

for some invertible matrix P. There are many possibilities for P, but we will choose the simplest, which is in fact a diagonal elementary matrix:

$$P \;=\; \begin{bmatrix} 1 & 0 \\ 0 & k^{-1} \end{bmatrix} : \begin{bmatrix} x \\ y \end{bmatrix} \;\mapsto\; \begin{bmatrix} x \\ k^{-1}y \end{bmatrix}$$

whose effect on vectors is to multiply y-coordinates by the scalar k^{-1}. Thus, in full,

$$S \;=\; \begin{bmatrix} 1 & k \\ 0 & 1 \end{bmatrix} \;=\; \begin{bmatrix} 1 & 0 \\ 0 & k^{-1} \end{bmatrix}\begin{bmatrix} 1 & 1 \\ 0 & 1 \end{bmatrix}\begin{bmatrix} 1 & 0 \\ 0 & k \end{bmatrix}.$$

If you read the right-hand side from right to left, in the order that the matrices multiply a column vector, it achieves a decomposition of the general "difficult" shear in terms of our standard "easy" shear using our template

expand – apply something easy – contract

which reinforces, of course, one of the main themes of this book!

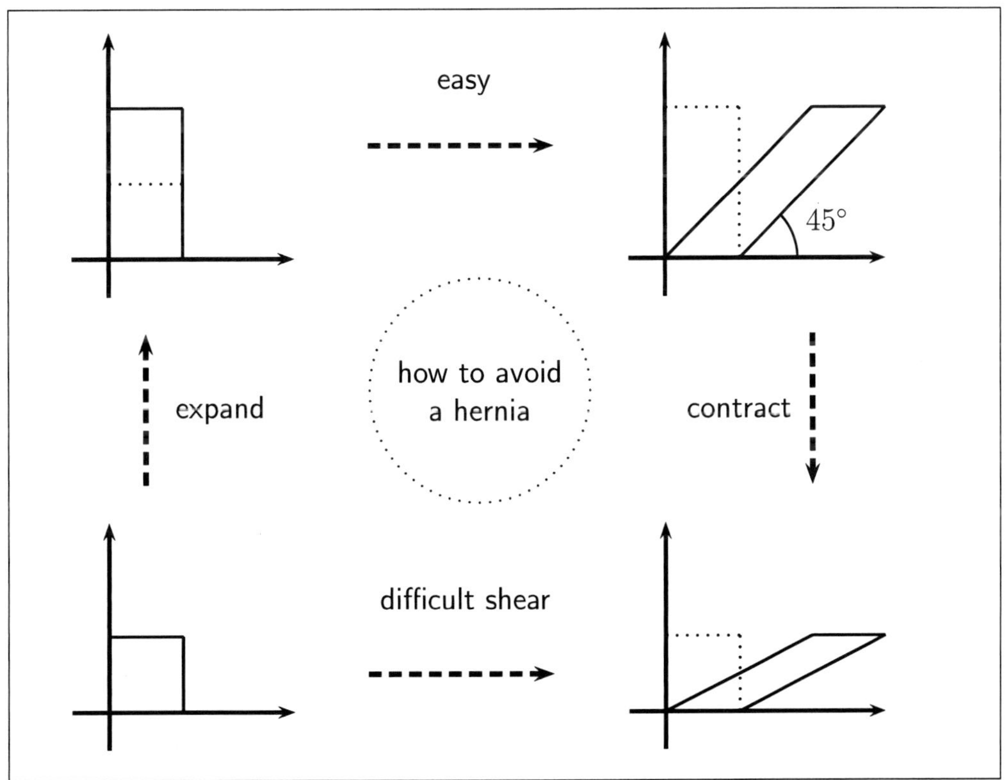

Chapter 12: Important Ideas and Useful Facts

12.1 A square matrix D is *diagonal* if all entries off the diagonal are zero. If D and E are diagonal then DE is also diagonal, and its diagonal entries are simply the products of corresponding diagonal entries of D and E. Thus the diagonal elements of D^n are just the nth powers of the diagonal elements of D.

12.2 Let M be a square $n \times n$ matrix with eigenvalues $\lambda_1, \ldots, \lambda_n$ and corresponding eigenvectors $\mathbf{v}_1, \ldots \mathbf{v}_n$. Then

$$MP = PD$$

where D is the diagonal matrix with eigenvalues down the diagonal and P the matrix with corresponding eigenvectors as columns. If P is invertible then

$$M = PDP^{-1} \qquad \text{and} \qquad D = P^{-1}MP\,.$$

In this case we say that M is *diagonalisable.*

12.3 In the preceding discussion, if the eigenvalues are all different then P is invertible and M is diagonalisable.

12.4 If M is diagonalisable then powers of M can be found easily by the formula

$$M^n = PD^nP^{-1}\,.$$

12.5 The matrix $\begin{bmatrix} 1 & 1 \\ 0 & 1 \end{bmatrix}$ cannot be diagonalised.

12.6 A square matrix M is *stochastic* if all the entries are nonnegative and the columns add to 1, and *regular* if further some positive power of M has all positive entries. A column matrix $\mathbf{v}$ is a *probability vector* if all of its entries are nonnegative and add to 1, and a *steady state vector* for M if $M\mathbf{v} = \mathbf{v}$.

12.7 If M is a regular stochastic matrix then there exists a unique steady state vector $\mathbf{v}$ for M and, for any probability vector $\mathbf{x}$,

$$\lim_{n\to\infty} M^n\mathbf{x} = \mathbf{v}\,.$$

12.8 Two square matrices A and B of the same size are *similar* if there exists an invertible matrix P such that

$$A = PBP^{-1}.$$

12.9 A *Jordan block* or *Jordan matrix* J is a square $n \times n$ matrix having the form

$$J = [\lambda]$$

for $n = 1$, and

$$J = \begin{bmatrix} \lambda & 1 & & & & \\ & \lambda & 1 & & & \\ & & \lambda & 1 & & \\ & & & \ddots & \ddots & \\ & & & & \lambda & 1 \\ & & & & & \lambda \end{bmatrix}$$

for $n > 1$, which has λ's down the diagonal, 1's sitting just above the diagonal and 0's everywhere else.

12.10 Jordan Normal Form Theorem: Every square matrix with complex entries is similar to a diagonal sum of Jordan blocks.

12.11 Complex Jordan Normal Form for 2×2 matrices: Every 2×2 matrix with complex number entries is similar either to a diagonal matrix, or to a matrix of the form

$$\begin{bmatrix} \lambda & 1 \\ 0 & \lambda \end{bmatrix}$$

for some $\lambda \in \mathbb{C}$.

12.12 Real Jordan Normal Form for 2×2 matrices: Every 2×2 matrix with real number entries is similar either to a diagonal matrix, to a matrix of the form

$$\begin{bmatrix} \lambda & 1 \\ 0 & \lambda \end{bmatrix}$$

for some $\lambda \in \mathbb{R}$, or to a matrix of the form

$$\begin{bmatrix} \lambda\cos\theta & -\lambda\sin\theta \\ \lambda\sin\theta & \lambda\cos\theta \end{bmatrix}$$

for some λ, $\theta \in \mathbb{R}$.

12.13 The matrix of a *shear* in the direction of the x-axis has the form

$$\begin{bmatrix} 1 & k \\ 0 & 1 \end{bmatrix},$$

which corresponds to a transformation of the plane which fixes the x-axis and shifts points sideways proportional to their y-coordinates.

Exercise 12.1 The matrix $A = \begin{bmatrix} 2 & 1 \\ 1 & 2 \end{bmatrix}$ has eigenvalues 1 and 3 with eigenvectors $\begin{bmatrix} -1 \\ 1 \end{bmatrix}$ and $\begin{bmatrix} 1 \\ 1 \end{bmatrix}$ respectively.

(i) Write down an invertible matrix P and a diagonal matrix D such that $A = PDP^{-1}$.

(ii) Find a formula for A^n and use it to find A^3 and A^4.

Exercise 12.2 The matrix $B = \begin{bmatrix} 0 & 1 & 1 \\ 1 & 2 & 1 \\ -1 & 1 & 2 \end{bmatrix}$ has eigenvalues 0, 1 and 3 with eigenvectors $\begin{bmatrix} 1 \\ -1 \\ 1 \end{bmatrix}$, $\begin{bmatrix} 0 \\ -1 \\ 1 \end{bmatrix}$ and $\begin{bmatrix} 1 \\ 2 \\ 1 \end{bmatrix}$ respectively.

(i) Write down an invertible matrix P and diagonal matrix D such that $B = PDP^{-1}$.

(ii) Find a formula for B^n and use it to find B^4.

Exercise 12.3 Let $C = \begin{bmatrix} 5/2 & -1/2 & 0 \\ -1/2 & 5/2 & 0 \\ -1/2 & 1/2 & 2 \end{bmatrix}$.

(i) Find the eigenvalues of C, an invertible matrix P and diagonal matrix D such that $C = PDP^{-1}$.

(ii) Find a formula for C^n and use it to find C^4.

Exercise 12.4 Diagonalise M and find M^k for any positive integer k where M is each of the following:

(i) $\begin{bmatrix} 0 & -1 \\ -1 & 0 \end{bmatrix}$ (ii) $\begin{bmatrix} 2 & 0 \\ 1 & 1 \end{bmatrix}$ (iii) $\begin{bmatrix} 1 & 1 \\ -2 & 4 \end{bmatrix}$

Exercise 12.5 Diagonalise M and find M^k for any positive integer k where M is each of the following:

(i) $\begin{bmatrix} 1 & 1 & 1 \\ 0 & 2 & -1 \\ 0 & 0 & 3 \end{bmatrix}$ (ii) $\begin{bmatrix} -7 & -2 & 6 \\ -2 & 1 & 2 \\ -10 & -2 & 9 \end{bmatrix}$ (iii) $\begin{bmatrix} 0 & 1 & 0 \\ -2 & 3 & 0 \\ 0 & 0 & 2 \end{bmatrix}$

Exercise 12.6 Verify that the matrix

$$M = \begin{bmatrix} 1/2 & 2/5 \\ 1/2 & 3/5 \end{bmatrix}$$

is regular stochastic and find its unique steady state vector. Find M^n and check the limiting behaviour as $n \to \infty$.

Exercise 12.7* Verify that some positive power of the following matrix M has all positive entries, so that M is regular stochastic:

$$M = \begin{bmatrix} 0 & 1/2 & 1/2 \\ 1/2 & 1/2 & 0 \\ 1/2 & 0 & 1/2 \end{bmatrix}$$

Find its unique steady state vector. Find M^n and check the limiting behaviour as $n \to \infty$.

Exercise 12.8* Find a stochastic matrix with more than one steady state vector. Find a stochastic matrix M with a unique steady state vector $\mathbf{v}$ and another probability vector $\mathbf{x}$ such that $\lim_{n\to\infty} M^n\mathbf{x} \neq \mathbf{v}$.

Exercise 12.9* Prove that $M = \begin{bmatrix} 2 & 1 \\ 0 & 2 \end{bmatrix}$ is not diagonalisable.

Exercise 12.10* The sequence of *Fibonacci numbers* is obtained by writing down 1 twice, and obtaining each successive number by adding the previous two numbers together:

$$1,\ 1,\ 2,\ 3,\ 5,\ 8,\ 13,\ 21,\ 34,\ 55,\ 89,\ 144,\ \ldots$$

If we let x_n denote the nth Fibonacci number then

$$x_1 = x_2 = 1\,, \quad x_n = x_{n-1} + x_{n-2} \quad \text{for } n \geq 3\,,$$

so that

$$\begin{bmatrix} x_n \\ x_{n-1} \end{bmatrix} = \begin{bmatrix} 1 & 1 \\ 1 & 0 \end{bmatrix} \begin{bmatrix} x_{n-1} \\ x_{n-2} \end{bmatrix} = \begin{bmatrix} 1 & 1 \\ 1 & 0 \end{bmatrix}^{n-2} \begin{bmatrix} 1 \\ 1 \end{bmatrix}.$$

Diagonalise $M = \begin{bmatrix} 1 & 1 \\ 1 & 0 \end{bmatrix}$ to find a general formula for the nth Fibonacci number.

Exercise 12.11* Suppose that $\mathbf{v}_1$ and $\mathbf{v}_2$ are eigenvectors for a 2×2 matrix M corresponding to different eigenvalues λ_1 and λ_2. Prove that $\mathbf{v}_1$ and $\mathbf{v}_2$ are not scalar multiples of each other (so are linearly independent). Show that

$$P = \begin{bmatrix} \mathbf{v}_1 & \mathbf{v}_2 \end{bmatrix}$$

is an invertible matrix.

Exercise 12.12* Prove that if $\mathbf{v}_1$, $\mathbf{v}_2$ and $\mathbf{v}_3$ are eigenvectors of a 3×3 matrix M corresponding to distinct eigenvalues, then they are linearly independent.

Exercise 12.13* Prove that if $\mathbf{v}_1, \ldots, \mathbf{v}_n$ are linearly independent column vectors of length n then the matrix

$$P = \begin{bmatrix} \mathbf{v}_1 & \ldots & \mathbf{v}_n \end{bmatrix}$$

is invertible.

Exercise 12.14** Prove that if $\mathbf{v}_1, \ldots, \mathbf{v}_n$ are eigenvectors of an $n \times n$ matrix M corresponding to distinct eigenvalues, then they are linearly independent. Deduce, from the previous exercise, that the matrix

$$P = \begin{bmatrix} \mathbf{v}_1 & \ldots & \mathbf{v}_n \end{bmatrix}$$

is invertible, so M is diagonalisable.

Exercise 12.15** Prove that every 2×2 complex matrix is similar to a diagonal matrix or to a matrix of the form

$$\begin{bmatrix} \lambda & 1 \\ 0 & \lambda \end{bmatrix}$$

for some $\lambda \in \mathbb{C}$. Deduce that every 2×2 real matrix is similar to a diagonal matrix or a matrix of the above form for some $\lambda \in \mathbb{R}$, or a scalar multiple of a rotation matrix.

Hints and Solutions

Hints and Short Solutions to Exercises

0 Introduction

0.1 Observe that $\frac{1}{2}(2x+4) - x \;=\; x + 2 - x \;=\; 2$.

0.2 The y-intercept is 2 and the x-intercept is 3. The slope of the line regarding the x-axis as horizontal is $-2/3$. Regarding the y-axis as horizontal the slope is $-3/2$.

0.3 The equation of the reflected line is $3x + 2y = 6$, with slope $-3/2$, which is the reciprocal of the slope of the original line.

0.4 The equation of the rotated line is $-3x + 2y = 6$, with slope $3/2$, which is the negative reciprocal of the slope of the original line.

0.5 The point of intersection is $(6/5, 6/5)$.

0.6* The equation of the reflected line is $bx + ay = c + k(a-b)$.

0.7* The equation of the rotated line is $-bx+ay = c+a(y_0-x_0)-b(x_0+y_0)$.

0.10** Verify that $1+2+\cdots+n = \dfrac{n(n+1)}{2}$ and play with inequalities.

1 Geometric Vectors

1.1 (i) $\mathbf{v}+\mathbf{u}$ (ii) $\mathbf{u}-\mathbf{v}$ (iii) $\mathbf{u}-\mathbf{v}$ (iv) $\mathbf{u}+\mathbf{v}-\mathbf{w}$ (v) $\mathbf{u}-\mathbf{v}-\mathbf{w}$

1.2 (i) $\mathbf{x} = \mathbf{a} - \mathbf{b}$ (ii) $\mathbf{x} = \mathbf{c} - \dfrac{12}{7}\mathbf{b}$ **1.3** $\frac{1}{2}\mathbf{v} = -\mathbf{a} + 6\mathbf{b} - 7\mathbf{c}$

1.5 $(\overrightarrow{PQ}+\overrightarrow{QR})+\overrightarrow{RS} = \overrightarrow{PR}+\overrightarrow{RS} = \overrightarrow{PS} = \overrightarrow{PQ}+\overrightarrow{QS} = \overrightarrow{PQ}+(\overrightarrow{QR}+\overrightarrow{RS})$

1.6 The rule works because ratios of corresponding sides of similar triangles are equal, and triangles which have been rotated 180° are congruent.

1.7 The shortest distance between two points is a straight line. The triangle inequality becomes equality when the vectors are parallel.

1.8 $\overrightarrow{OR} = \overrightarrow{OP} + \frac{1}{2}(\overrightarrow{PO}+\overrightarrow{OQ}) = \overrightarrow{OP} - \frac{1}{2}\overrightarrow{OP} + \frac{1}{2}\overrightarrow{OQ} = \frac{1}{2}(\overrightarrow{OP}+\overrightarrow{OQ})$

1.9* Use the previous exercise. Check that the sum of vectors representing the medians, emanating from the vertices and pointing in the direction of midpoints of opposite sides, is the zero vector.

1.10* Let P, Q, R, S be the respective midpoints of the edges AB, BC, CD, DA of a quadrilateral $ABCD$. Apply Exercise 1.8 twice to get $\overrightarrow{PQ} = \frac{1}{2}\overrightarrow{AC} = \overrightarrow{SR}$.

1.11* Let $PQRS$ be a parallelogram and T the midpoint of PR. Verify that $\overrightarrow{QT} = \frac{1}{2}\overrightarrow{QS}$ by first expanding $\overrightarrow{QT} = \overrightarrow{QP} + \overrightarrow{PT} = \overrightarrow{QP} + \frac{1}{2}\overrightarrow{PR}$.

1.12** Suppose that $\overrightarrow{PU} = \alpha\overrightarrow{PR}$ and $\overrightarrow{QU} = \beta\overrightarrow{QT}$. Use the fact that $\overrightarrow{PS}$ and $\overrightarrow{PQ}$ are not parallel to deduce that $\alpha = \beta = \dfrac{r+s}{r+2s}$.

2 Position Vectors and Components

2.1 (i) $5\mathbf{i} + 9\mathbf{j} + 4\mathbf{k}$ (ii) $3\mathbf{j} - 4\mathbf{k}$ (iii) $5\mathbf{i} + 15\mathbf{j} - 4\mathbf{k}$ (iv) $3\mathbf{i} + 6\mathbf{j} - 6\mathbf{k}$ (v) $\dfrac{\mathbf{i}}{2} - \dfrac{3\mathbf{j}}{2} - 4\mathbf{k}$ (vi) 13 (vii) 5 (viii) 3 (ix) 9 (x) 21 (xi) $\dfrac{5\mathbf{i}}{13} - \dfrac{12\mathbf{j}}{13}$ (xii) $-\dfrac{3\mathbf{j}}{5} + \dfrac{4\mathbf{k}}{5}$ (xiii) $\dfrac{\mathbf{i}}{3} + \dfrac{2\mathbf{j}}{3} - \dfrac{2\mathbf{k}}{3}$ (xiv) $\sqrt{6}$ (xv) $\sqrt{62}$

2.2 (i) $\overrightarrow{OP} = \mathbf{i}+\mathbf{j}+\mathbf{k}$, $\overrightarrow{OQ} = -\mathbf{i}-\mathbf{j}$, $\overrightarrow{OR} = \mathbf{j}+2\mathbf{k}$, $\overrightarrow{OS} = 2\mathbf{i}+3\mathbf{j}+3\mathbf{k}$, $\overrightarrow{PQ} = -2\mathbf{i} - 2\mathbf{j} - \mathbf{k}$, $\overrightarrow{QP} = 2\mathbf{i} + 2\mathbf{j} + \mathbf{k}$, $\overrightarrow{QR} = \mathbf{i} + 2\mathbf{j} + 2\mathbf{k}$, $\overrightarrow{RS} = 2\mathbf{i} + 2\mathbf{j} + \mathbf{k}$, $\overrightarrow{SP} = -\mathbf{i} - 2\mathbf{j} - 2\mathbf{k}$

(ii) Observe that $\overrightarrow{PQ} = \overrightarrow{SR}$ so $PQRS$ is a parallelogram. Further, $|\overrightarrow{PQ}| = |\overrightarrow{QR}| = 3$, so $PQRS$ is a rhombus. But the diagonals may be represented by the vectors $\overrightarrow{PR} = -\mathbf{i} + \mathbf{k}$ of length $\sqrt{2}$ and $\overrightarrow{QS} = 3\mathbf{i} + 4\mathbf{j} + 3\mathbf{k}$ of length $\sqrt{34} \neq \sqrt{2}$, so the rhombus cannot be a square.

2.4 (i) -2, 6 (ii) $4/3$

2.6* $$\overrightarrow{OR} = \overrightarrow{OP} + \overrightarrow{PR} = \overrightarrow{OP} + \frac{\lambda}{\lambda+\mu}\left(\overrightarrow{PO} + \overrightarrow{OQ}\right) = \frac{\mu\overrightarrow{OP} + \lambda\overrightarrow{OQ}}{\lambda+\mu}$$

2.7* $\overrightarrow{QT} = \dfrac{1}{3}(\mathbf{v} - 2\mathbf{u})$, $\overrightarrow{QB} = \dfrac{1}{2}(\mathbf{v} - 2\mathbf{u})$ **2.8*** $\alpha = \frac{1}{3}$, $\beta = \frac{2}{3}$

2.9* Either check your answer directly after conjecturing (say from a diagram) the value of the ratio, or use Exercise 2.8, exploiting the fact that adjacent sides of the parallelogram are not parallel. The latter is guaranteed to work but takes longer.

2.10* If $\lambda_1\mathbf{v}_1 + \cdots + \lambda_n\mathbf{v}_n = \mathbf{0}$ where some $\lambda_i \neq 0$ then

$$\mathbf{v}_i = \left(\frac{-\lambda_1}{\lambda_i}\right)\mathbf{v}_1 + \cdots + \left(\frac{-\lambda_{i-1}}{\lambda_i}\right)\mathbf{v}_{i-1} + \left(\frac{-\lambda_{i+1}}{\lambda_i}\right)\mathbf{v}_{i+1} + \cdots + \left(\frac{-\lambda_n}{\lambda_i}\right)\mathbf{v}_n .$$

Conversely, if some $\mathbf{v}_i$ is a linear combination of the other vectors, say

$$\mathbf{v}_i = a_1\mathbf{v}_1 + \cdots + a_{i-1}\mathbf{v}_{i-1} + a_{i+1}\mathbf{v}_{i+1} + \cdots + a_n\mathbf{v}_n ,$$

then

$$a_1\mathbf{v}_1 + \cdots + a_{i-1}\mathbf{v}_{i-1} + (-1)\mathbf{v}_i + a_{i+1}\mathbf{v}_{i+1} + \cdots + a_n\mathbf{v}_n = \mathbf{0},$$

so that the implication in the definition of linear independence fails.

2.12** Suppose that

$$\lambda_0 f_0 + \cdots + \lambda_n f_n = \mathbf{0},$$

where $\lambda_0, \ldots, \lambda_n$ are real numbers and $\mathbf{0}$ denotes the zero function. Interpret this as a polynomial equation. Prove that a nonzero polynomial of degree n has at most n roots. (An elegant proof, for example, is given in Exercise 10.20, using Vandermonde's determinant.) Use the fact that there are infinitely many real numbers to deduce that $\lambda_0 = \ldots = \lambda_n = 0$.

3 Dot Products and Projections

3.1 $2\mathbf{i} + \mathbf{j} + 3\mathbf{k}$, $\sqrt{14}$, $\mathbf{i} + 2\mathbf{j} - \mathbf{k}$, $\sqrt{6}$, 1, $\dfrac{1}{2\sqrt{21}}$

3.2 (i) -4 (ii) $\frac{1}{\sqrt{5}}(\mathbf{i} - 2\mathbf{j})$ (iii) $\frac{1}{\sqrt{5}}(-2\mathbf{i} + \mathbf{j})$ (iv) $-\frac{4}{\sqrt{5}}$ (v) $-\frac{4}{\sqrt{5}}$ (vi) $-\frac{4}{5}$
(vii) $-\frac{4}{5}(\mathbf{i} - 2\mathbf{j})$ (viii) $\frac{4}{5}(2\mathbf{i} - \mathbf{j})$ (ix) $-\frac{3}{5}(2\mathbf{i} + \mathbf{j})$ (x) $-\frac{3}{5}(\mathbf{i} + 2\mathbf{j})$
(xi) $-\frac{4}{5}$ (xii) $-\frac{4}{\sqrt{5}}$ (xiii) $-\frac{4}{\sqrt{5}}$ (xiv) $\frac{4}{5}(2\mathbf{i} - \mathbf{j})$ (xv) $-\frac{4}{5}(\mathbf{i} - 2\mathbf{j})$
(xvi) $-\frac{3}{5}(\mathbf{i} + 2\mathbf{j})$ (xvii) $-\frac{3}{5}(2\mathbf{i} + \mathbf{j})$

3.3 (i) -36, 29, -14, 15, 100 (ii) obtuse, obtuse, acute
(iii) $-\dfrac{36}{5}$, $\dfrac{36}{25}(3\mathbf{j} - 4\mathbf{k})$, $5\mathbf{i} + \dfrac{192}{25}\mathbf{j} + \dfrac{144}{25}\mathbf{k}$
(iv) $\dfrac{5}{3}(\mathbf{i} + 2\mathbf{j} - 2\mathbf{k})$, $\dfrac{1}{3}(10\mathbf{i} + 17\mathbf{j} + 22\mathbf{k})$, $\dfrac{15}{122}(5\mathbf{i} + 9\mathbf{j} + 4\mathbf{k})$, $\dfrac{1}{122}(47\mathbf{i} + 109\mathbf{j} - 304\mathbf{k})$

3.4 (i) acute, obtuse (ii) both equal to $\left(\frac{1}{2}, 1, \frac{3}{2}\right)$ (iii) $\overrightarrow{PR} \cdot \overrightarrow{QS} = 0$

3.5 If $|\mathbf{a}| = |\mathbf{b}|$ then $(\mathbf{a}+\mathbf{b})\cdot(\mathbf{a}-\mathbf{b}) = |\mathbf{a}|^2 - \mathbf{a}\cdot\mathbf{b} + \mathbf{b}\cdot\mathbf{a} - |\mathbf{b}|^2 = 0$. If $PQRS$ is a rhombus then $|\overrightarrow{PQ}| = |\overrightarrow{QR}|$ and the result follows since the diagonals are $\overrightarrow{PQ} + \overrightarrow{QR}$ and $\overrightarrow{PQ} - \overrightarrow{QR}$.

3.6 If PR is a diameter of a semicircle with centre O, and Q is located on the semicircle, then $\overrightarrow{QP} = \overrightarrow{QO} + \overrightarrow{OP}$ and $\overrightarrow{QR} = \overrightarrow{QO} - \overrightarrow{OP}$, so the result follows by the previous exercise, since $|\overrightarrow{QO}| = |\overrightarrow{OP}|$.

3.7 If $\mathbf{a}\cdot\mathbf{b} = 0$ then $|\mathbf{a}+\mathbf{b}|^2 = (\mathbf{a}+\mathbf{b})\cdot(\mathbf{a}+\mathbf{b}) = \mathbf{a}\cdot\mathbf{a}+\mathbf{a}\cdot\mathbf{b}+\mathbf{b}\cdot\mathbf{a}+\mathbf{b}\cdot\mathbf{b} = |\mathbf{a}|^2+|\mathbf{b}|^2$, which is the Theorem of Pythagoras where the vectors label edges of a right-angled triangle.

3.8 (i) $\mathbf{v}\cdot(a\mathbf{x}+b\mathbf{y}) \;=\; a(\mathbf{v}\cdot\mathbf{x}) + b(\mathbf{v}\cdot\mathbf{y}) \;=\; a(0)+b(0) \;=\; 0$

(ii) $\mathbf{v}\cdot\mathbf{y} \;=\; \mathbf{v}\cdot\left(\frac{1}{b}(a\mathbf{x}+b\mathbf{y}-a\mathbf{x})\right) \;=\; \frac{1}{b}\left(\mathbf{v}\cdot(a\mathbf{x}+b\mathbf{y}) - a(\mathbf{v}\cdot\mathbf{x})\right) \;=\; \frac{1}{b}(0-a(0)) \;=\; 0$

3.9 Expand brackets, bring scalars to the front, and evaluate.

3.10 Drop a perpendicular to create a right-angled triangle.

3.11* Given $\mathbf{v} = a\mathbf{i}+b\mathbf{j}+c\mathbf{k}$ and $\mathbf{w} = d\mathbf{i}+e\mathbf{j}+f\mathbf{k}$ and the geometric definition, we have, applying the Cosine Rule and the length formula,

$$\begin{aligned}\mathbf{v}\cdot\mathbf{w} &= |\mathbf{v}||\mathbf{w}|\cos\theta = \frac{1}{2}\left(|\mathbf{v}|^2+|\mathbf{w}|^2-|\mathbf{v}-\mathbf{w}|^2\right)\\ &= \frac{1}{2}\left(a^2+b^2+c^2+d^2+e^2+f^2-\left((a-d)^2+(b-e)^2+(c-f)^2\right)\right)\\ &= \frac{1}{2}\left(2ad+2be+2cf\right) = ad+be+cf\,.\end{aligned}$$

3.12* To the nearest degree, answers are 71°, 55° and 35° respectively.

3.13* The identity follows by properties of the dot product, expanding and simplifying. Use the identity with **a**, **b**, **c**, **d** as position vectors of points A, B, C, D respectively, where D is the intersection of the altitudes through A and B.

3.14* Let PQR be a triangle and A, B, C be the midpoints of QR, RP, PQ respectively. Let D be the intersection of the perpendicular bisectors of RP and QR. Observe that

$$\overrightarrow{DC}\cdot\overrightarrow{PQ} \;=\; (\overrightarrow{DB}+\overrightarrow{BC})\cdot(\overrightarrow{PR}+\overrightarrow{RQ})\,,$$

expand brackets, and evaluate to 0.

3.15* With the notation of the previous exercise, use Pythagoras to show that $|PD| = |RD|$.

3.16** Proving that the geometric dot product distributes over vector addition is the difficult part. Draw a good diagram and carefully label angles. Apply the Sine Rule for triangles and manipulate expressions using trigonometric formulae.

4 Cross Products

4.1 (i) $\mathbf{k}$ (ii) $6\mathbf{k}$ (iii) $-4\mathbf{k}$ (iv) $-\mathbf{k}$ (v) $4\mathbf{k}$ (vi) $\mathbf{0}$
(vii) $\mathbf{0}$ (viii) $-\mathbf{j}$ (ix) $-\mathbf{i}$ (x) $\mathbf{j}$ (xi) $2\mathbf{i}$ (xii) $2\mathbf{i}$

4.2 (i) $48\mathbf{i} - 20\mathbf{j} - 15\mathbf{k}$ (ii) $-24\mathbf{i} + 10\mathbf{j} - 2\mathbf{k}$ (iii) $-2\mathbf{i} + 4\mathbf{j} + 3\mathbf{k}$
(iv) $-26\mathbf{i} + 14\mathbf{j} + \mathbf{k}$ (v) $70\mathbf{i} + 81\mathbf{j} + 116\mathbf{k}$ (vi) 38 (vii) 38 (viii) $38\mathbf{v}$

4.3 (i) 3 (ii) $\sqrt{3}$ (iii) $-\mathbf{i} + 4\mathbf{j} + 3\mathbf{k}$ (iv) $\sqrt{26}$ (v) $\dfrac{\sqrt{78}}{9}$

4.4 $-28\mathbf{i} - 11\mathbf{j} - 10\mathbf{k}\,,\ \dfrac{\sqrt{1005}}{2}$ **4.5** $20\sqrt{5}$

4.6 (i) (c) (ii) (d) (iii) (e) (iv) (b) (v) (a)

4.7 (i) $\dfrac{1}{\sqrt{2}}(\mathbf{i} + \mathbf{k})$ (ii) $-\dfrac{1}{\sqrt{2}}(\mathbf{i} + \mathbf{k})$

4.9 (i) Both are $\dfrac{\sqrt{17}}{2}$. There is no surprise since $PQRS$ is a rhombus.

(ii) $d_1 = \sqrt{2}$, $d_2 = \sqrt{34}$, $\dfrac{d_1 d_2}{4} = \dfrac{\sqrt{17}}{2}$, which coincides with the areas of the triangles in the first part, which is no surprise since the diagonals of a rhombus are mutually perpendicular.

4.10* Label three directed edges of the tetrahedron emanating from a given vertex by $\mathbf{u}$, $\mathbf{v}$ and $\mathbf{w}$ in a clockwise orientation. By the Right-Hand Rule we may take

$$\mathbf{a} = \frac{1}{2}(\mathbf{v} \times \mathbf{u})\,, \quad \mathbf{b} = \frac{1}{2}(\mathbf{w} \times \mathbf{v})\,, \quad \mathbf{c} = \frac{1}{2}(\mathbf{u} \times \mathbf{w})$$

and $\mathbf{d} = \dfrac{1}{2}\big[(\mathbf{w} - \mathbf{v}) \times (\mathbf{u} - \mathbf{v})\big]$. These add to $\mathbf{0}$.

4.14* Show that $\mathbf{u}$ is both parallel and perpendicular to $\mathbf{v} - \mathbf{w}$.

4.15* Having verified the first two equations, then $(\mathbf{u}\times\mathbf{v})\times\mathbf{w} = \mathbf{u}\times(\mathbf{v}\times\mathbf{w})$ if and only if $(\mathbf{v}\cdot\mathbf{w})\,\mathbf{u} = (\mathbf{u}\cdot\mathbf{v})\,\mathbf{w}$ which occurs if and only if $\mathbf{u}$ and $\mathbf{w}$ are parallel, or $\mathbf{v}$ is perpendicular to both $\mathbf{u}$ and $\mathbf{w}$.

4.16** In proving distributivity, project the vector $\mathbf{a}+\mathbf{b}$ into the plane perpendicular to $\mathbf{c}$, and then rotate the projected image in this plane until it is perpendicular to $\mathbf{a}+\mathbf{b}$. Compare with rotated projections of $\mathbf{a}$ and $\mathbf{b}$.

5 Lines in Space

5.1 (i) $\mathbf{r} = 3\mathbf{i}+2\mathbf{j}+\mathbf{k}+t(\mathbf{i}+2\mathbf{j}+3\mathbf{k})$, $\left.\begin{aligned} x &= 3+t \\ y &= 2+2t \\ z &= 1+3t \end{aligned}\right\}$ $t\in\mathbb{R}$,

$$x-3 = \frac{y-2}{2} = \frac{z-1}{3}$$

(ii) $\mathbf{r} = \mathbf{k}+t\mathbf{k} = (1+t)\mathbf{k} = s\mathbf{k}$, $\left.\begin{aligned} x &= 0 \\ y &= 0 \\ z &= s \end{aligned}\right\}$ $s\in\mathbb{R}$, $x=y=0$

(iii) $\mathbf{r} = 2\mathbf{i}-\mathbf{k}+t(-\mathbf{i}+2\mathbf{j}-3\mathbf{k})$, $\left.\begin{aligned} x &= 2-t \\ y &= 2t \\ z &= -1-3t \end{aligned}\right\}$ $t\in\mathbb{R}$,

$$\frac{x-2}{-1} = \frac{y}{2} = \frac{z+1}{-3}$$

(iv) $\mathbf{r} = t(\mathbf{j}+\mathbf{k})$, $\left.\begin{aligned} x &= 0 \\ y &= t \\ z &= t \end{aligned}\right\}$ $t\in\mathbb{R}$, $x = 0,\ y = z$

(v) $\mathbf{r} = -4\mathbf{i}-3\mathbf{j}+t(\mathbf{j}-\mathbf{k})$, $\left.\begin{aligned} x &= -4 \\ y &= -3+t \\ z &= -t \end{aligned}\right\}$ $t\in\mathbb{R}$,

$x=-4,\ y+3=-z$

(vi) $\mathbf{r} = \mathbf{i}+\mathbf{k}+t(\mathbf{i}-\mathbf{j})$, $\left.\begin{aligned} x &= 1+t \\ y &= t \\ z &= 1 \end{aligned}\right\}$ $t\in\mathbb{R}$, $x-1=-y,\ z=1$

5.2 (i) $\mathbf{r} = 3\mathbf{i}+2\mathbf{j}+\mathbf{k}+t(\mathbf{i}-\mathbf{k})$, $\left.\begin{aligned} x &= 3+t \\ y &= 2 \\ z &= 1-t \end{aligned}\right\}$ $t\in\mathbb{R}$,

$x-3 = 1-z,\ y = 2$

(ii) $\mathbf{r} = \mathbf{k} + t(\mathbf{i} - \mathbf{k})$, $\left.\begin{aligned} x &= t \\ y &= 0 \\ z &= 1 - t \end{aligned}\right\}$ $t \in \mathbb{R}$, $x = 1 - z$, $y = 0$

(iii) $\mathbf{r} = 2\mathbf{i}+\mathbf{k}+t(2\mathbf{i}+7\mathbf{j}-4\mathbf{k})$, $\left.\begin{aligned} x &= 2 + 2t \\ y &= 7t \\ z &= 1 - 4t \end{aligned}\right\}$ $t \in \mathbb{R}$,

$$\frac{x-2}{2} = \frac{y}{7} = \frac{z-1}{-4}$$

(iv) $\mathbf{r} = \mathbf{i}+\mathbf{j}+\mathbf{k}+t(\mathbf{i}+\mathbf{j})$, $\left.\begin{aligned} x &= 1 + t \\ y &= 1 + t \\ z &= 1 \end{aligned}\right\}$ $t \in \mathbb{R}$, $x = y$, $z = 1$

5.3 The lines are identical since they are both parallel to $4\mathbf{i} + 3\mathbf{j} + 5\mathbf{k}$ and contain $(1, 2, 10)$.

5.4 Vectors in the directions of the lines are $2\mathbf{i} + 3\mathbf{j} + 5\mathbf{k}$ and $3\mathbf{i} - 2\mathbf{j} - \mathbf{k}$ respectively, which are not parallel. If the lines intersect then

$$\frac{1 + 3s + 3}{2} = \frac{6 - 2s}{3} = \frac{15 - s - 4}{5}$$

for some s. But the first equation implies $s = 0$ and the second implies $s = -3/7$, which is a contradiction. This proves that the lines are skew.

5.5 The point of intersection is $(-8, 4, -11)$.

5.6 (i) $\mathcal{L}_1 :\ \mathbf{r} = \mathbf{i}+\mathbf{j}+\mathbf{k}+t(5\mathbf{i}-4\mathbf{j}-2\mathbf{k})$, $\left.\begin{aligned} x &= 1 + 5t \\ y &= 1 - 4t \\ z &= 1 - 2t \end{aligned}\right\}$ $t \in \mathbb{R}$,

$$\frac{x-1}{5} = \frac{y-1}{-4} = \frac{z-1}{-2}$$

$\mathcal{L}_2 : \mathbf{r} = 5\mathbf{i}-5\mathbf{j}-3\mathbf{k}+s(-3\mathbf{i}+8\mathbf{j}+6\mathbf{k})$, $\left.\begin{aligned} x &= 5 - 3s \\ y &= -5 + 8s \\ z &= -3 + 6s \end{aligned}\right\}$ $s \in \mathbb{R}$,

$$\frac{x-5}{-3} = \frac{y+5}{8} = \frac{z+3}{6}$$

(ii) $T = (7/2, -1, 0)$

(iii) The coordinates of T are the averages of those of P and R and also of Q and S, so T is the midpoint of PR and QS. This is not surprising since $PQRS$ is a parallelogram.

5.7* $\sqrt{270}/\sqrt{11}$, $(7/11, -1/11, -40/11)$ **5.8*** $\sqrt{2275}/13$

5.9* Observe that $\overrightarrow{OR} = \lambda\overrightarrow{OP} + (1-\lambda)\overrightarrow{OQ} = \overrightarrow{OQ} + \lambda\overrightarrow{QP}$.

(i) $0 \le \lambda \le 1$ (ii) $\lambda > 1$ (iii) $\lambda < 0$ (iv) $\lambda = 1/3$ or $\lambda = -1$

5.10* (i) same line, zero rotation (ii) $90°$ rotation (iii) $180°$ rotation (iv) $90° - \theta$ rotation where θ is the angle between $\mathbf{v}$ and $\mathbf{w}$.

5.11** $4/\sqrt{3}$, $(2, 25/3, -22/3)$, $(2/3, 29/3, -6)$

5.12* Interpret $\mathbf{r}'$ as the velocity of the particle on $\mathcal{C}$ at time t. Split the limit into components.

6 Planes in Space

6.1 $-x + y + 3z = -25$

6.2 (i) $-4\mathbf{i} - 3\mathbf{j} - 4\mathbf{k}$, $\mathbf{i} - 5\mathbf{j} - 2\mathbf{k}$ (ii) $-14\mathbf{i} - 12\mathbf{j} + 23\mathbf{k}$

(iii) $-14x - 12y + 23z = 29$

6.3 (i) $x + y + z = 0$ (ii) $101x + 22y - 8z = 256$

6.4 (i) (b)(n) (ii) (d)(i) (iii) (f)(m) (iv) (a)(k) (v) (g)(l)

(vi) (e)(h) (vii) (c)(j)

6.5 $x - 3 = \dfrac{y+1}{-3} = \dfrac{z-2}{2}$ **6.6** $(3\mathbf{i} + 3\mathbf{j} - \mathbf{k}) \cdot (-3\mathbf{i} + 4\mathbf{j} + 3\mathbf{k}) = 0$

6.7
$$\left.\begin{aligned} x &= 3t \\ y &= -1 + t \\ z &= -7 + 7t \end{aligned}\right\} \quad t \in \mathbb{R}, \qquad \frac{x}{3} = y + 1 = \frac{z+7}{7}$$

6.8 $11x + 31y + z = -6$ **6.9** $-\dfrac{\sqrt{2}}{3}$

6.10 Take $Q = (d/a, 0, 0), (0, d/b, 0), (0, 0, d/c)$ according to whether $a \neq 0$, $b \neq 0$, $c \neq 0$ respectively. Take $\mathbf{n} = a\mathbf{i} + b\mathbf{j} + c\mathbf{k}$ in all cases. If $a = b = c = d = 0$ then the equation is satisfied by all points in space. If $a = b = c = 0 \neq d$ then no points satisfy the equation.

6.11 If $Q = R$ then $\overrightarrow{PQ} \cdot \overrightarrow{PR} = |\overrightarrow{PQ}|^2 > 0$. If $Q \neq R$ then as one travels from Q to R in a straight line, the distance to P is decreasing, so the angle QPR must be acute, so $\overrightarrow{PQ} \cdot \overrightarrow{PR} > 0$.

6.12* $32\sqrt{3}/9$, $\left(-\dfrac{113}{27}, \dfrac{86}{27}, \dfrac{29}{27}\right)$

6.13* $\left(\mathbf{i}+\frac{\partial f}{\partial x}\mathbf{k}\right)\times\left(\mathbf{j}+\frac{\partial f}{\partial y}\mathbf{k}\right) = \mathbf{k}-\frac{\partial f}{\partial y}\mathbf{j}-\frac{\partial f}{\partial x}\mathbf{i}$

6.14* The equation describes a sphere of radius r centred at the origin. The position vector of (x_0, y_0, z_0) is normal to the tangent plane to the sphere at that point. It follows quickly that the equation of the tangent plane is

$$x_0x + y_0y + z_0z = r^2 .$$

6.15* Intersection points are $\left(0, \frac{\pm 100}{t-10}, \frac{t\sqrt{101}}{t-10}\right)$. As $t \to -\infty$, $z \to \sqrt{101}$.

7 Systems of Linear Equations

7.1 $x = 1,\ y = 2,\ z = -3$ **7.2** $x_1 = -\frac{4t}{3},\ x_2 = 1 + \frac{t}{3},\ x_3 = t$

7.3 $x_1 = 1 - s - \frac{t}{2},\ x_2 = \frac{1}{2} - \frac{t}{2},\ x_3 = s,\ x_4 = t,\ x_5 = \frac{1}{2}$

7.4 $x_1 = -2s,\ x_2 = \frac{1}{2}(3+3s-2t),\ x_3 = \frac{1}{2}(1+9s-4t),\ x_4 = s,\ x_5 = t$

7.5 $x = y = z = 0$ **7.6** $x_1 = 4t/3,\ x_2 = -5t/3,\ x_3 = 7t/3,\ x_4 = t$

7.7 $x = -1 - 17t,\ y = -1 - 29t,\ z = t$ **7.8** (i) (c) (ii) (a) (iii) (b) (iv) (a)

7.9 $x = 1,\ y = 3,\ z = 2$ **7.10** $p(x) = x^3 + 2x^2 - 3x + 4$

7.11 $A = 1,\ B = 3,\ C = 3,\ D = 1$ **7.12** $x = 2,\ y = 25,\ z = 16,\ w = 18$

7.13* Take $\lambda = -5$ for no solutions, $\lambda = 2$ for infinitely many solutions, and $\lambda \neq -5, 2$ for a unique solution.

7.15* The performance of these operations simultaneously cannot be invertible because it yields a process (function) that is not one-one on augmented matrices (so that inversion is ambiguous).

7.17* The asterisks are intended to be arbitrary real numbers chosen independently:

$$\begin{bmatrix} 0&0&0&0\\0&0&0&0\\0&0&0&0 \end{bmatrix}, \begin{bmatrix} 0&0&0&1\\0&0&0&0\\0&0&0&0 \end{bmatrix}, \begin{bmatrix} 0&0&1&*\\0&0&0&0\\0&0&0&0 \end{bmatrix}, \begin{bmatrix} 0&1&*&*\\0&0&0&0\\0&0&0&0 \end{bmatrix},$$

$$\begin{bmatrix} 1&*&*&*\\0&0&0&0\\0&0&0&0 \end{bmatrix}, \begin{bmatrix} 0&0&1&0\\0&0&0&1\\0&0&0&0 \end{bmatrix}, \begin{bmatrix} 0&1&*&0\\0&0&0&1\\0&0&0&0 \end{bmatrix}, \begin{bmatrix} 0&1&0&*\\0&0&1&*\\0&0&0&0 \end{bmatrix},$$

$$\begin{bmatrix} 1&*&*&0\\0&0&0&1\\0&0&0&0 \end{bmatrix}, \begin{bmatrix} 1&*&0&*\\0&0&1&*\\0&0&0&0 \end{bmatrix}, \begin{bmatrix} 1&0&*&*\\0&1&*&*\\0&0&0&0 \end{bmatrix}, \begin{bmatrix} 0&1&0&0\\0&0&1&0\\0&0&0&1 \end{bmatrix},$$

$$\begin{bmatrix} 1&*&0&0\\0&0&1&0\\0&0&0&1 \end{bmatrix}, \begin{bmatrix} 1&0&*&0\\0&1&*&0\\0&0&0&1 \end{bmatrix}, \begin{bmatrix} 1&0&0&*\\0&1&0&*\\0&0&1&* \end{bmatrix}$$

The system must have a nontrivial solution, since there are more variables than equations (so that there is at least one nonleading variable to which a parameter may be assigned). If we have four vectors in space, say $\mathbf{u} = u_1\mathbf{i} + u_2\mathbf{j} + u_3\mathbf{k}$, $\mathbf{v} = v_1\mathbf{i} + v_2\mathbf{j} + v_3\mathbf{k}$, $\mathbf{w} = w_1\mathbf{i} + w_2\mathbf{j} + w_3\mathbf{k}$, $\mathbf{t} = t_1\mathbf{i} + t_2\mathbf{j} + t_3\mathbf{k}$, then the vector equation $\alpha\mathbf{u} + \beta\mathbf{v} + \gamma\mathbf{w} + \delta\mathbf{t} = \mathbf{0}$ is equivalent to the given homogeneous system, so has a nontrivial solution, which proves $\mathbf{u}$, $\mathbf{v}$, $\mathbf{w}$, $\mathbf{t}$ are linearly dependent.

7.18* (i) Different systems produce different augmented matrices, and every augmented matrix arises from some system. (ii) To operate on the system and then form the augmented matrix has the same effect as forming the augmented matrix and then performing the corresponding elementary row operation.

7.19* (i) $f(n) = n^2$ (ii) $g(n) = n^2(n+1)$ including operations with zeros

7.20** Each reduced row echelon matrix corresponds to a homogeneous system. As the matrix varies, so does the solution set of the corresponding homogeneous system. Since the solution sets are invariant under elementary row operations, the reduced echelon forms cannot be row equivalent.

8 Matrix Operations

8.1 2×2, 1×3, 4×2 **8.2** (i) 0 (ii) -7 (iii) 8 (iv) -3

8.3 (i) $\begin{bmatrix} 2 & 4 \\ 0 & 6 \end{bmatrix}$ (ii) $\begin{bmatrix} 6 & -3 \\ -4 & -1 \end{bmatrix}$ (iii) $\begin{bmatrix} -5 & 5 \\ 4 & 4 \end{bmatrix}$ (iv) $\begin{bmatrix} 7 & -1 \\ -4 & 2 \end{bmatrix}$

(v) $\begin{bmatrix} 1 & 8 \\ 0 & 9 \end{bmatrix}$ (vi) $\begin{bmatrix} 2 & 5 \\ 12 & 3 \end{bmatrix}$ (vii) $\begin{bmatrix} -6 & -3 \\ 4 & 11 \end{bmatrix}$ (viii) -3 (ix) $\begin{bmatrix} -13 \\ -6 \\ -1 \end{bmatrix}$

(x) -64

8.4 $\begin{bmatrix} \cos(\alpha+\beta) & -\sin(\alpha+\beta) \\ \sin(\alpha+\beta) & \cos(\alpha+\beta) \end{bmatrix}$

8.5
$$\begin{aligned} & X^2+XY+XY+Y^2=(X+Y)^2=X^2+2XY+Y^2 \\ \iff\ & YX=XY \quad\iff\quad -XY+YX=0 \\ \iff\ & X^2-Y^2=X^2-XY+YX-Y^2=(X+Y)(X-Y) \end{aligned}$$

8.6 $A0_{n\times n} = 0_{m\times m}A = 0_{m\times n}$ and $0_{n\times n} \neq 0_{m\times m} \neq 0_{m\times n}$

8.7 (i) $c=3(2x+3y)-5(x-4y)=x+29y\,,$

$d=2(2x+3y)+3(x-4y)=7x-6y\,.$

(ii) $\begin{bmatrix} c \\ d \end{bmatrix} = \begin{bmatrix} 3 & -5 \\ 2 & 3 \end{bmatrix}\begin{bmatrix} 2 & 3 \\ 1 & -4 \end{bmatrix}\begin{bmatrix} x \\ y \end{bmatrix} = \begin{bmatrix} 1 & 29 \\ 7 & -6 \end{bmatrix}\begin{bmatrix} x \\ y \end{bmatrix} = \begin{bmatrix} x+29y \\ 7x-6y \end{bmatrix}$

8.8 (i) $x=5,\ y=-1,\ z=4$ (ii) $x=-t/2,\ y=2t/3,\ z=t$

8.9 By associativity of multiplication, $AA^2=A(AA)=(AA)A=A^2A\,.$

8.10 All are true except for (ii) and (iii). **8.11** $\begin{bmatrix} 1 & -1 \\ 1 & -1 \end{bmatrix}$

8.12 Multiply each (i) row, (ii) column, by the corresponding diagonal entry.

8.13* (ii) $M^n=nM-(n-1)I$ (iii) $\begin{bmatrix} 11 & -5 \\ 20 & -9 \end{bmatrix}, \begin{bmatrix} 21 & -10 \\ 40 & -19 \end{bmatrix}, \begin{bmatrix} 201 & -100 \\ 400 & -199 \end{bmatrix}$

8.14* (i) 1 (ii) -5 **8.15*** (i) $b=c=0$ (ii) $a=d,\ b=c$ (iii) $a=d,\ c=0$

8.16* (i) $\begin{bmatrix} 1 & nk \\ 0 & 1 \end{bmatrix}$ (ii) $\begin{bmatrix} k^n & nk^{n-1} \\ 0 & k^n \end{bmatrix}$ (iii) $\begin{bmatrix} k^n & nk^{n-1} & \frac{n(n-1)}{2}k^{n-2} \\ 0 & k^n & nk^{n-1} \\ 0 & 0 & k^n \end{bmatrix}$

8.17* Suppose $\mathbf{x}_1+\lambda(\mathbf{x}_1-\mathbf{x}_2)=\mathbf{x}_1+\mu(\mathbf{x}_1-\mathbf{x}_2)$. Then $(\lambda-\mu)(\mathbf{x}_1-\mathbf{x}_2)=\mathbf{0}$ so that $(\lambda-\mu)z=0$ for all entries z in $\mathbf{x}_1-\mathbf{x}_2$. But at least one value of z is nonzero, since $\mathbf{x}_1\neq\mathbf{x}_2$. It follows that $\lambda-\mu=0$, that is, $\lambda=\mu$. Hence, distinct expressions evaluate to different vectors, so there must be infinitely many.

8.18* $(YZ)^T = \left[\sum_j y_{ij}z_{jk}\right]^T = \left[\sum_j y_{kj}z_{ji}\right] = \left[\sum_j z_{ji}y_{kj}\right] = Z^TY^T$

8.19* $A(B+C) = \left[\sum_j a_{ij}(b_{jk}+c_{jk})\right] = \left[\sum_j a_{ij}b_{jk}\right] + \left[\sum_j a_{ij}c_{jk})\right] = AB + AC$, and so $(A+B)C = (C^T(A^T+B^T))^T = (C^TA^T + C^TB^T)^T = (C^TA^T)^T + (C^TB^T)^T = (A^T)^T(C^T)^T + (B^T)^T(C^T)^T = AC + BC$

8.20* $(AB)C = \left[\sum_j a_{ij}b_{jk}\right]\left[c_{kl}\right] = \left[\sum_{jk} a_{ij}b_{jk}c_{kl}\right] = \left[a_{ij}\right]\left[\sum_k b_{jk}c_{kl}\right] = A(BC)$

8.21* The matrix product simplifies to $\begin{bmatrix} rs\cos(\alpha+\beta) & -rs\sin(\alpha+\beta) \\ rs\sin(\alpha+\beta) & rs\cos(\alpha+\beta) \end{bmatrix}$.

If we identify $r \operatorname{cis} \theta$ (in polar form) with $\begin{bmatrix} r\cos\theta & -r\sin\theta \\ r\sin\theta & r\cos\theta \end{bmatrix}$ then the above matrix product becomes the usual multiplication of complex numbers: $(r \operatorname{cis} \alpha)(s \operatorname{cis} \beta) = rs \operatorname{cis}(\alpha+\beta)$. But $r \operatorname{cis} \theta = x + iy$ where $x = r\cos\theta$ and $y = r\sin\theta$, which then gets identified with the matrix $\begin{bmatrix} x & -y \\ y & x \end{bmatrix}$ so that the usual addition of complex numbers corresponds to addition of matrices. Thus there is a copy of complex number arithmetic within the arithmetic of 2×2 real matrices, where the real number 1 corresponds to the identity matrix $\begin{bmatrix} 1 & 0 \\ 0 & 1 \end{bmatrix}$ and the imaginary number i corresponds to $\begin{bmatrix} 0 & -1 \\ 1 & 0 \end{bmatrix}$.

8.22** The *trace* of a matrix is the sum of the diagonal entries. Show that the trace of $AB - BA$ is zero.

9 Matrix Inverses

9.1 Let A be $p \times q$ and B be $r \times s$. Since AB and BA are defined, $q = r$ and $s = p$. But $p = n$, since $AB = I_n$, and $r = n$ since $BA = I_n$, whence $p = q = r = s = n$.

9.2 (i) $\frac{1}{3}\begin{bmatrix} 3 & -2 \\ 0 & 1 \end{bmatrix}$ (ii) $\frac{1}{18}\begin{bmatrix} -1 & 3 \\ 4 & 6 \end{bmatrix}$ (iv) $\begin{bmatrix} \cos\alpha & \sin\alpha \\ -\sin\alpha & \cos\alpha \end{bmatrix}$

(v) $\begin{bmatrix} 2 & 2 & -1 \\ -1 & -2 & 1 \\ -1 & -1 & 1 \end{bmatrix}$ (vi) $\begin{bmatrix} 33 & -17 & 9 \\ -11 & 6 & -3 \\ -4 & 2 & -1 \end{bmatrix}$

(viii) $\frac{1}{2}\begin{bmatrix} -1 & 1 & 1 \\ 1 & -1 & 1 \\ 1 & 1 & -1 \end{bmatrix}$ (ix) $\begin{bmatrix} 1 & -1 & 0 & 0 \\ 0 & 1 & -1 & 0 \\ 0 & 0 & 1 & -1 \\ 0 & 0 & 0 & 1 \end{bmatrix}$

9.3 $\begin{bmatrix} -4 & 3 \\ -7 & 5 \end{bmatrix}$, $x = 1$, $y = -1$, $z = -2$, $w = -3$

9.4 Observe that $(AB)(B^{-1}A^{-1}) = A(BB^{-1})A^{-1} = AIA^{-1} = AA^{-1} = I$ and similarly $(B^{-1}A^{-1})(AB) = I$.

9.5 Suppose that A is $m \times m$ and D is $n \times n$. For $ABD = ACD$ to be sensibly defined, both B and C are $m \times n$. It does not matter if m and n are different: since $A^{-1}A = I_m$ and $DD^{-1} = I_n$, we have

$$B = I_m B I_n = A^{-1}ABDD^{-1} = A^{-1}ACDD^{-1} = I_m C I_n = C\,.$$

9.6 Let A, B be $n \times n$. If A has a row of zeros then AB also has a row of zeros. If A has a column of zeros then BA also has a column of zeros. In either case it is impossible to have $AB = BA = I_n$.

9.7 Let $M = \begin{bmatrix} a & b \\ c & d \end{bmatrix}$. We consider the case $a \neq 0$. (The case $c \neq 0$ is similar.) Then, in two steps,

$$\left[\begin{array}{cc|cc} a & b & 1 & 0 \\ c & d & 0 & 1 \end{array}\right] \sim \left[\begin{array}{cc|cc} 1 & b/a & 1/a & 0 \\ 0 & (ad-bc)/a & -c/a & 1 \end{array}\right].$$

If $ad - bc = 0$ then the left-hand side has a row of zeros so M is not invertible. If $ad - bc \neq 0$ then we proceed further:

$$\left[\begin{array}{cc|cc} a & b & 1 & 0 \\ c & d & 0 & 1 \end{array}\right] \sim \left[\begin{array}{cc|cc} 1 & b/a & 1/a & 0 \\ 0 & 1 & -c/(ad-bc) & a/(ad-bc) \end{array}\right]$$
$$\sim \left[\begin{array}{cc|cc} 1 & 0 & d/(ad-bc) & -b/(ad-bc) \\ 0 & 1 & -c/(ad-bc) & a/(ad-bc) \end{array}\right]$$

so that M^{-1} exists and equals $\dfrac{1}{ad-bc}\begin{bmatrix} d & -b \\ -c & a \end{bmatrix}$.

9.8 A diagonal matrix is invertible if and only if the diagonal entries are nonzero, in which case the inverse matrix is formed by inverting the diagonal entries.

9.9 Matching dimensions as in Exercise 9.1, we see that B is $n \times n$ and C is $m \times m$. Let i be an integer from 1 to n. Let A be any $m \times n$ matrix with the first row consisting of 0 everywhere except for 1 in the ith place. Then the first row of $A = AB$ is also $\begin{bmatrix} b_{i1} & b_{i2} & \cdots & b_{in} \end{bmatrix}$, so that $b_{ij} = \begin{cases} 0 & \text{if } i \neq j \\ 1 & \text{if } i = j \end{cases}$, which proves $B = I_n$. A similar argument using the first column of A shows $C = I_m$.

9.10* All are true except for (i), (iv) and (viii).

9.11* $(I - M)(I - \frac{1}{n-1}M) = I - \frac{n}{n-1}M + \frac{1}{n-1}M^2 = I - \frac{n}{n-1}M + \frac{n}{n-1}M = I$

9.12* Draw diagrams of matrices for each type of elementary row operation, and carefully keep track of labels of rows and columns.

9.13* The first part is immediate by inspection and Exercise 9.12. Suppose A is invertible. By the recipe for finding A^{-1} we have $I = E_m \dots E_1 A$ for some elementary matrices $E_1, \dots, E_m$. Then, by Exercise 9.4,

$$E_1^{-1} \dots E_m^{-1} = (E_m \dots E_1)^{-1} E_m \dots E_1 A = IA = A,$$

which proves, by the first part, that A is a product of elementary matrices.

9.14* Suppose $E_1, \dots, E_m$ are elementary such that $E_m \dots E_1 A$ has a row of zeros. If A is invertible then $I = (E_m \dots E_1 A) A^{-1} E_1^{-1} \dots E_m^{-1}$ has a row of zeros, which is nonsense.

9.15* Suppose $AB = I$. By the recipe for finding inverses, if a row of zeros appears then a contradiction is reached as in the previous exercise. Hence, there must be elementary matrices $E_1, \dots, E_m$ such that $E_m \dots E_1 A = I$. Hence, $B = IB = E_m \dots E_1 AB = E_m \dots E_1 I = E_m \dots E_1$, yielding $AB = BA = I$.

9.16* Suppose $m < n$. (The case $n < m$ is similar.) Inspecting dimensions shows that A is $n \times m$ and B is $m \times n$. Since $n > m$, row reducing A must produce a row of zeros. Hence, there are elementary $n \times n$ matrices $E_1, \dots, E_k$ such that $E_k \dots E_1 A$ has a row of zeros. But then the invertible matrix $E_k \dots E_1 = E_k \dots E_1 I_n = E_k \dots E_1 AB$ must have a row of zeros, contradicting Exercise 9.6.

9.17* If $A = A^T$ and A^{-1} exists then $I = I^T = (AA^{-1})^T = \left(A^{-1}\right)^T A^T = \left(A^{-1}\right)^T A$, so that $A^{-1} = \left(A^{-1}\right)^T$. If $A = -A^T$ and A^{-1} exists then

$I = I^T = (AA^{-1})^T = \left(A^{-1}\right)^T A^T = \left(A^{-1}\right)^T(-A) = \left(-\left(A^{-1}\right)^T\right)A$, so that $A^{-1} = -\left(A^{-1}\right)^T$.

9.18* If M is any square matrix then $M = \frac{1}{2}(M + M^T) + \frac{1}{2}(M - M^T)$, the sum of a symmetric matrix and a skew-symmetric matrix.

9.19** Yes, for example, let

$$A = \begin{bmatrix} 0 & 1 & 0 & 0 & 0 & \cdots \\ 0 & 0 & 1 & 0 & 0 & \cdots \\ 0 & 0 & 0 & 1 & 0 & \cdots \\ 0 & 0 & 0 & 0 & 1 & \cdots \\ 0 & 0 & 0 & 0 & 0 & \cdots \\ \vdots & \vdots & \vdots & \vdots & \vdots & \ddots \end{bmatrix},$$

the result of adding a column of zeros to the front of I, moving all the other columns along one space, and

$$B = \begin{bmatrix} 0 & 0 & 0 & 0 & 0 & \cdots \\ 1 & 0 & 0 & 0 & 0 & \cdots \\ 0 & 1 & 0 & 0 & 0 & \cdots \\ 0 & 0 & 1 & 0 & 0 & \cdots \\ 0 & 0 & 0 & 1 & 0 & \cdots \\ \vdots & \vdots & \vdots & \vdots & \vdots & \ddots \end{bmatrix},$$

the result of adding a row of zeros to the top of I, moving all the other rows down one space. Then $AB = I$, but $BA \neq I$.

10 Determinants

10.1 2, 3, 1, 1, −1 **10.2** 78

10.3 (i) 10 (ii) −36 (iii) 0 (iv) 2 (v) −90 (vi) 16

10.4 (i) −14 (ii) 0 (iii) 32

10.5 (i) −1 (ii) 1 (iii) 14 (iv) 126 (v) 82 (vi) $bd(c-d)$ (vii) $(t^2-4)(t+4)$ (viii) −4

10.6 By subtracting one row or column from its identical row or column, a matrix is obtained with a row or column of zeros, which has zero determinant (by expanding along that row or down that column).

10.7 (i) true (ii) true (iii) false (iv) false

10.8 (i) $-3\mathbf{i}+18\mathbf{j}+13\mathbf{k}$ (ii) $\mathbf{i}-\mathbf{j}+\mathbf{k}$ (iii) $-3\mathbf{i}+\mathbf{k}$ (iv) $\mathbf{i}+\mathbf{j}+\mathbf{k}$

10.9 (i) anticlockwise (ii) clockwise

10.10 (i) inside (ii) outside (iii) on the boundary

10.11 The linear equations have a unique solution if and only if the equation

$$\begin{bmatrix} a & b \\ c & d \end{bmatrix}\begin{bmatrix} x \\ y \end{bmatrix} = \begin{bmatrix} k \\ \ell \end{bmatrix}$$

has a unique solution, which occurs if and only if the matrix $\begin{bmatrix} a & b \\ c & d \end{bmatrix}$ is invertible, that is, if and only if its determinant $ad-bc$ is nonzero.

10.12 $\mathbf{u}\times\mathbf{v}\cdot\mathbf{w}=(\mathbf{u}\times\mathbf{v})\cdot\mathbf{w}$ (i) -5 (ii) 21

10.13* Interchanging the first and second rows and then the second and third rows leaves the determinant unchanged, giving the following:

$$\begin{vmatrix} \mathbf{i} & \mathbf{j} & \mathbf{k} \\ u_1 & u_2 & u_3 \\ v_1 & v_2 & v_3 \end{vmatrix} \cdot (w_1\mathbf{i}+w_2\mathbf{j}+w_3\mathbf{k}) = \begin{vmatrix} w_1 & w_2 & w_3 \\ u_1 & u_2 & u_3 \\ v_1 & v_2 & v_3 \end{vmatrix} = \begin{vmatrix} u_1 & u_2 & u_3 \\ v_1 & v_2 & v_3 \\ w_1 & w_2 & w_3 \end{vmatrix}$$

10.14* (i) Suppose that A has a row of zeros. The induction starts trivially. Suppose that the result holds for square matrices of size less than n where $n \geq 2$. If the first row of A is zero then $\det A = \sum(-1)^{1+j}\,0\,\det A_{1j} = 0$. If the ith row of A is zero where $i>1$ then, by the inductive hypothesis,

$$\det A = \sum(-1)^{1+j}\,a_{1j}\,\det A_{1j} = \sum(-1)^{1+j}\,a_{1j}\,0 = 0,$$

and the result follows by induction. Suppose now that A has a column of zeros. Again the induction starts trivially and we make a corresponding inductive hypothesis. If it is the kth column of A which is zero then

$$\begin{aligned}\det A &= \sum_{j\neq k}(-1)^{1+j}\,a_{1j}\,\det A_{1j} + (-1)^{1+k}\,a_{1k}\,\det A_{1k} \\ &= \sum_{j\neq k}(-1)^{1+j}\,a_{1j}\,0 + (-1)^{1+k}\,0\,\det A_{1k} = 0,\end{aligned}$$

and again we are done by induction.

(ii) The lower triangular case is a simple induction. In the case that A is upper triangular, A_{11} is also upper triangular and A_{1j} has its first column all zero for $j > 1$, so that, by part (i), $\det A = a_{11} \det A_{11}$, and again the result follows by a simple induction.

(iii) The proof is by induction on the size of B.

10.15* This is a simple induction using the fact that A_{12} is another matrix of the same shape.

10.16* The first two cases follow immediately from **10.14**(ii), since the elementary matrices are upper or lower triangular. The last case follows by **10.14**(iii) and **10.15** and a simple induction.

10.17** If B is the result of interchanging two rows of A then $\det B = -\det A$ by an elaborate induction that splits up into cases depending on whether the first row is involved in the interchange. It follows quickly that if two rows of A are identical then $\det A = 0$. If B is the result of multiplying a row of A by the scalar λ then $\det B = \lambda \det A$ by a simple induction. Finally, if B is the result of adding a scalar multiple of one row of A to another row then $\det B = \det A$, using the previous results of this exercise and the fact that if C, D, E are identical square matrices except that for some i the ith row of E is the sum of the ith rows of C and D then $\det E = \det C + \det D$, which itself follows by a simple induction.

10.18** (i) The first step is to prove that A is invertible if and only if $\det A \neq 0$, and this follows from the previous exercise and the fact that a matrix is invertible if and only if it can be row reduced to the identity matrix. If A is not invertible then it is easy to show AB also is not invertible, so $\det AB = 0 = \det A \det B$. If A is invertible then A is a product of elementary matrices and the equality $\det AB = \det A \det B$ now follows by the previous exercise and a simple induction.

(ii) If A is not invertible then A^T also is not invertible, so from the proof of part (i), $\det A = 0 = \det A^T$. If A is invertible then it is a product of elementary matrices and the result follows by a simple induction after first checking that $\det E = \det E^T$ where E is elementary by **10.16**.

(iii) The result about expanding along the ith row follows by using the matrix B obtained from A by interchanging rows to bring the ith row to the top and the fact that interchanging rows multiplies the determinant by -1, so that $\det B = (-1)^{i-1} \det A$. The result about expanding down any column follows from the result about rows and part (ii).

10.19* Suppose A is a square matrix with nonzero determinant. Row reduce A, so that $E_k \dots E_1 A = J$ where $E_1, \dots, E_k$ are elementary and J is in reduced row echelon form. If $J \neq I$ then J has a row of zeros so has zero determinant by **10.14**(ii), so, by the multiplicative property and the fact that elementary matrices are invertible,

$$\det A \;=\; \det(E_k \dots E_1)^{-1} \det J \;=\; \det(E_k \dots E_1)^{-1} 0 \;=\; 0\,,$$

which is a contradiction. Hence, $J = I$ and $A = E_k \dots E_1$ is invertible.

10.20** When $n = 2$, $\det V = \begin{vmatrix} 1 & x_1 \\ 1 & x_2 \end{vmatrix} = x_2 - x_1$, which starts an induction. For $n > 2$, we apply column operations and an inductive hypthesis to get

$$\begin{aligned}
\det V \;&=\; \begin{vmatrix} 1 & x_1 & x_1^2 & \dots & x_1^{n-1} \\ 1 & x_2 & x_2^2 & \dots & x_2^{n-1} \\ 1 & x_3 & x_3^2 & \dots & x_3^{n-1} \\ \vdots & \vdots & \vdots & \ddots & \vdots \\ 1 & x_n & x_n^2 & \dots & x_n^{n-1} \end{vmatrix} \\
&=\; \begin{vmatrix} 1 & 0 & 0 & \dots & 0 \\ 1 & x_2 - x_1 & x_2^2 - x_2 x_1 & \dots & x_2^{n-1} - x_2^{n-2} x_1 \\ 1 & x_3 - x_1 & x_3^2 - x_3 x_1 & \dots & x_3^{n-1} - x_3^{n-2} x_1 \\ \vdots & \vdots & \vdots & \ddots & \vdots \\ 1 & x_n - x_1 & x_n^2 - x_n x_1 & \dots & x_n^{n-1} - x_n^{n-2} x_1 \end{vmatrix} \\
&=\; (x_2 - x_1)(x_3 - x_1) \dots (x_n - x_1) \begin{vmatrix} 1 & x_2 & \dots & x_2^{n-2} \\ 1 & x_3 & \dots & x_3^{n-2} \\ \vdots & \vdots & \ddots & \vdots \\ 1 & x_n & \dots & x_n^{n-2} \end{vmatrix} \\
&=\; (x_2 - x_1)(x_3 - x_1) \dots (x_n - x_1) \prod_{2 \le i < j \le n} x_j - x_i \\
&=\; \prod_{1 \le i < j \le n} x_j - x_i\,,
\end{aligned}$$

completing the proof of the formula for Vandermonde's determinant. Suppose p has more than $n-1$ roots, so there exist distinct $x_1, \dots, x_n$

such that $p(x_1) = \ldots = p(x_n) = 0$. Hence, the matrix equation

$$V\mathbf{a} = \mathbf{0}$$

holds where $\mathbf{a} = \begin{bmatrix} a_0 \\ a_1 \\ \vdots \\ a_{n-1} \end{bmatrix}$. But $x_j - x_i \neq 0$ whenever $j > i$, since the roots are distinct. Hence, $\det V \neq 0$, being a product of nonzero elements, so V is invertible. Hence,

$$\mathbf{a} = V^{-1}V\mathbf{a} = V^{-1}\mathbf{0} = \mathbf{0},$$

contradicting that $a_{n-1} \neq 0$. Hence, p has at most $n-1$ roots.

10.21** Suppose that f is a permutation that is both even and odd, so

$$f = \sigma_1 \ldots \sigma_k = \tau_1 \ldots \tau_\ell$$

for some even integer k and odd integer ℓ and transpositions $\sigma_1, \ldots, \sigma_k$ and $\tau_1, \ldots, \tau_\ell$. A transposition that interchanges elements i and j corresponds to an elementary matrix that interchanges the ith and jth rows of an $n \times n$ matrix by pre-multiplication. Denote by E_i and F_j the elementary matrices corresponding to σ_i and τ_j respectively, for each i and j. Then the two decompositions of f yield, by applying elementary row operations successively,

$$E_1 \ldots E_k = E_1 \ldots E_k I_n = F_1 \ldots F_\ell I_n = F_1 \ldots F_\ell,$$

so that

$$\begin{aligned} (-1)^k &= (\det E_1) \ldots (\det E_k) = \det(E_1 \ldots E_k) \\ &= \det(F_1 \ldots F_\ell) = (\det F_1) \ldots (\det F_\ell) = (-1)^\ell. \end{aligned}$$

Hence, $1 = (-1)^k(-1)^\ell = (-1)^{k+\ell} = -1$, since $k + \ell$ is odd, which is a contradiction.

11 Eigenvalues and Eigenvectors

11.1 $\begin{bmatrix} 3 \\ 3 \end{bmatrix}, \begin{bmatrix} 1 \\ -1 \end{bmatrix}, \ 3, \ -1$

11.2 $(\lambda - 3)(\lambda + 1)$; the roots 3 and -1 are the eigenvalues of A

11.3 (i) $(\lambda-1)(\lambda-2), 1, 2$ (ii) $(\lambda-1)(\lambda+1), 1, -1$ (iii) $(\lambda+3)(\lambda-2), -3, 2$

11.4 (i) $1,\ 2,\ \left\{\begin{bmatrix} t \\ 0 \end{bmatrix} \middle| t \in \mathbb{R}\right\},\ \left\{\begin{bmatrix} 0 \\ t \end{bmatrix} \middle| t \in \mathbb{R}\right\}$

(ii) $1,\ -1,\ \left\{\begin{bmatrix} -t \\ t \end{bmatrix} \middle| t \in \mathbb{R}\right\},\ \left\{\begin{bmatrix} t \\ t \end{bmatrix} \middle| t \in \mathbb{R}\right\}$

(iii) $-3,\ 2,\ \left\{\begin{bmatrix} -t \\ t \end{bmatrix} \middle| t \in \mathbb{R}\right\},\ \left\{\begin{bmatrix} 2t \\ 3t \end{bmatrix} \middle| t \in \mathbb{R}\right\}$

11.5 $\begin{bmatrix} 0 \\ 0 \\ 0 \end{bmatrix},\ \begin{bmatrix} 0 \\ -1 \\ 1 \end{bmatrix},\ \begin{bmatrix} 3 \\ 6 \\ 3 \end{bmatrix},\ 0,\ 1,\ 3$

11.6 $\lambda(\lambda-1)(3-\lambda)$; the roots 0, 1 and 3 are the eigenvalues of B

11.7 (i) $1,\ \left\{\begin{bmatrix} t \\ 0 \end{bmatrix} \middle| t \in \mathbb{R}\right\}$

(ii) $2,\ -1,\ \left\{\begin{bmatrix} -3t \\ t \end{bmatrix} \middle| t \in \mathbb{R}\right\},\ \left\{\begin{bmatrix} 0 \\ t \end{bmatrix} \middle| t \in \mathbb{R}\right\}$

(iii) $1,\ 7,\ \left\{\begin{bmatrix} t \\ 0 \\ 0 \end{bmatrix} \middle| t \in \mathbb{R}\right\},\ \left\{\begin{bmatrix} 6t \\ t \\ 0 \end{bmatrix} \middle| t \in \mathbb{R}\right\}$

11.8 (i) $1,\ 2,\ \left\{\begin{bmatrix} 0 \\ t \end{bmatrix} \middle| t \in \mathbb{R}\right\},\ \left\{\begin{bmatrix} t \\ t \end{bmatrix} \middle| t \in \mathbb{R}\right\}$

(ii) $2,\ 3,\ \left\{\begin{bmatrix} t \\ t \end{bmatrix} \middle| t \in \mathbb{R}\right\},\ \left\{\begin{bmatrix} \frac{t}{2} \\ t \end{bmatrix} \middle| t \in \mathbb{R}\right\}$

(iii) $1, 2, 3,\ \left\{\begin{bmatrix} t \\ 0 \\ 0 \end{bmatrix} \middle| t \in \mathbb{R}\right\},\ \left\{\begin{bmatrix} t \\ t \\ 0 \end{bmatrix} \middle| t \in \mathbb{R}\right\},\ \left\{\begin{bmatrix} 0 \\ -t \\ t \end{bmatrix} \middle| t \in \mathbb{R}\right\}$

(iv) $2,\ \left\{\begin{bmatrix} t \\ t \end{bmatrix} \middle| t \in \mathbb{R}\right\}$

(v) $\pm 1, 3,\ \left\{\begin{bmatrix} t \\ 0 \\ t \end{bmatrix} \middle| t \in \mathbb{R}\right\},\ \left\{\begin{bmatrix} t \\ -t \\ t \end{bmatrix} \middle| t \in \mathbb{R}\right\},\ \left\{\begin{bmatrix} t/2 \\ t/2 \\ t \end{bmatrix} \middle| t \in \mathbb{R}\right\}$

(vi) $1,\ 2,\ \left\{\begin{bmatrix} t \\ t \\ 0 \end{bmatrix} \middle| t \in \mathbb{R}\right\},\ \left\{\begin{bmatrix} t/2 \\ t \\ s \end{bmatrix} \middle| s, t \in \mathbb{R}\right\}$

11.9 $M\mathbf{v} = \lambda\mathbf{v} \implies M^k\mathbf{v} = M^{k-1}M\mathbf{v} = M^{k-1}\lambda\mathbf{v} = \lambda M^{k-1}\mathbf{v} = \cdots = \lambda^k\mathbf{v}$

11.10 Suppose $M\mathbf{v} = \lambda\mathbf{v}$ where $\mathbf{v}$ is a nonzero vector. If $\lambda = 0$ then $M\mathbf{v} = 0\mathbf{v} = \mathbf{0}$, so $\mathbf{v} = M^{-1}M\mathbf{v} = M^{-1}\mathbf{0} = \mathbf{0}$, which is a contradiction. Hence, $\lambda \neq 0$ and

$$\mathbf{v} = I\mathbf{v} = M^{-1}M\mathbf{v} = M^{-1}\lambda\mathbf{v} = \lambda M^{-1}\mathbf{v},$$

so that $M^{-1}\mathbf{v} = \lambda^{-1}\mathbf{v}$, proving λ^{-1} is an eigenvalue of M^{-1} with eigenvector $\mathbf{v}$.

11.11 The formula still holds for $n \leq 0$ since $\begin{bmatrix} 2 & 0 \\ 0 & 3 \end{bmatrix}^n = \begin{bmatrix} 2^n & 0 \\ 0 & 3^n \end{bmatrix}$ always.

11.12* $\det(M^T - \lambda I) = \det(M^T - \lambda I^T) = \det(M - \lambda I)^T = \det(M - \lambda I)$

11.13* $\det(B^{-1}AB - \lambda I) = \det B^{-1}\det(A - \lambda I)\det B = \det(A - \lambda I)$

11.14* The Conjugation Principle can be expressed by $Z = XYX^{-1}$ where Z is difficult, Y is easy, and X changes the conditions to facilitate the performance of Y. In the proof template, Z represents the final proof, X the expansion step, X^{-1} the contraction step, and Y the "easy" steps in between. In the train example, Z represents getting from Bondi Junction to Redfern, X hopping on the train, X^{-1} hopping off the train, and Y sitting on the train until the appropriate time to alight. In the cake example, $Z = XYWX^{-1}W^{-1}$ where Z is baking the cake, Y is mixing the ingredients, X putting the cake in the oven and W putting on oven mitts so as not to get burnt.

11.15* Let $\mathbf{v}$ be an eigenvector of A corresponding to λ.

(i) If $A^2 = 0$ and $\lambda \neq 0$ then $\mathbf{v} = \lambda^{-2}\lambda^2\mathbf{v} = \lambda^{-2}A^2\mathbf{v} = \lambda^{-2}0\mathbf{v} = \mathbf{0}$, a contradiction.

(ii) If $A^2 = A$ and $\lambda \neq 0$ then $\mathbf{v} = \lambda^{-1}\lambda\mathbf{v} = \lambda^{-1}A\mathbf{v} = \lambda^{-2}A^2\mathbf{v} = \lambda^{-1}\lambda^2\mathbf{v} = \lambda\mathbf{v}$, so that $(1-\lambda)\mathbf{v} = \mathbf{0}$, yielding $1 - \lambda = 0$, so that $\lambda = 1$.

(iii) If $A^2 = I$ then $\mathbf{v} = A^2\mathbf{v} = \lambda^2\mathbf{v}$, so that $(1-\lambda^2)\mathbf{v} = \mathbf{0}$, yielding $1 - \lambda^2 = 0$, so that $\lambda = 1$ or -1.

11.16* This is a straightforward evaluation using the definitions.

11.17* This is a simple induction using as inductive hypothesis that the determinant of a square matrix with entries which are polynomials in λ is also a polynomial in λ. The determinant expansion along the first row then becomes a sum of polynomials, which is itself a polynomial.

11.18* This follows by row reducing the noninvertible square matrix A, which guarantees a row of zeros. Solving the associated homogeneous system yields a parametric solution. Picking one nonzero instance gives a nonzero vector $\mathbf{v}$ such that $A\mathbf{v} = \mathbf{0}$.

11.19* By a simple induction, $M^n = \begin{bmatrix} \cos n\theta & -\sin n\theta \\ \sin n\theta & \cos n\theta \end{bmatrix}$. If $M^n = I$ then $n\theta = 2\pi k$ for some integer k, so that $\pi = n\theta/(2k) \in \mathbb{Q}$, which is absurd, since π is irrational.

11.20* The matrix $M = \begin{bmatrix} \cos\theta & -\sin\theta \\ \sin\theta & \cos\theta \end{bmatrix}$ has eigenvalues $\cos\theta + i\sin\theta$ and $\cos\theta - i\sin\theta$ with eigenspaces $\left\{ \begin{bmatrix} it \\ t \end{bmatrix} \middle| t \in \mathbb{C} \right\}$ and $\left\{ \begin{bmatrix} -it \\ t \end{bmatrix} \middle| t \in \mathbb{C} \right\}$ respectively.

11.21* $\det(\lambda I - M) = \lambda^3 - 3\lambda^2 - \lambda + 3$, so $M^3 - 3M^2 - M + 3I = 0$ and

$$M^{-1} = -\frac{1}{3}(M^2 - 3M - I) = \frac{1}{3}\begin{bmatrix} -13 & -6 & 10 \\ 2 & 3 & -2 \\ -14 & -6 & 11 \end{bmatrix}$$

11.22* Making the substitution before taking the determinant is invalid and produces the zero scalar, not the zero matrix, except for 1×1 matrices, the only cases when the reasoning becomes valid.

11.23* The (i, ℓ)-entry of $A(\operatorname{adj} A)$ is $\sum_{j=1}^{n}(-1)^{\ell+j}\, a_{ij} \det A_{\ell j}$. If $i = \ell$ then this is the expansion along the ℓth row of A, yielding $\det A$. If $i \neq \ell$ then this is the expansion along the ℓth row of the matrix obtained by replacing the ℓth row of A by the ith row, yielding the determinant of a matrix with two identical rows, which is zero. This is enough to prove $A(\operatorname{adj} A) = I$, and the rest follows.

11.25* Given $M\mathbf{x} = \mathbf{c}$, we have $\mathbf{x} = M^{-1}\mathbf{c} = \frac{1}{\det M}(\operatorname{adj} M)\,\mathbf{c}$, so that

$$x_i = \frac{1}{\det M}\sum_{j=1}^{n}(-1)^{j+i}\det(M_{ji})c_j = \frac{\det M_i}{\det M},$$

since $\det M_i = \sum_{j=1}^{n}(-1)^{j+i}c_j \det M_{ji}$, expanding down the ith column. In the given system, $M = \begin{bmatrix} 2 & 3 & 4 \\ 5 & 5 & 6 \\ 3 & 1 & 2 \end{bmatrix}$ and $\mathbf{c} = \begin{bmatrix} -4 \\ -3 \\ -1 \end{bmatrix}$, so

$$\det M = \begin{vmatrix} 2 & 3 & 4 \\ 5 & 5 & 6 \\ 3 & 1 & 2 \end{vmatrix} = 2\begin{vmatrix} 5 & 6 \\ 1 & 2 \end{vmatrix} - 3\begin{vmatrix} 5 & 6 \\ 3 & 2 \end{vmatrix} + 4\begin{vmatrix} 5 & 5 \\ 3 & 1 \end{vmatrix} = -8\,,$$

$$\det M_1 = \begin{vmatrix} -4 & 3 & 4 \\ -3 & 5 & 6 \\ -1 & 1 & 2 \end{vmatrix} = -4\begin{vmatrix} 5 & 6 \\ 1 & 2 \end{vmatrix} - 3\begin{vmatrix} -3 & 6 \\ -1 & 2 \end{vmatrix} + 4\begin{vmatrix} -3 & 5 \\ -1 & 1 \end{vmatrix} = -8\,,$$

$$\det M_2 = \begin{vmatrix} 2 & -4 & 4 \\ 5 & -3 & 6 \\ 3 & -1 & 2 \end{vmatrix} = 2\begin{vmatrix} -3 & 6 \\ -1 & 2 \end{vmatrix} + 4\begin{vmatrix} 5 & 6 \\ 3 & 2 \end{vmatrix} + 4\begin{vmatrix} 5 & -3 \\ 3 & -1 \end{vmatrix} = -16\,,$$

$$\det M_3 = \begin{vmatrix} 2 & 3 & -4 \\ 5 & 5 & -3 \\ 3 & 1 & -1 \end{vmatrix} = 2\begin{vmatrix} 5 & -3 \\ 1 & -1 \end{vmatrix} - 3\begin{vmatrix} 5 & -3 \\ 3 & -1 \end{vmatrix} - 4\begin{vmatrix} 5 & 5 \\ 3 & 1 \end{vmatrix} = 24\,,$$

yielding $x = \frac{-8}{-8} = 1\,,\ y = \frac{-16}{-8} = 2\,,\ z = \frac{24}{-8} = -3\,.$

12 Diagonalising a Matrix

12.1 (i) $P = \begin{bmatrix} -1 & 1 \\ 1 & 1 \end{bmatrix},\ D = \begin{bmatrix} 1 & 0 \\ 0 & 3 \end{bmatrix}$

(ii) $\begin{bmatrix} \frac{1}{2}(1+3^n) & \frac{1}{2}(-1+3^n) \\ \frac{1}{2}(-1+3^n) & \frac{1}{2}(1+3^n) \end{bmatrix},\ \begin{bmatrix} 14 & 13 \\ 13 & 14 \end{bmatrix},\ \begin{bmatrix} 41 & 40 \\ 40 & 41 \end{bmatrix}$

12.2 (i) $P = \begin{bmatrix} 1 & 0 & 1 \\ -1 & -1 & 2 \\ 1 & 1 & 1 \end{bmatrix},\ D = \begin{bmatrix} 0 & 0 & 0 \\ 0 & 1 & 0 \\ 0 & 0 & 3 \end{bmatrix}$

(ii) $\begin{bmatrix} 0 & 3^{n-1} & 3^{n-1} \\ 1 & 2(3^{n-1}) & -1+2(3^{n-1}) \\ -1 & 3^{n-1} & 1+3^{n-1} \end{bmatrix},\ \begin{bmatrix} 0 & 27 & 27 \\ 1 & 54 & 53 \\ -1 & 27 & 28 \end{bmatrix}$

12.3 (i) eigenvalues 2 and 3, $P = \begin{bmatrix} 1 & 0 & -1 \\ 1 & 0 & 1 \\ 0 & 1 & 1 \end{bmatrix},\ D = \begin{bmatrix} 2 & 0 & 0 \\ 0 & 2 & 0 \\ 0 & 0 & 3 \end{bmatrix}$

(ii) $\frac{1}{2}\begin{bmatrix} 2^n+3^n & 2^n-3^n & 0 \\ 2^n-3^n & 2^n+3^n & 0 \\ 2^n-3^n & -2^n+3^n & 2^{n+1} \end{bmatrix},\ \begin{bmatrix} 97/2 & -65/2 & 0 \\ -65/2 & 97/2 & 0 \\ -65/2 & 65/2 & 16 \end{bmatrix}$

12.4 (i) $\begin{bmatrix} \frac{1+(-1)^k}{2} & \frac{-1+(-1)^k}{2} \\ \frac{-1+(-1)^k}{2} & \frac{1+(-1)^k}{2} \end{bmatrix}$ (ii) $\begin{bmatrix} 2^k & 0 \\ 2^k - 1 & 1 \end{bmatrix}$

(iii) $\begin{bmatrix} 2^{k+1} - 3^k & -2^k + 3^k \\ 2^{k+1} - 2(3^k) & -2^k + 2(3^k) \end{bmatrix}$

12.5 (i) $\begin{bmatrix} 1 & 2^k - 1 & 2^k - 1 \\ 0 & 2^k & 2^k - 3^k \\ 0 & 0 & 3^k \end{bmatrix}$ (iii) $\begin{bmatrix} 2 - 2^k & 2^k - 1 & 0 \\ 2 - 2^{k+1} & 2^{k+1} - 1 & 0 \\ 0 & 0 & 2^k \end{bmatrix}$

(ii) $\begin{bmatrix} 3(-1)^k - 3^k - 1 & (-1)^k - 1 & 1 - 2(-1)^k + 3^k \\ 1 - 3^k & 1 & 3^k - 1 \\ 3(-1)^k - 2(3^k) - 1 & (-1)^k - 1 & 1 - 2(-1)^k + 2(3^k) \end{bmatrix}$

12.6 The columns add to 1 so the matrix is regular stochastic. The steady state vector is

$$\begin{bmatrix} \frac{4}{9} \\ \frac{5}{9} \end{bmatrix} \text{ and } M^n = \begin{bmatrix} \frac{4}{9} + \frac{5}{9}\left(\frac{1}{10}\right)^n & \frac{4}{9} - \frac{4}{9}\left(\frac{1}{10}\right)^n \\ \frac{5}{9} - \frac{5}{9}\left(\frac{1}{10}\right)^n & \frac{5}{9} + \frac{4}{9}\left(\frac{1}{10}\right)^n \end{bmatrix} \to \begin{bmatrix} \frac{4}{9} & \frac{4}{9} \\ \frac{5}{9} & \frac{5}{9} \end{bmatrix}.$$

12.7* The entries of M^2 are all positive. The steady state vector is $\begin{bmatrix} \frac{1}{3} \\ \frac{1}{3} \\ \frac{1}{3} \end{bmatrix}$ and $M^n =$

$$\begin{bmatrix} \frac{1}{3} + \frac{2}{3}\left(-\frac{1}{2}\right)^n & \frac{1}{3} - \frac{1}{3}\left(-\frac{1}{2}\right)^n & \frac{1}{3} - \frac{1}{3}\left(-\frac{1}{2}\right)^n \\ \frac{1}{3} - \frac{1}{3}\left(-\frac{1}{2}\right)^n & \frac{1}{3} + \left(\frac{1}{2}\right)^{n+1} + \frac{1}{6}\left(-\frac{1}{2}\right)^n & \frac{1}{3} - \left(\frac{1}{2}\right)^{n+1} + \frac{1}{6}\left(-\frac{1}{2}\right)^n \\ \frac{1}{3} - \frac{1}{3}\left(-\frac{1}{2}\right)^n & \frac{1}{3} - \left(\frac{1}{2}\right)^{n+1} + \frac{1}{6}\left(-\frac{1}{2}\right)^n & \frac{1}{3} + \left(\frac{1}{2}\right)^{n+1} + \frac{1}{6}\left(-\frac{1}{2}\right)^n \end{bmatrix}$$
$$\to \begin{bmatrix} \frac{1}{3} & \frac{1}{3} & \frac{1}{3} \\ \frac{1}{3} & \frac{1}{3} & \frac{1}{3} \\ \frac{1}{3} & \frac{1}{3} & \frac{1}{3} \end{bmatrix}.$$

12.8* The identity matrix $\begin{bmatrix} 1 & 0 \\ 0 & 1 \end{bmatrix}$ is stochastic with steady state vectors $\begin{bmatrix} 1 \\ 0 \end{bmatrix}$ and $\begin{bmatrix} 0 \\ 1 \end{bmatrix}$. The matrix $M = \begin{bmatrix} 0 & 1 \\ 1 & 0 \end{bmatrix}$ is stochastic with a unique steady state vector $\mathbf{v} = \begin{bmatrix} \frac{1}{2} \\ \frac{1}{2} \end{bmatrix}$. The vector $\mathbf{x} = \begin{bmatrix} -1 \\ 1 \end{bmatrix}$ has the property that $\lim_{n\to\infty} M^n\mathbf{x}$ does not exist.

12.9* Suppose $P^{-1}MP$ is diagonal where $P = \begin{bmatrix} a & b \\ c & d \end{bmatrix}$. Deduce that the determinant $ad - bc$ is zero, contradicting that P is invertible.

12.10* The eigenvalues of M are $\lambda_1 = \dfrac{1+\sqrt{5}}{2}$ and $\lambda_2 = \dfrac{1-\sqrt{5}}{2}$,

$$M^n = \frac{1}{\sqrt{5}}\begin{bmatrix} \lambda_1^{n+1} - \lambda_2^{n+1} & \lambda_1\lambda_2^{n+1} - \lambda_2\lambda_1^{n+1} \\ \lambda_1^n - \lambda_2^n & \lambda_1\lambda_2^n - \lambda_2\lambda_1^n \end{bmatrix},$$

$$x_n = \frac{1}{\sqrt{5}}\left[\left(\frac{1+\sqrt{5}}{2}\right)^n - \left(\frac{1-\sqrt{5}}{2}\right)^n\right].$$

12.11* and **12.12*** These are special cases of **12.14****.

12.13* If P^T can be row reduced to a matrix with a row of zeros then the associated homogeneous system has a nonzero solution, which implies that $\mathbf{v}_1^T, \ldots, \mathbf{v}_n^T$ are linearly dependent, contradicting that $\mathbf{v}_1, \ldots, \mathbf{v}_n$ are linearly independent. Hence, P^T can be row reduced to the identity matrix, so that P^T and P are invertible.

12.14** Suppose $\mu_1\mathbf{v}_1 + \ldots + \mu_n\mathbf{v}_n = \mathbf{0}$. Apply M and the definition of eigenvector to get another equation. Combine the two equations to eliminate one of the vectors, and then apply an inductive hypothesis, exploiting the assumption that the eigenvalues are distinct, to deduce that $\mu_1 = \ldots = \mu_n = 0$.

12.15** Let M be a complex 2×2 matrix. If the eigenvalues are distinct, or if M has one eigenvalue whose eigenspace contains two nonparallel vectors, then M is diagonalisable so M is similar to a diagonal matrix. It remains to assume that M has just one eigenvalue λ such that the eigenspace is one-dimensional (that is, consists precisely of all scalar multiples of some fixed nonzero vector). The characteristic equation is a perfect square, so, by the Cayley-Hamilton Theorem, $(M-\lambda I)^2 = 0$. Consider

$$\mathbf{w}_1 = (M-\lambda I)\begin{bmatrix} 1 \\ 0 \end{bmatrix} \quad \text{and} \quad \mathbf{w}_2 = (M-\lambda I)\begin{bmatrix} 0 \\ 1 \end{bmatrix},$$

the first and second columns of $M - \lambda I$ respectively. If both are zero then $M - \lambda I = 0$, so that $M = \lambda I$ is diagonal, contradicting that the eigenspace is one-dimensional. Without loss of generality, we may suppose $\mathbf{w}_1 = \begin{bmatrix} a \\ b \end{bmatrix}$ is nonzero. But

$$(M-\lambda I)\mathbf{w}_1 = (M-\lambda I)^2\begin{bmatrix} 1 \\ 0 \end{bmatrix} = 0\begin{bmatrix} 1 \\ 0 \end{bmatrix} = \begin{bmatrix} 0 \\ 0 \end{bmatrix},$$

so that $\mathbf{w}_1$ is an eigenvector. Certainly $\begin{bmatrix} 1 \\ 0 \end{bmatrix}$ is not an eigenvector, so $\mathbf{w}_1$ and $\begin{bmatrix} 1 \\ 0 \end{bmatrix}$ are not scalar multiples of each other. It follows quickly that $b \neq 0$ and so $P = \begin{bmatrix} a & 1 \\ b & 0 \end{bmatrix}$ is invertible. Note that

$$M\mathbf{w}_1 = \lambda \mathbf{w}_1 \qquad \text{and} \qquad M\begin{bmatrix} 1 \\ 0 \end{bmatrix} = \mathbf{w}_1 + \begin{bmatrix} \lambda \\ 0 \end{bmatrix} = \begin{bmatrix} a+\lambda \\ b \end{bmatrix},$$

so $MP = [M\mathbf{w}_1 \;\; \mathbf{w}_1] = \begin{bmatrix} \lambda a & a+\lambda \\ \lambda b & b \end{bmatrix} = P\begin{bmatrix} \lambda & 1 \\ 0 & \lambda \end{bmatrix}$, whence

$$M = P\begin{bmatrix} \lambda & 1 \\ 0 & \lambda \end{bmatrix}P^{-1}.$$

Thus M is similar to $\begin{bmatrix} \lambda & 1 \\ 0 & \lambda \end{bmatrix}$. The real case is identical except when the eigenvalues form a complex conjugate pair $\lambda = r \operatorname{cis}(\pm\theta)$, in which case there is an invertible complex matrix P such that (suppressing some straightforward calculations)

$$M = P\begin{bmatrix} r \operatorname{cis}\theta & 0 \\ 0 & r \operatorname{cis}(-\theta) \end{bmatrix}P^{-1} = Q\begin{bmatrix} r\cos\theta & -r\sin\theta \\ r\sin\theta & r\cos\theta \end{bmatrix}Q^{-1}$$

where $Q = P\begin{bmatrix} 1 & i \\ 1 & -i \end{bmatrix}$, so that M is similar to a scalar multiple of a rotation matrix. In fact, if we are careful to choose complex conjugate eigenvectors to form $P = \begin{bmatrix} \alpha & \overline{\alpha} \\ \beta & \overline{\beta} \end{bmatrix}$ (as we may since M is unchanged by conjugating all of its elements, and complex conjugation clearly preserves matrix multiplication) then $Q = \begin{bmatrix} \alpha+\overline{\alpha} & i\alpha - i\overline{\alpha} \\ \beta+\overline{\beta} & i\beta - i\overline{\beta} \end{bmatrix}$ will be real (as is readily checked because the entries are fixed by conjugation).

Appendix 1 The Theorem of Pythagoras

- **Railway tracks** Time: 7.03

Appendix 1 The Theorem of Pythagoras

The Theorem of Pythagoras is an important foundation stone from geometry and is a key to the development of the theory of vectors.

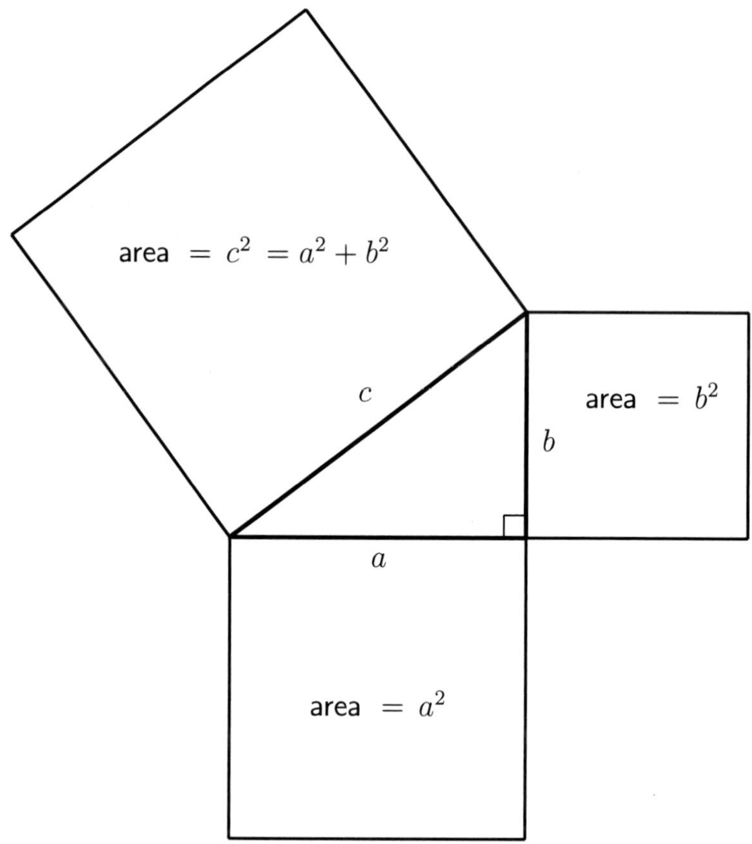

It says simply that if you form squares on the sides of a right-angled triangle, then the area of the larger square is the sum of the areas of the two smaller squares. The longest side-length is called the *hypotenuse*.

Theorem of Pythagoras: If a, b and c are side-lengths of a right-angled triangle and c is the length of the hypotenuse then

$$a^2 + b^2 = c^2 .$$

Though attributed to Pythagoras (569–475 BC), the result was known long before his lifetime. A Babylonian clay tablet (pictured on the cover of the *Notices of the American Mathematical Society*, Volume 29, Number 1, 2002) dating from about 1800 BC appears to depict a proof for isosceles triangles. The argument goes as follows, in the special case where the common side-length is 1 and the length of the hypotenuse is denoted by x:

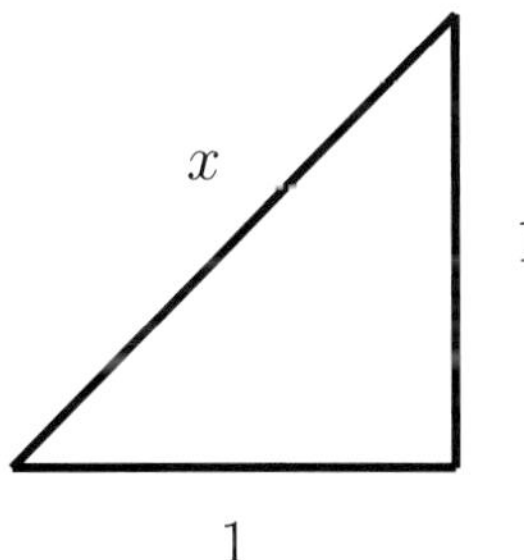

Rotate the triangle quarter turns about the point opposite the hypotenuse, duplicating the triangle three more times to get the following figure, which is a square with side-length x:

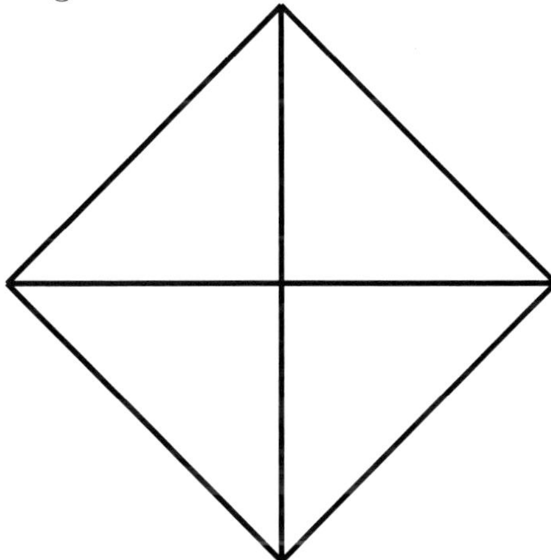

This larger figure has area four times the area of the original triangle, which itself is half the area of a square of side-length one. This gives $x^2 = 4 \times \frac{1}{2} = 2$ and so $x = \sqrt{2}$. If the side-length 1 is replaced with k then the same argument gives $x^2 = 2k^2$, which is the isosceles special case of Pythagoras.

It is not difficult to adjust this proof to get the general argument. Suppose our right-angled triangle has hypotenuse c and other side-lengths $a \geq b$.

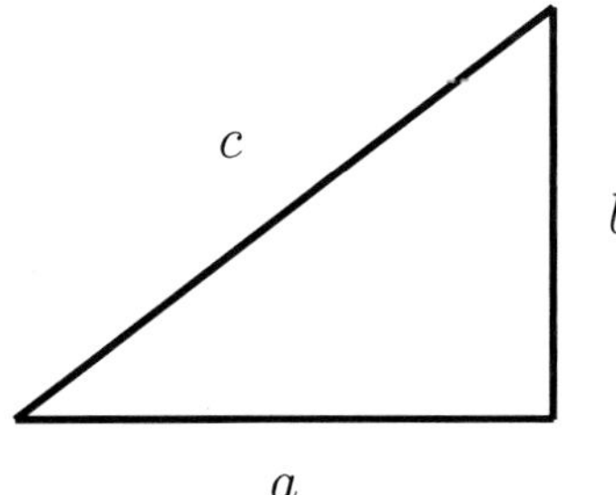

We again duplicate the triangle after rotating quarter turns about the centre of a middle square of side-length $a - b$. (In the isosceles case this middle square is just a single point.)

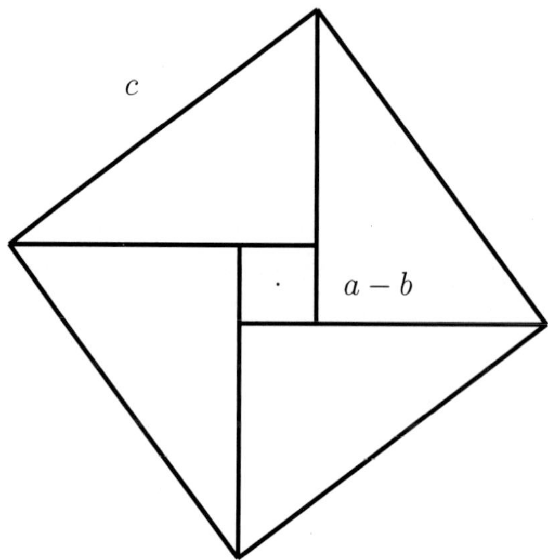

The larger square has area c^2 and the original triangle has area $\frac{1}{2}ab$. The middle square has area $(a-b)^2$. Adding up gives

$$c^2 = 4 \times \frac{1}{2}ab + (a-b)^2 = 2ab + a^2 - 2ab + b^2 = a^2 + b^2$$

and the Theorem of Pythagoras is proved.

The Cosine Rule, described in Section 3.2, follows easily by dropping a perpendicular from one vertex of the triangle to its opposite side, creating two right-angled triangles to which the Theorem of Pythagoras applies. Adding up the expressions and simplifying yields the Cosine Rule (Exercise 3.10). Clearly, the Theorem of Pythagoras is a special case of the Cosine Rule where the angle is 90°. Thus, the Cosine Rule and the Theorem of Pythagoras are mathematically equivalent, in the sense described in Appendix 2.

We complete Appendix 1 by discussing an amusing problem due to Paul Halmos. It can be regarded as a demonstration of the sensitivity of matrix multiplication to perturbations, and appears in his book *Problems for Mathematicians, Young and Old*, published by the Mathematical Association of America, 1991.

> **The Railway Track Problem:** You have built a railway line 20 kilometres long in a straight line on a flat plain, fixed at one end. You discover at the other end that you are 1 metre over, so push it in, to evenly distribute the extra metre to form an arc. How high does the arc rise above the plain in the middle of the track?

Here is a drawing, not to scale, and the task is to estimate the height h:

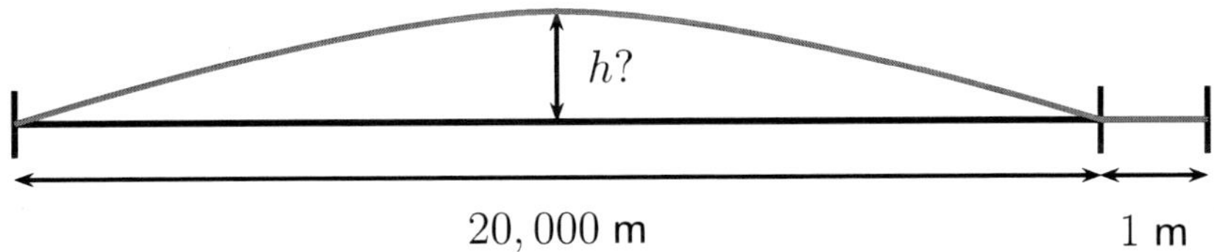

The reaction of most people is that h should be very small, since we are distributing 1 metre over 20,000 metres. It is a common experience in life that small perturbations have small effects. However, in this problem, the perturbation is orthogonal to the effect, so that the Theorem of Pythagoras kicks in, up to an approximation:

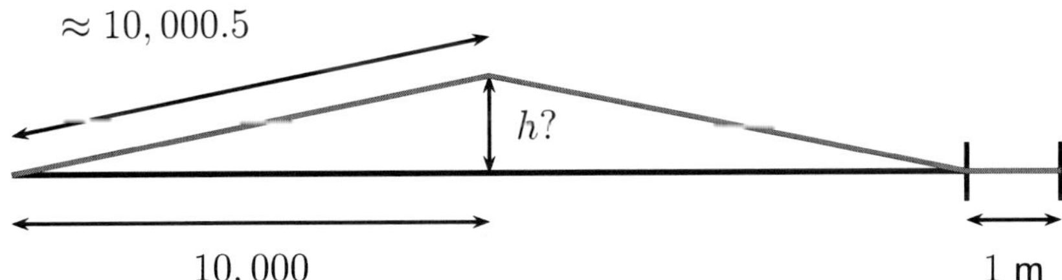

We have

$$(10,00.5)^2 \approx h^2 + (10,000)^2 ,$$

so

$$\begin{aligned} h^2 &\approx 10,000.5^2 - 10,000^2 \\ &= (10,000.5 + 10,000)(10,000.5 - 10,000) \\ &= (20,000.5)(0.5) \\ &\approx 10,000 , \end{aligned}$$

giving $h \approx 100$. The railway track springs about 100 metres into the air!

The connection with matrix multiplication is in the factorisation of the difference of two squares when approximating h^2 above. Consider the matrix product

$$M = \begin{bmatrix} x & y \end{bmatrix} \begin{bmatrix} 1 & 0 \\ 0 & -1 \end{bmatrix} \begin{bmatrix} x \\ y \end{bmatrix} = \begin{bmatrix} x & -y \end{bmatrix} \begin{bmatrix} x \\ y \end{bmatrix} = x^2 - y^2 .$$

For example, if $x = y = 10,000$ then $M = 0$. However, perturbing x or y slightly (changing x from $10,000$ to $10,000.5$ in the above) results in huge changes in M when x and y are themselves large.

Matrix multiplication, described in Section 8.2, is just a cascade of dot products of vectors. Where there are dot products there are notions of angles and orthogonality. Where there are notions of orthogonality, there is lurking the Theorem of Pythagoras!

Appendix 2 Mathematical Implication

Appendix 2 Mathematical Implication

Throughout this book, implication has been used extensively, and most noticeably in proofs and statements which involve the words

"if ... then" and "if and only if"

and corresponding symbols

$$\Longrightarrow \quad \text{and} \quad \Longleftrightarrow .$$

It is the purpose of this appendix to explain precisely what we mean by *mathematical implication* and provide the student with an opportunity to develop some fluency in using it.

In mathematics there are two *truth values*, abbreviated T for *TRUE* and F for *FALSE*. A *proposition* is a statement which can meaningfully be assigned a truth value. It is a basic premise of mathematical logic that

Propositions are either *TRUE* or *FALSE*, but never both simultaneously.

For example, the following is a true proposition:

Sydney is in Australia.

The following is false:

Sydney is not in Australia.

Of course, the words are subject to interpretation, and by "Sydney" we mean the city of Sydney, not some particular person whose name happens to be Sydney, who may or may not happen to be in Australia. The following are grammatical statements in English, but they are not propositions because they cannot be assigned truth values. The first is a question, and the second is the basis of the so-called *liar paradox*, violating our basic premise above by trying to be both true and false simultaneously:

Why are we here?

This sentence is false.

Here are two familiar mathematical propositions, the first of which is true:

$$2 + 2 = 4$$

and the second false:

$$2 + 2 \neq 4$$

The only difference is that "equals" in the first is replaced by "not equals" in the second, using the device of a stroke through the equality symbol. This is very common in mathematics: a stroke through a symbol can flip the truth value by representing the word "not". This leads to the first and simplest logical connective:

> **Negation:** If P is a proposition then $\sim P$ (read "not P") is the *negation* of P and is
>
> $$\begin{cases} \textit{TRUE} & \text{when } P \text{ is } \textit{FALSE} \\ \textit{FALSE} & \text{when } P \text{ is } \textit{TRUE}. \end{cases}$$

Thus $2+2 \neq 4$ is the negation of $2+2=4$, and the statement "Sydney is not in Australia" is the negation of "Sydney is in Australia".

The next two connectives combine two propositions into one:

> **Conjunction:** If P and Q are propositions then $P \wedge Q$ (read "P and Q") is the *conjunction* of P and Q and is
>
> $$\begin{cases} \textit{TRUE} & \text{when } P \text{ and } Q \text{ are both } \textit{TRUE} \\ \textit{FALSE} & \text{when } \begin{cases} P \text{ is } \textit{TRUE} \text{ and } Q \text{ is } \textit{FALSE}, \text{ or} \\ P \text{ is } \textit{FALSE} \text{ and } Q \text{ is } \textit{TRUE}, \text{ or} \\ P \text{ is } \textit{FALSE} \text{ and } Q \text{ is } \textit{FALSE}. \end{cases} \end{cases}$$

Thus a conjunction can only be true if each of the component propositions is true, and is false otherwise.

Disjunction: If P and Q are propositions then $P \vee Q$ (read "P or Q") is the *disjunction* of P and Q and is

$$\begin{cases} \mathit{TRUE} \quad \text{when} \begin{cases} P \text{ is } \mathit{TRUE} \text{ and } Q \text{ is } \mathit{TRUE}, \text{ or} \\ P \text{ is } \mathit{TRUE} \text{ and } Q \text{ is } \mathit{FALSE}, \text{ or} \\ P \text{ is } \mathit{FALSE} \text{ and } Q \text{ is } \mathit{TRUE} \end{cases} \\ \mathit{FALSE} \quad \text{when } P \text{ and } Q \text{ are both } \mathit{FALSE}. \end{cases}$$

Thus a disjunction can only be false if each of the component propositions is false. A disjunction will be true provided at least one of the component propositions is true.

(There is another connective called "exclusive or" which is only true if *exactly one* of the components is true. "Exclusive or" is rarely used in mathematics arguments without explicit mention. It is, however, common in ordinary language. For example, if you tell a child to choose an apple *or* an orange, the child will almost certainly interpret your instruction to take one or the other of the pieces of fruit but not both.)

A **truth table** displays the truth values for compound propositions given truth values for components. Here is the truth table for negation:

P	$\sim P$
T	F
F	T

and a combined truth table for both conjunction and disjunction:

P	Q	$P \wedge Q$	$P \vee Q$
T	T	T	T
T	F	F	T
F	T	F	T
F	F	F	F

Combinations of connectives can be built up in steps. For example, the compound proposition

$$\sim P \vee Q$$

can be interpreted as asserting either P is false or Q is true (or both). The following truth table lists its truth values, after an intermediate step:

P	Q	$\sim P$	$\sim P \vee Q$
T	T	F	T
T	F	F	F
F	T	T	T
F	F	T	T

Thus the only way $\sim P \vee Q$ can be false is if P is true and Q is false. In a sense we can make precise, the truth of P "entails" the truth of Q. Shortly we will discuss *implication* and the expression $P \implies Q$ and explain why this is naturally defined as an abbreviation for $\sim P \vee Q$.

We say that (compound) propositions P and Q are *logically equivalent*, and write

$$P \equiv Q$$

if they produce the same truth values, that is, if they have identical columns in a truth table.

For example, for any proposition P we have

$$P \equiv \sim (\sim P)$$

because P is true precisely when $\sim P$ is false, which occurs precisely when $\sim (\sim P)$ is true:

P	$\sim P$	$\sim (\sim P)$
T	F	T
F	T	F

A more elaborate example is one of *De Morgan's Laws*:

$$\sim (P \vee Q) \equiv \sim P \wedge \sim Q$$

which follows from the fact that the third and last columns of truth values in the next table are identical:

P	Q	$P \vee Q$	$\sim (P \vee Q)$	$\sim P$	$\sim Q$	$\sim P \wedge \sim Q$
T	T	T	F	F	F	F
T	F	T	F	F	T	F
F	T	T	F	T	F	F
F	F	F	T	T	T	T

The other *De Morgan's Law*

$$\sim (P \wedge Q) \equiv \sim P \vee \sim Q$$

can be verified similarly.

Let's think now informally about the meaning of the proposition

if P then Q ,

which can be expressed symbolically by

$P \implies Q$.

In the above we call P the *hypothesis*, the arrow symbol *implies* and Q the *conclusion*. The hypothesis being true should somehow "force" the conclusion to be true. We would like the following

Rule of inference: If P is $TRUE$ and $P \Longrightarrow Q$ is $TRUE$ then Q is $TRUE$.

Logicians call this rule of inference *Modus Ponens*, and it can be displayed as follows (though this is rather old-fashioned):

$$\begin{array}{rl} & P \\ & P \Longrightarrow Q \\ \hline \therefore & Q \end{array}$$

This diagram is called a *syllogism* and the three dots to the left are an abbreviation of "therefore". Probably the most famous example is attributed to Aristotle:

$$\begin{array}{rl} & \text{Socrates is a man} \\ & \text{All men are mortal} \\ \hline \therefore & \text{Socrates is mortal} \end{array}$$

We can express this symbolically by putting

$$P(X) \equiv X \text{ is a man}$$

$$Q(X) \equiv X \text{ is mortal}$$

and letting the upside-down symbol $\forall$ represent "for all":

$$\begin{array}{ll} & P(\text{Socrates}) \\ & (\forall X)\ P(X) \implies Q(X) \\ \hline \therefore & Q(\text{Socrates}) \end{array}$$

This is a variation of Modus Ponens involving the "universal quantifier" $\forall$. If we remove it by taking the special case $X \equiv \text{Socrates}$ at the second line, then this becomes the usual form of Modus Ponens:

$$\begin{array}{ll} & P(\text{Socrates}) \\ & P(\text{Socrates}) \implies Q(\text{Socrates}) \\ \hline \therefore & Q(\text{Socrates}) \end{array}$$

Now, the expression $P \Longrightarrow Q$ should be some kind of compound proposition built from P and Q. The connective $\Longrightarrow$ must have a truth table. We will find this table in a couple of steps, by thinking about properties implication should have. For example, the following implication is evidently true:

$$\textit{student is listening attentively} \implies \textit{student is awake}$$

However, a related implication, where we interchange the hypothesis and conclusion, need not be true:

$$\textit{student is awake} \implies \textit{student is listening attentively}$$

> The *converse* of the implication $P \Longrightarrow Q$ is the implication $Q \Longrightarrow P$ obtained by interchanging the hypothesis and conclusion.

The previous example relating the attentiveness and wakefulness of a student demonstrates that an implication and its converse should not be logically equivalent. Thus

$$\boxed{P \Longrightarrow Q \ \not\equiv \ Q \Longrightarrow P}$$

and so

The truth tables for $P \Longrightarrow Q$ and $Q \Longrightarrow P$ **must** be different!

We have yet another variation for transforming an implication:

The *contrapositive* of the implication $P \Longrightarrow Q$ is the implication $\sim Q \Longrightarrow \sim P$ obtained by negating and then interchanging the hypothesis and the conclusion.

If you think about the contrapositive in familiar settings you quickly become convinced that it should be logically equivalent to the original implication. For example, expressed as an "if ... then" statement, the following implication is evidently true:

If I am in Sydney then I am in Australia.

The contrapositive is also evidently true and carries the same information:

If I am not in Australia then I am not in Sydney.

If you agree that our logic requires

$$P \Longrightarrow Q \ \equiv \ \sim Q \Longrightarrow \sim P$$

then you are forced to conclude the following:

The truth tables for $P \Longrightarrow Q$ and $\sim Q \Longrightarrow \sim P$ **must** be the same!

We now set about constructing the truth table for $P \Longrightarrow Q$. There are two cases where there should be no controversy about the truth values. If P and Q are both true then $P \Longrightarrow Q$ should be true. If P is true whilst Q is false then $P \Longrightarrow Q$ should be false, for otherwise we have not captured the idea of the truth of P entailing the truth of Q.

The less intuitively obvious truth values for $P \Longrightarrow Q$ occur when P is false. In the table below we have called them X, for when Q is true, and Y, for when Q is false:

P	Q	$P \Longrightarrow Q$
T	T	T
T	F	F
F	T	X
F	F	Y

As a first step, these values now imply the following table for the contrapositive $\sim Q \Longrightarrow \sim P$, expressed in terms of X and Y:

P	Q	$\sim Q$	$\sim P$	$\sim Q \Longrightarrow \sim P$
T	T	F	F	Y
T	F	T	F	F
F	T	F	T	X
F	F	T	T	T

The last column of truth values has to coincide with the values for the table for $P \Longrightarrow Q$, which forces

$$Y = T.$$

We can revise our truth table further and add a column for the converse $Q \Longrightarrow P$:

P	Q	$P \Longrightarrow Q$	$Q \Longrightarrow P$
T	T	T	T
T	F	F	X
F	T	X	F
F	F	T	T

If $X = F$ then the last two columns have identical truth values, which will contradict that implication and its converse are not logically equivalent. Hence, we are forced to conclude

$$X = T.$$

Here then is the final truth table for implication:

P	Q	$P \Longrightarrow Q$
T	T	T
T	F	F
F	T	T
F	F	T

Notice then, from the truth table of an earlier example, that

$$P \Longrightarrow Q \;\equiv\; \sim P \vee Q .$$

If you are ever in doubt, simply think of an implication as asserting either that the hypothesis is false or the conclusion is true (or both).

Intuitively, it may seem strange that $P \Longrightarrow Q$ has a truth value at all when P is false, since there seems to be no "connection" with the truth value of Q. However, mathematical logic is coherent and requires assigning truth values to every proposition. The above analysis demonstrates that the truth values are forced, even in these less intuitive cases.

When P is false, mathematicians say that the implication $P \Longrightarrow Q$ is *trivially true* or *trivially satisfied.*

For example, both of the following implications are trivially true, regardless of the conclusion, since the hypothesis is obviously false:

$$0 = 1 \quad \Longrightarrow \quad \textit{the author of this book is Superman}$$

$$0 = 1 \quad \Longrightarrow \quad \textit{the author of this book is not Superman}$$

Typical examples in mathematics use a *quantifier*, such as the following:

$$(\forall x \in \mathbb{R}) \quad x^2 = 1 \quad \Longrightarrow \quad x = \pm 1$$

For a particular $x \in \mathbb{R}$ the implication holds, either trivially, when $x \neq \pm 1$, or by actual properties of the hypothesis, when $x = \pm 1$. By altering the hypothesis slightly, so that it is always false, we can make the statement trivially true with any conclusion we like:

$$(\forall x \in \mathbb{R}) \quad x^2 = -1 \quad \Longrightarrow \quad x \text{ is a cow jumping over the moon}$$

Notice in the previous two examples that the quantifier is linked to membership of the set $\mathbb{R}$ of real numbers. The overriding membership class is called the *universe of discourse*. If we change the universe of discourse to the set $\mathbb{C}$ of complex numbers, then the previous statement becomes false:

$$(\forall x \in \mathbb{C}) \quad x^2 = -1 \quad \Longrightarrow \quad x \text{ is a cow jumping over the moon}$$

We can make it true by adjusting the conclusion:

$$(\forall x \in \mathbb{C}) \quad x^2 = -1 \quad \Longrightarrow \quad x = \pm i$$

There is one final matter to discuss:

Double implication: Define $P \iff Q$ to be an abbreviation for the conjunction $(P \Longrightarrow Q) \wedge (Q \Longrightarrow P)$ and we say or write

"P if and only if Q",

which in turn may get abbreviated to

"P iff Q".

The words "only if" take a little getting used to. To say

P only if Q

means

$$P \Longrightarrow Q\,.$$

Removing the word "only" changes the meaning dramatically. To say

$$P \text{ if } Q$$

means

$$Q \Longrightarrow P,$$

which may also be written

$$P \Longleftarrow Q.$$

Putting the two arrows $\Longrightarrow$ and $\Longleftarrow$ together gives the double arrow or double implication $\Longleftrightarrow$. Using the truth table for conjunction, we can work out its truth table in steps:

P	Q	$P \Longrightarrow Q$	$Q \Longrightarrow P$	$(P \Longrightarrow Q) \wedge (Q \Longrightarrow P)$
T	T	T	T	T
T	F	F	T	F
F	T	T	F	F
F	F	T	T	T

More compactly, we have:

P	Q	$P \Longleftrightarrow Q$
T	T	T
T	F	F
F	T	F
F	F	T

This can be expressed in words:

> The double implication $P \Longleftrightarrow Q$ is true precisely when P and Q have identical truth values, that is, P and Q are logically equivalent propositions.

Mathematicians also commonly refer to "necessary" and/or "sufficient" conditions when analysing mathematical relationships. The idea is that for

$$P \Longrightarrow Q$$

to be true, the truth of P should be *sufficient* to guarantee the truth of Q, and we say

$$P \textit{ is a sufficient condition for } Q\,.$$

But here, also, the truth of Q is *necessary* given that we know the truth of P, and we say

$$Q \textit{ is a necessary condition for } P.$$

From the point of view of mathematical logic, both of the preceding wordings carry the same information. The choice of words is often a question of emphasis in a given problem.

An ideal situation sought by mathematicians is when one phenomenon is *characterised* in terms of another. If the phenomena are P and Q, then to say P *is characterised by* Q means precisely that P is a necessary and sufficient condition for Q (and automatically vice versa). In terms of the above discussion, this happens when the double implication

$$P \iff Q$$

holds. There are many examples in this book. One of the most striking is the characterisation of invertibility of a square matrix M:

$$M \textit{ is invertible if and only if } \det M \neq 0\,,$$

or, symbolically,

$$M^{-1} \textit{ exists} \iff \det M \neq 0\,.$$

If one says that Q *is a necessary but not sufficient condition for* P, then technically this means that $P \Longrightarrow Q$ holds, but $Q \Longrightarrow P$ fails. An example of this would be

$$P \textit{ is the property that a square matrix } M \textit{ has a row of zeros}$$

and

$$Q \textit{ is the property that a square matrix } M \textit{ has zero deteminant.}$$

Here, the implication $Q \implies P$ can be made to fail by choosing, say,

$$M = \begin{bmatrix} 1 & -1 \\ 1 & -1 \end{bmatrix}.$$

Then M has zero determinant but does not possess a row of zeros. We call M a *counterexample* to the implication $Q \implies P$. It provides an instance where Q is true whilst P is false. We can also say that P *is a sufficient but not necessary condition for* Q.

As a final illustration, consider the following conditions:

(1) Students who study linear algebra have a truly perceptive view of the world.

(2) No person leading an unfulfilling life can be happy.

(3) People who are unhappy are unable to grasp a truly perceptive view of the world.

(4) Students who study linear algebra lead fulfilling lives.

We will explain why condition (4) is necessary for conditions (1), (2) and (3). Therefore, if you agree that (1), (2) and (3) are plausible, and you have studied this book on linear algebra, then you are guaranteed to lead a fulfilling life!

To analyse the morass of information in these statements using our knowledge of implication, we introduce the following symbols:

$$\begin{aligned} S(X) &\equiv X \text{ is a student studying linear algebra} \\ P(X) &\equiv X \text{ has a truly perceptive view of the world} \\ F(X) &\equiv X \text{ leads a fulfilling life} \\ H(X) &\equiv X \text{ is happy} \end{aligned}$$

Earlier we introduced the *universal quantifier* $\forall$ which means "for all". We also now need the *existential quantifier* $\exists$ which means "for some" or "there exists". Universally quantified statements are like gigantic conjunctions. Existentially quantified statements are like disjunctions. Analogues of De Morgan's Laws discussed earlier apply where $Q(X)$ is a proposition about X:

$$\sim [(\forall X)\ Q(X)] \equiv (\exists X)\ \sim Q(X)$$

and similarly

$$\sim [(\exists X)\ Q(X)] \equiv (\forall X)\ \sim Q(X)$$

The earlier statements can now be translated into symbols, ready then for processing:

(1)	$(\forall X)\ S(X) \Longrightarrow P(X)$
(2)	$\sim [(\exists X)\ \sim F(X) \wedge H(X)]$
(3)	$(\forall X)\ \sim H(X) \Longrightarrow \sim P(X)$
(4)	$(\forall X)\ S(X) \Longrightarrow F(X)$

Proof that (4) follows from (1), (2) and (3): By applying De Morgan's Laws to (2), we get

$$(\forall X)\ \sim(\sim F(X)) \vee \ \sim H(X),$$

which simplifies to

$$(\forall X)\ \sim H(X) \vee F(X),$$

which in turn can be reformulated using implication as

(5) $(\forall X)\ H(X) \Longrightarrow F(X)$.

Using the contrapositive, we can reformulate (3) as

(6) $(\forall X)\ P(X) \Longrightarrow H(X)$.

Now let X be any given person. Suppose X is a student studying linear alegebra, that is, $S(X)$ holds. By (1), (5) and (6) we have the following chain of implications:

$$S(X) \Longrightarrow P(X) \Longrightarrow H(X) \Longrightarrow F(X)$$

By three applications of Modus Ponens, we conclude that $F(X)$ holds. Thus the truth of $S(X)$ entails the truth of $F(X)$, so that we can conclude

$$(\forall X)\ S(X) \Longrightarrow F(X).$$

Thus (4) holds and the proof is complete. May students of linear algebra indeed have fulfilling lives!

Appendix 3 Complex Numbers

Appendix 3 Complex Numbers

There are several places in this book, especially in the last chapter, when the arithmetic of the real numbers $\mathbb{R}$ is just not quite powerful enough to fully understand or adequately describe the behaviour of a square matrix M. The problem arises when we look for the eigenvalues of M, which are the roots of its *characteristic polynomial*

$$p(\lambda) = \det(M - \lambda I) .$$

Such roots may or may not exist in $\mathbb{R}$. It is the purpose of this appendix to introduce the complex number arithmetic $\mathbb{C}$, which extends $\mathbb{R}$, and explain why it is rich enough to always guarantee the existence of eigenvalues. This is a direct consequence of the following seminal fact, which leads to the phenomenon of *algebraic closure*:

The Fundamental Theorem of Algebra: Every nonconstant polynomial with coefficients from $\mathbb{C}$ has a root in $\mathbb{C}$.

In particular, the characteristic polynomial $p(\lambda)$ of a real or complex $n \times n$ matrix M is nonconstant (having leading term $\pm\lambda^n$), so has roots in $\mathbb{C}$, the eigenvalues of M.

Corollary: Every complex (including real) square matrix has complex eigenvalues.

There are many proofs of the Fundamental Theorem in the literature, but one of the most appealing, in the author's opinion, is by Stephen Smale, expressed in a brief paragraph in his paper "The Fundamental Theorem of Algebra and Complexity Theory" in the *Bulletin of the American Mathematical Society* 4 (1981), 1–36. This proof, given towards the end of this appendix, adapts Newton's Method from ordinary calculus and relies on some facts about continuity. But first we introduce and illustrate the basic arithmetic of $\mathbb{C}$.

Right from the start, in this book, we have been rearranging linear equations. If a and b are real numbers and x is an unknown such that

$$ax + b = 0\,,$$

then, provided $a \neq 0$, we can find x quickly, using a subtraction followed by a division:

$$x = -b/a\,.$$

Suppose now that a, b, c are real numbers such that $a \neq 0$ and

$$ax^2 + bx + c \;=\; 0\,.$$

This is called a *quadratic* equation, and the unknown x appears twice on the left-hand side. It is not immediately obvious how to manipulate the expression to isolate x. There is a trick, however, called *completing the square*, that leads to the so-called *quadratic formula*:

$$\begin{aligned}
ax^2 + bx + c = 0 \quad &\Longrightarrow \quad x^2 + \frac{b}{a}x = -\frac{c}{a} \\
&\Longrightarrow \quad x^2 + \frac{b}{a}x + \left(\frac{b}{2a}\right)^2 = -\frac{c}{a} + \left(\frac{b}{2a}\right)^2 \\
&\Longrightarrow \quad \left(x + \frac{b}{2a}\right)^2 = \frac{b^2 - 4ac}{4a^2} \\
&\Longrightarrow \quad x + \frac{b}{2a} = \pm\sqrt{\frac{b^2 - 4ac}{4a^2}} = \frac{\pm\sqrt{b^2 - 4ac}}{2a} \\
&\Longrightarrow \quad x = \frac{-b \pm \sqrt{b^2 - 4ac}}{2a}
\end{aligned}$$

But in the real number system it only makes sense to take square roots of nonnegative numbers. So, for x above to be a solution in $\mathbb{R}$ we require the *discriminant* $\Delta = b^2 - 4ac$ to be nonnegative. To be meaningful in general (including when $\Delta < 0$), we need to be able to take square roots of negative numbers. Just for this purpose alone, it suffices to "toss in" or "invent" the square root of -1, and we have a special name for it:

$$i = \sqrt{-1}$$

But then a miracle occurs. By combining i with all real numbers, in a way that we can make precise shortly, not only are we able to solve quadratic equations, but, in principle (because of the Fundamental Theorem of Algebra), we are able to solve all *polynomial equations*:

$$a_0 + a_1 x + \ldots + a_n x^n = 0$$

where n is a positive integer, $a_0, \ldots, a_n \in \mathbb{R}$ and $a_n \neq 0$. Moreover, we can go the whole hog and allow the coefficients $a_0, \ldots, a_n$ to be any complex numbers (not necessarily real), and we study such an example at the end of this appendix. Polynomial equations are also referred to as *algebraic equations*, because the left-hand sides involve only repeated use of the algebra of addition and multiplication. The fact that they always have solutions in $\mathbb{C}$ means that this new arithmetic is self-contained and "closed". That is why we say that $\mathbb{C}$ is *algebraically closed.*

Extending real number arithmetic to include $i = \sqrt{-1}$ turns out, in fact, to be natural and uncomplicated. (The consequences of doing this, however, turn out to be highly nontrivial!) We want to be able to perform the usual operations $+$, $-$, $\times$, $\div$, so we need at least expressions of the form

$a + bi$ where a and b are arbitrary real numbers

that result by first formally "multiplying" i by b, and then "adding" a. We use usual conventions of arithmetic (such as reading multiplication before addition, and writing multiplication as juxtaposition). These formal expressions $a + bi$, such as $7 + 4i$ and $3 + 2i$, are called *complex numbers*, gathered together collectively in a set denoted by $\mathbb{C}$:

$$\mathbb{C} = \{ a + bi \mid a, b \in \mathbb{R} \}$$

Complex numbers can be combined in natural ways that simplify. They "close up" very easily under addition and subtraction:

Rule for addition and subtraction:

$$(a+bi)\pm(c+di) \;=\; (a\pm c)+(b\pm d)i$$

For example,

$$(7+4i)+(3+2i) \;=\; 10+6i \quad \text{and} \quad (7+4i)-(3+2i) \;=\; 4+2i\,.$$

Combining real number multiples of i, and real numbers on their own, is like gathering and sorting apples and oranges. Also, we write $a - bi$ for $a + (-b)i$. For example, the following simplifies quickly to the right-hand side:

$$(7-4i)-(-3-2i) \;=\; 10-2i\,.$$

Multiplication is a bit more work, but remember that

$$i=\sqrt{-1}\,,\ \text{so that}\ \ i^2=-1\,.$$

For example, expanding the brackets and sorting in a natural way yields

$$\begin{aligned}(7-4i)(-3-2i) \;&=\; -21-14i+12i+8i^2 \;=\; -21-2i+8(-1)\\ &=\; -29-2i\,.\end{aligned}$$

More generally,

$$\begin{aligned}(a+bi)(c+di) \;&=\; ac+adi+bci+bdi^2 \;=\; ac+(ad+bc)i+bd(-1)\\ &=\; (ac-bd)+(ad+bc)i\,,\end{aligned}$$

yielding the following:

Rule for multiplication:

$$(a+bi)(c+di) \;=\; (ac-bd)+(ad+bc)i$$

All of these manipulations can be formally justified. We are working in an arithmetic called a *field*, which satisfies certain properties (called *field axioms*) that tell us, for example, that addition and multiplication are *associative* (brackets don't matter) and *commutative* (order doesn't matter), and multiplication *distributes* over addition (we can expand brackets under multiplication). Fields are studied, from an abstract axiomatic point of view, in more advanced algebra courses.

So far we have *addition*, *subtraction* and *multiplication* of complex numbers. But we also want to explore *division*. Before doing so, however, we introduce some geometry closely related to vectors in the plane. The key is to identify a complex number $a+bi$ with the point (a,b) in the usual Cartesian plane, where the x-axis becomes the usual real line $\mathbb{R}$, now called the *real axis*. Points on the perpendicular y-axis are identified with all real multiples of i and comprise the so-called *imaginary axis*, denoted $i\,\mathbb{R}$. Together the real and imaginary axes span the so-called *complex* or *Argand plane* (named after the Swiss-French mathematician Jean-Robert Argand, 1768–1822).

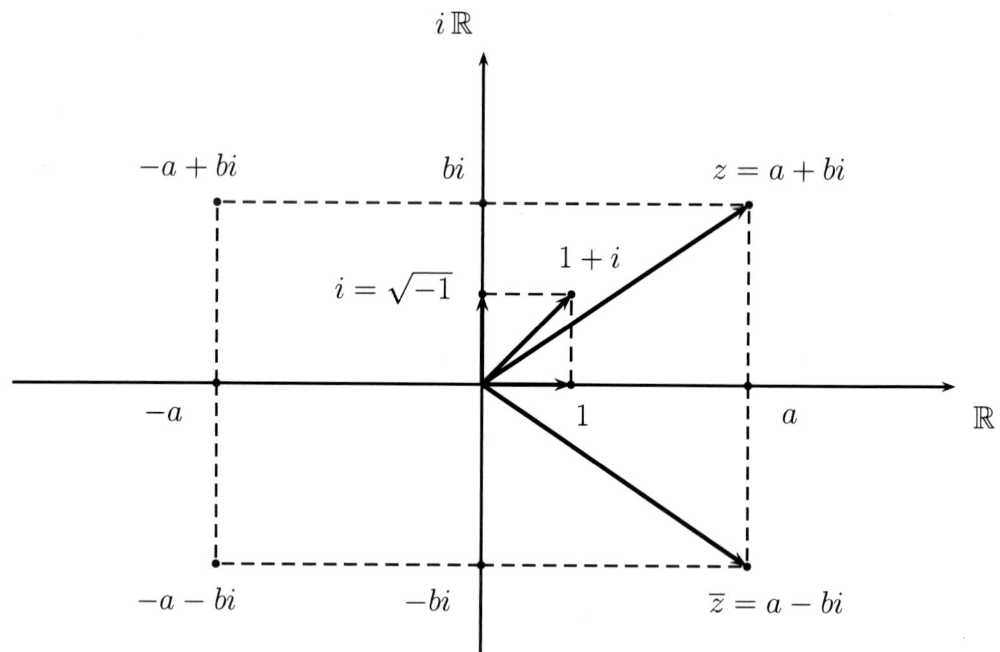

The numbers a and b are referred to as the *real* and *imaginary parts* respectively of the complex number $a+bi$. We think of 1 and i as unit vectors in the positive x- and y-directions respectively and then $a+bi$ becomes just a linear combination of vectors. Addition and subtraction of complex numbers, in turn, becomes usual addition and subtraction of vectors. These amount to collecting real and imaginary parts:

$$(a+bi) \pm (c+di) \;=\; (a \pm c) + (b \pm d)i$$

If we reflect the complex number $z = a+bi$ in the real axis we get $\overline{z} = a-bi$, called the *complex conjugate* of z. Observe that

$$z\overline{z} \;=\; (a+bi)(a-bi) \;=\; a^2 - b^2 i^2 \;=\; a^2 - b^2(-1) \;=\; a^2 + b^2 \,.$$

This is a very important property, because $a^2 + b^2$ is the familiar formula for the square of the length of the position vector of the point (a, b) in the Cartesian plane. If we adopt the usual notation $|z|$ for length or magnitude of the vector $z = a + bi$, then we have the following:

$$z\overline{z} \;=\; a^2 + b^2 \;=\; |z|^2$$

This enables us to perform division of complex numbers easily, by multiplying the numerator and denominator by the complex conjugate of the denominator. This is called *realising the denominator* (because $z\overline{z}$ is always a real number). For example,

$$\frac{3+2i}{1+2i} = \frac{(3+2i)(1-2i)}{(1+2i)(1-2i)} = \frac{3+2i-6i-4i^2}{1^2+2^2} = \frac{7-4i}{5} = \frac{7}{5} - \frac{4}{5}i\,.$$

In general we have the following:

Rule for division:

$$\frac{a+bi}{c+di} = \frac{(a+bi)(c-di)}{c^2+d^2}$$

Note that this rule only makes sense if the denominator is nonzero, that is, if at least one of c or d is nonzero. In particular, we have the following elegant rule for finding the reciprocal of a nonzero complex number z:

$$z^{-1} \;=\; \frac{1}{z} \;=\; \frac{\overline{z}}{z\overline{z}} \;=\; \frac{\overline{z}}{|z|^2}$$

There are other useful formulations of multiplication and division of complex numbers, but these require *polar forms*, which we describe now. Consider the complex number $z = a + bi$ and put

$$r \;=\; |z| \;=\; \sqrt{a^2+b^2}\,.$$

Think of r as a "radius", the distance from the point z to the origin. (Shortly we will be sweeping out circles centred at the origin to justify this terminology.) Alternatively, r is the length of z regarded as a vector. Suppose that this vector makes an angle θ, called the *argument* of z, measured anticlockwise (in radians) from the positive real axis:

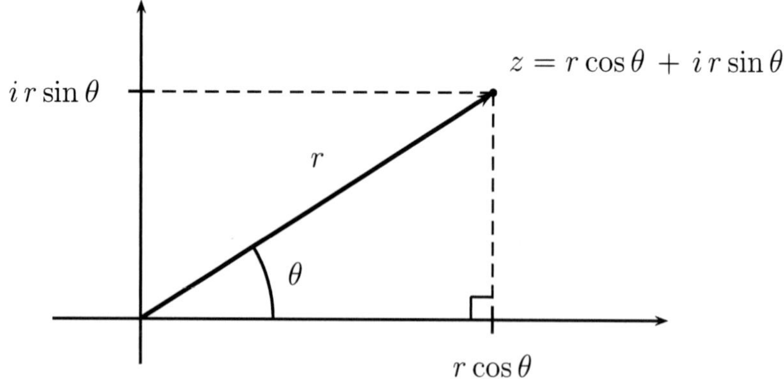

Because we deliberately placed the imaginary axis perpendicular to the real axis, elementary trigonometry yields

$$z \;=\; r\cos\theta + i\,r\sin\theta \;=\; r(\cos\theta + i\,\sin\theta)\,,$$

commonly abbreviated to

$$\boxed{z \;=\; r\,\mathrm{cis}\,\theta\,,}$$

known as the *polar form*. The real numbers r and θ are referred to as the *polar coordinates* of z. For example, in the following diagrams we can easily read off the polar coordinates and convert from Cartesian to polar forms:

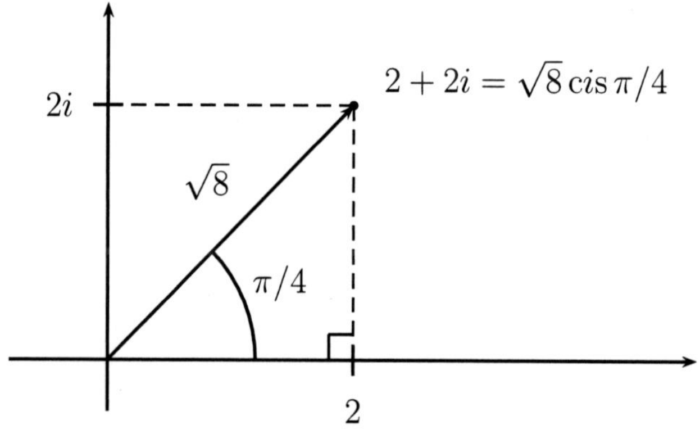

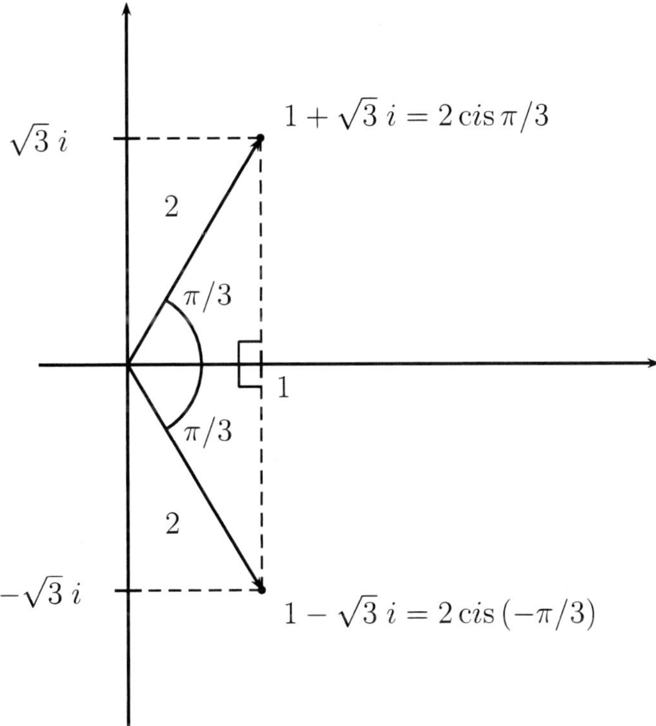

The rule for multiplication takes on a particularly elegant form using polar coordinates, by simplifying expressions using trigonometric identities:

$$\begin{aligned} r_1\, cis\, \theta_1\, r_2\, cis\, \theta_2 &= r_1(\cos\theta_1 + i\, \sin\theta_1)r_2(\cos\theta_2 + i\, \sin\theta_2) \\ &= r_1r_2(\cos\theta_1\cos\theta_2 - \sin\theta_1\sin\theta_2 \\ &\qquad + i\,(\cos\theta_1\sin\theta_2 + \sin\theta_1\cos\theta_2)) \\ &= r_1r_2(\cos(\theta_1+\theta_2) + i\,(\sin\theta_1+\theta_2)) \\ &= r_1r_2\, cis\, (\theta_1+\theta_2)\ . \end{aligned}$$

Geometrically this may be interpreted as a mixture of dilation and rotation of vectors. We thus have the following rule for multiplying polar forms:

Multiply moduli and add arguments:

$$r_1\, cis\, \theta_1 r_2\, cis\, \theta_2 \quad = \quad r_1r_2\, cis\, (\theta_1+\theta_2)$$

For example, if

$$z \;=\; 2\sqrt{3} - 2i \;=\; 4\, cis\, (-\pi/6)$$

and

$$w = -1 + \sqrt{3}\,i = 2\,cis\,(2\pi/3)$$

then

$$zw = 8\,cis\,(-\pi/6 + 2\pi/3) = 8\,cis\,\pi/2 = 8\,i\,.$$

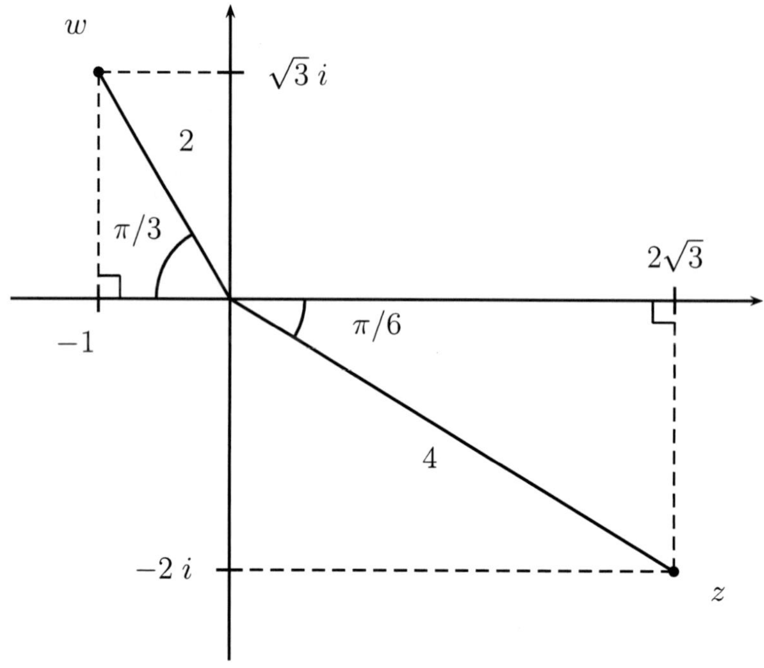

Division of complex numbers in polar form also becomes straightforward. First suppose that $z = r\,cis\,\theta$ is a nonzero complex number. Certainly its length r is a nonzero real number, so that r^{-1} makes sense. We claim that

$$\boxed{z^{-1} = r^{-1}\,cis\,(-\theta)\,.}$$

To see why, just check that multiplying z by this formula gets us back to 1:

$$r\,cis\,\theta\,r^{-1}\,cis\,(-\theta) = rr^{-1}\,cis\,(\theta - \theta) = 1\,cis\,0 = 1 \qquad \surd$$

There is also a connection with complex conjugation, because clearly $\overline{z} = r\,cis\,(-\theta)$. An earlier formula gives us another check:

$$z^{-1} = \frac{\overline{z}}{|z|^2} = \frac{r\,cis\,(-\theta)}{r^2} = r^{-1}\,cis\,(-\theta) \qquad \surd$$

Thus we get the following rule for division in polar form:

Divide moduli and subtract arguments:

$$r_1\, cis\,\theta_1 / r_2\, cis\,\theta_2 \;=\; r_1\, cis\,(\theta_1)\, r_2^{-1}\, cis\,(-\theta_2) \;=\; r_1 r_2^{-1}\, cis\,(\theta_1 - \theta_2)$$

For example, passing between Cartesian and polar forms, we have

$$\frac{2\sqrt{3}-2i}{-1+\sqrt{3}\,i} \;=\; \frac{4\,cis\,(-\pi/6)}{2\,cis\,(2\pi/3)} \;=\; 2\,cis\,(-5\pi/6) \;=\; -\sqrt{3}-i\,.$$

Finding nth powers of complex numbers is easy in polar form, where n is an integer: raise length to the nth power and multiply the argument by n.

Finding integral powers:

$$(r\,cis\,\theta)^n \;=\; r^n\,cis\,(n\theta)$$

This result follows easily from the rules for multiplication and division and is known as *De Moivre's Theorem* (named after the French mathematician Abraham de Moivre, 1667–1754). For example, the following calculations would be tedious in Cartesian form, but simplify quickly using polar form:

$$(1+i)^{10} \;=\; \sqrt{2}^{10}\;cis\,10\pi/4 \;=\; 2^5\;cis\,5\pi/2 \;=\; 32\,i\,,$$

and

$$(1+i)^{-7} \;=\; \sqrt{2}^{\,-7}\;cis\,(-7\pi/4) \;=\; \frac{1}{\sqrt{2}^7}\;cis\,\pi/4 \;=\; \frac{1}{16}(1+i)\,.$$

The inverse process to taking nth powers is to extract nth roots.

We call w an **nth root** of z if $w^n = z$, and write $w = z^{1/n}$.

Here, the situation is delicate because nth roots turn out not to be unique for integers n greater than 1. In real number arithmetic this is familiar: there are two square roots of every positive real number. In fact, there are exactly n distinct nth roots of any nonzero complex number. It is easy to find at least one, by reversing the rule for taking nth powers: if $z = r\,cis\,\theta$ is a nonzero complex number then $w = r^{1/n}\,cis\,(\theta/n)$ should be an nth root. We can check this using the rule for finding powers:

$$w^n \;=\; (r^{1/n}\,cis\,(\theta/n))^n \;=\; (r^{1/n})^n\,cis\,(n\theta/n) \;=\; r\,cis\,\theta \;=\; z \qquad \surd$$

To find the other nth roots we exploit the fact that arguments of complex numbers are not unique. Note that, if θ is any angle then

$$\text{cis}\,\theta \;=\; \cos\theta + i\,\sin\theta \;=\; \cos(\theta+2\pi k) + i\,\sin(\theta+2\pi k) \;=\; \text{cis}\,(\theta+2\pi k)$$

for any integer k. Consider a general polar form

$$w \;=\; s\,\text{cis}\,\phi$$

for an nth root $w = z^{1/n}$ of $z = r\,\text{cis}\,\theta$, where s and ϕ are unknowns. Then

$$r\,\text{cis}\,\theta \;=\; z \;=\; w^n \;=\; (s\,\text{cis}\,\phi)^n \;=\; s^n\,\text{cis}\,(n\phi)\,.$$

The lengths of these polar forms match:

$$s^n \;=\; r\,,$$

so that

$$s \;=\; r^{1/n} \qquad \text{(nth real root)}\,.$$

The arguments of these polar forms match, up to addition by an integer multiple of 2π:

$$n\phi = \theta + 2\pi k\,,$$

so that

$$\phi \;=\; \frac{\theta+2\pi k}{n} \qquad \text{for some integer } k\,.$$

But the integer k can be restricted to the values $0,\ 1,\ \ldots,\ n-1$, because any k outside these values yields duplicates for $\text{cis}\,\phi$. In summary:

Finding nth roots: If $z = r\,\text{cis}\,\theta$ is a nonzero complex number then

$$z^{1/n} \;=\; r^{1/n}\,\text{cis}\,\frac{\theta+2\pi k}{n} \qquad \text{for } k = 0,\ldots,n-1\,.$$

Example: Let's find all three cube roots of -8. There is only one real cube root, namely -2, so there must be two more lurking somewhere in $\mathbb{C}$. We have the polar form

$$-8 \;=\; 8\,\text{cis}\,\pi\,,$$

so that

$$
\begin{aligned}
(-8)^{1/3} &= 8^{1/3}\,cis\left(\frac{\pi+2\pi k}{3}\right) \qquad \text{for } k=0,1,2 \\
&= 2\,cis\,\pi/3\,, \quad 2\,cis\,\pi\,, \quad 2\,cis\,5\pi/3 \\
&= 1+\sqrt{3}\,i\,, \quad -2, \quad 1-\sqrt{3}\,i\,.
\end{aligned}
$$

All three roots have the same length and may be visualised as equally spaced around a circle centred at the origin:

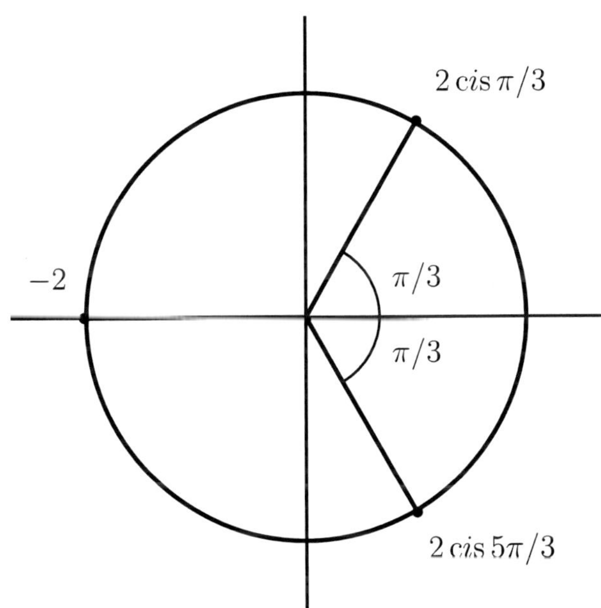

Example: Now let's find all 8th roots w of 1. There are only two in $\mathbb{R}$, namely ± 1, so there are six more somewhere in $\mathbb{C}$. Even though the polar form of 1 is not very interesting, we exploit it to get interesting solutions! We have

$$w^8 = 1 = 1\,cis\,0\,,$$

so that

$$
\begin{aligned}
w &= 1^{1/8}\,cis\left(\frac{0+2\pi k}{8}\right) \qquad \text{for } k = 0, 1, \ldots, 7 \\
&= cis\,(\pi/4)\,,\ cis\,(2\pi/4)\,, cis\,(3\pi/4)\,, cis\,(4\pi/4)\,,\ cis\,(5\pi/4)\,, \\
&\qquad\qquad cis\,(6\pi/4)\,,\ cis\,(7\pi/4) \\
&= \pm 1\,,\ \pm i\,,\ \pm\frac{1}{\sqrt{2}}(1+i)\,,\ \pm\frac{1}{\sqrt{2}}(1-i)\,.
\end{aligned}
$$

These are equally spaced around the circle of radius 1 centred at the origin:

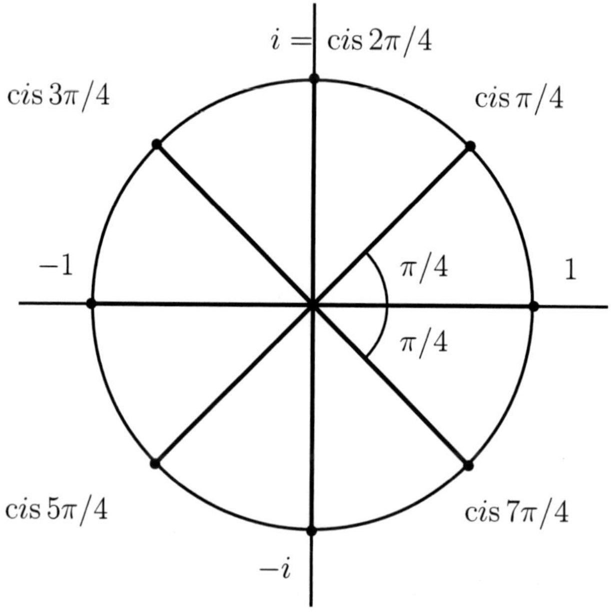

Example: Solve the equation

$$z^8 - 2z^4 + 2 = 0\,.$$

Solution: Observe that the left-hand side is a quadratic expression in z^4. By the quadratic formula,

$$z^4 = \frac{2 \pm \sqrt{4-8}}{2} = 1 \pm i = \sqrt{2}\,cis\,(\pm\pi/4)\,.$$

Taking fourth roots yields

$$z = (\sqrt{2})^{1/4}\,cis\,(\pm\pi/16 + 2\pi k/4) \qquad \text{for } k = 0, 1, 2, 3\,,$$

that is,

$$z = 2^{1/8}\, cis\,(\pm\pi/16)\,,\ 2^{1/8}\, cis\,(\pm 7\pi/16)\,,\ 2^{1/8}\, cis\,(\pm 9\pi/16)\,,\ 2^{1/8}\, cis\,(\pm 15\pi/16)\,,$$

depicted on the following diagram:

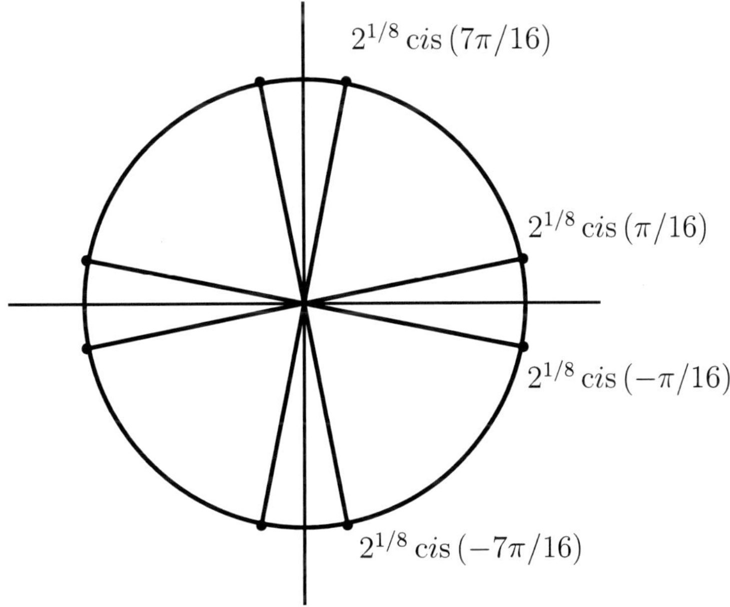

Notice, in this example, that the roots come in complex conjugate pairs (reflections of each other in the horizontal real axis). This is no accident, as the coefficients of the polynomial $z^8 - 2z^4 + 2$ are real numbers. This can be explained by the following general useful facts:

Complex conjugation respects addition and multiplication:

$$\overline{z_1 + z_2} = \overline{z_1} + \overline{z_2} \qquad \text{and} \qquad \overline{z_1 z_2} = \overline{z_1}\,\overline{z_2}\,.$$

Proof: To see why these hold, we use both Cartesian and polar forms, say $z_1 = a + bi = r\,\mathrm{cis}\,\theta\,,\ z_2 = c + di = s\,\mathrm{cis}\,\phi$. Then $\overline{z_1} = a - bi = r\,\mathrm{cis}\,(-\theta)\,,$ $\overline{z_2} = c - di = s\,\mathrm{cis}\,(-\phi)\,,$ so that

$$\begin{aligned}\overline{z_1 + z_2} &= \overline{(a+c)+(b+d)i} = (a+c)-(b+d)i \\ &= a - bi + c - di = \overline{z_1} + \overline{z_2}\end{aligned}$$

and

$$\begin{aligned} \overline{z_1 z_2} &= \overline{r\,cis\,\theta\ s\,cis\,\phi} = \overline{rs\,cis\,(\theta+\phi)} = rs\,cis\,(-\theta-\phi) \\ &= r\,cis\,(-\theta) s\,cis\,(-\phi) = \overline{z_1}\,\overline{z_2}\,. \end{aligned}$$

Corollary: The complex roots of any polynomial with real coefficients come in complex conjugate pairs.

Proof: Suppose that

$$p(z) = a_0 + a_1 z + \cdots + a_n z^n\,,$$

where $a_0, a_1, \ldots, a_n \in \mathbb{R}$ ($a_n \neq 0$). Suppose that w is a root of p so that $p(w) = 0$. Note that conjugation leaves real numbers unchanged. Hence $\overline{0} = 0$ and $\overline{a}_j = a_j$ for each j. Now, since conjugation respects addition and multiplication, we have

$$\begin{aligned} p(\overline{w}) &= a_0 + a_1\overline{w} + \cdots + a_n(\overline{w})^n = \overline{a_0} + \overline{a_1}\overline{w} + \cdots + \overline{a_n}(\overline{w})^n \\ &= \overline{a_0} + \overline{a_1 w} + \cdots + \overline{a_n w^n} = \overline{a_0 + a_1 w + \cdots + a_n w^n} \\ &= \overline{p(z)} = \overline{0} = 0\,. \end{aligned}$$

Hence $\overline{w}$ is also a root of p, and the corollary is proved.

We have just exploited the fact that complex conjugation is well-behaved. But there are other useful relationships between arithmetic operations. For example, from our discussion of polar forms, the *cis* operator "converts" real number addition into complex number multiplication, in the following sense:

For any real numbers θ_1 and θ_2,

$$cis\,(\theta_1 + \theta_2) = cis\,(\theta_1)\,cis\,(\theta_2)\,.$$

The real exponential function has this property also, and that motivates the following definition:

Complex exponential function: For any complex number $a + bi$ where a and b are real, put

$$e^{a+bi} = e^a\,cis\,b\,.$$

In particular, considering purely imaginary complex numbers, we have

> **Euler's formula:** for any real number θ ,
>
> $$e^{i\theta} \;=\; \text{cis}\,\theta\;.$$

In the way we have set things up, Euler's formula is immediate from the definition. However, it becomes a theorem if one defines exponentials differently, using say power series expansions from mathematical analysis, though we will have no need for them here. The property of turning sums into products now lifts to the complex exponential function: for any complex numbers $z = a + bi$ and $w = c + di$, where $a, b, c, d \in \mathbb{R}$,

$$e^{z+w} \;=\; e^a\,\text{cis}\;b\;e^c\,\text{cis}\;d \;=\; e^a e^b(\,\text{cis}\;b)(\,\text{cis}\;d) \;=\; e^{a+b}\,\text{cis}\,(b+d) \;=\; e^{z+w}\;,$$

proving the general

> **Exponential Law:** for any complex numbers z and w ,
>
> $$e^z e^w \;=\; e^{z+w}\;.$$

We now have an alternative notation for polar forms of complex numbers:

$$z \;=\; r\,\text{cis}\,\theta \;=\; re^{i\theta}\;.$$

When $r = 1$ the point z falls on the unit circle centred at the origin:

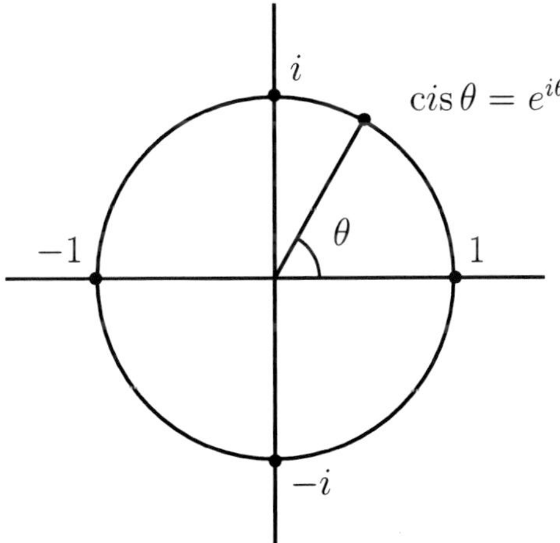

We are now well placed to amplify the proof of the Fundamental Theorem of Algebra that appears in Smale's paper. The underlying idea is Newton's

Method from ordinary calculus. Suppose that we are given some function f of a real variable and want to approximate an exact root α that yields the equation $f(\alpha) = 0$. Suppose also that we have made a "stab in the dark" and have an initial approximation α_k. The idea is to move from α_k on the x-axis vertically up to the point $(\alpha_k, f(\alpha_k))$ on the curve, and then travel down the tangent line until one hits the x-axis at a new value α_{k+1}. If the tangent is a good approximation to the curve in this region, then we expect α_{k+1} to be a better approximation than α_k to α.

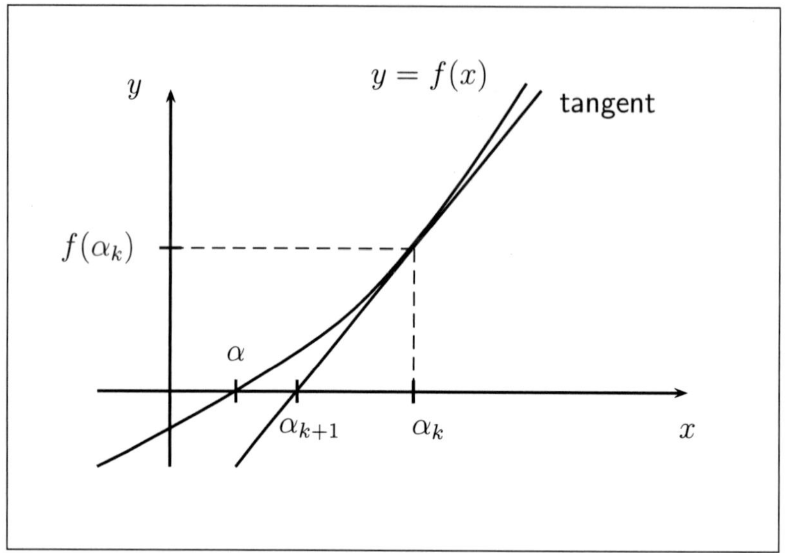

To find a formula for α_{k+1} in terms of α_k that we can subsequently iterate, observe that the derivative of f at $x = \alpha_k$ is the slope of the tangent line:

$$f'(\alpha_k) = \frac{\text{rise}}{\text{run}} = \frac{f(\alpha_k)}{\alpha_k - \alpha_{k+1}}$$

After rearranging, we get the following formula:

$$\boxed{\alpha_{k+1} = \alpha_k - \frac{f(\alpha_k)}{f'(\alpha_k)}}$$

How can we adapt this idea to deal with functions involving complex numbers? We can't even graph a complex function. The complex plane is two dimensional, so a graph would require four dimensions: two for inputs and another two for outputs. The key heuristic idea in the following narrative is to separate the domain and codomain into twin copies of the complex plane: one copy of $\mathbb{C}$ is a *source* of inputs and the other a *target* copy of possible outputs.

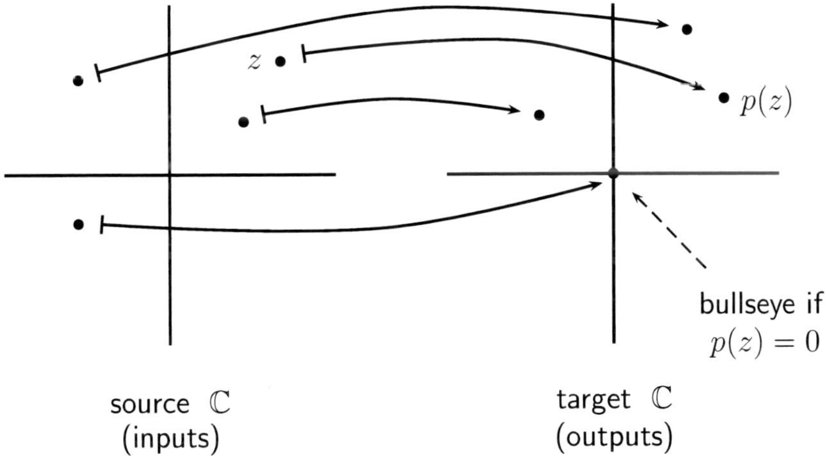

Proof of the Fundamental Theorem of Algebra: Throughout, consider the polynomial function

$$p : \mathbb{C} \to \mathbb{C} \quad \text{where} \quad p(z) = a_0 + a_1 z + a_2 z^2 + \ldots + a_n z^n$$

where $a_0, \ldots, a_n$ are complex number constants. The variable z can take any complex number as input. The function p returns the output $p(z)$, evaluated in complex number arithmetic. We are supposing here that n is a positive integer and the coefficient a_n is nonzero. We call n the *degree* of the polynomial. The theorem asserts that there exists a complex number z such that $p(z) = 0$. The proof of existence is by contradiction.

Suppose to the contrary that $p(z) \neq 0$ for all $z \in \mathbb{C}$. This means that, no matter how hard we try,

we always miss the bullseye.

Step (i). Form the derivative $p'(z)$, a polynomial of degree $n - 1$. List the critical points, that is, the complex numbers $c_1, \ldots, c_k$ (for some $k \geq 0$, including the possibility that there are none) in the source copy of $\mathbb{C}$ such that

$$p'(c_1) = \cdots = p'(c_k) = 0\,.$$

(Note that k is finite because $k \leq n-1$ by **Exercise 10.20**.) Plot the critical values $p(c_1), \ldots, p(c_k)$, wherever they fall, in the target copy of $\mathbb{C}$, and join them to the origin, forming "spokes". Then extend the spokes, as rays, indefinitely towards infinity.

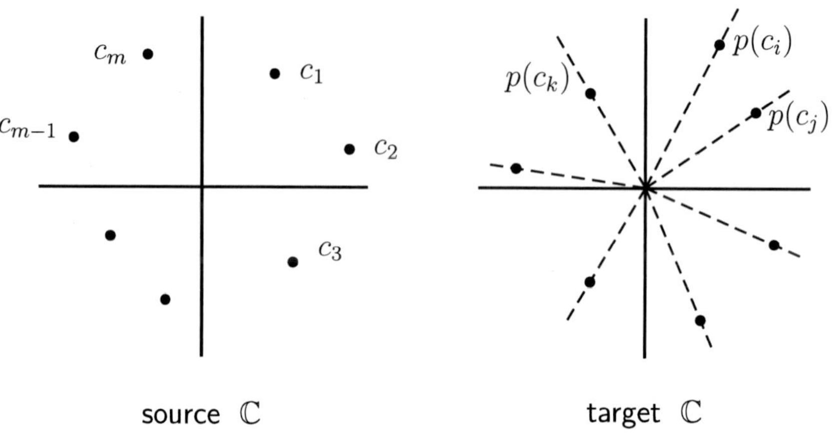

Step (ii). We claim that

> we can choose z in the source copy of $\mathbb{C}$ such that $p(z)$ falls in a wedge between two of these spokes as we move clockwise around the origin.

We are implicitly assuming that there are at least two spokes (so certainly $k \geq 2$) in what follows, but the reader will have no difficulty adapting the subsequent steps in the argument if there are no spokes ($k = 0$), or if all of the critical values lie on a single spoke (including the case $k = 1$).

To see why this claim holds, suppose $p(c_i)$ and $p(c_j)$ lie on adjacent spokes as we move clockwise around the origin in the target $\mathbb{C}$. In the source copy of $\mathbb{C}$, we may draw a straight line between two closest distinct critical points c_i and c_j (which of course is possible since there are only finitely many critical points). As we move z along this line, the target value $p(z)$ must trace a continuous path joining $p(c_i)$ to $p(c_j)$. (This relies on the fact that polynomial functions of a complex variable are continuous. We used a similar fact concerning real polynomial functions, in discussing the Right-Hand

Rule in Section 4.5.) This path avoids the origin (because we always miss the bullseye), so must leave the spokes and traverse the interior of one of the wedges between successive spokes.

There is no loss in generality then in supposing that, for some input z, the target $p(z)$ lies in a wedge formed by the origin and the spokes passing through $p(c_i)$ and $p(c_j)$ as in the following diagram:

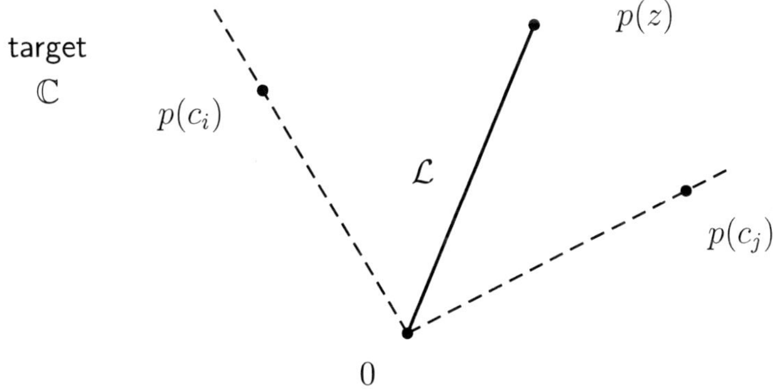

Step (iii). We now join $p(z)$ to the origin (bullseye) with a line segment $\mathcal{L}$. Because $p(z)$ lies in the interior of a wedge between spokes, there are no critical values on $\mathcal{L}$. Thus, most importantly,

> the derivative $p'(z) \neq 0$ for all target values $p(z)$ that fall on $\mathcal{L}$.

Imagine standing at the point $p(z)$ and looking in the distance towards the origin, which falls outside the range of p. Now imagine walking carefully along $\mathcal{L}$ towards the origin, but remaining in the range of p. Heuristically, think of yourself on dry land at $p(z)$ (within the range of the function p) and the origin is out at sea, attached to a spot by an anchor. You want to get as close as possible to the origin without getting wet. If the boundary between the sea and land is a cliff that meanders along the shore-line, then the closest point on land to the origin, along this straight line path $\mathcal{L}$, will be on the cliff-face. (If you move any closer you tumble into the sea.) Thus, there is no loss of generality in supposing in the above diagram that

> of all possible target values, $p(z)$ is the closest point on $\mathcal{L}$ to the origin.

The curious reader may wonder what guarantee there is, as we move towards the origin along $\mathcal{L}$, that we actually stop at the cliff-face or even stop at all.

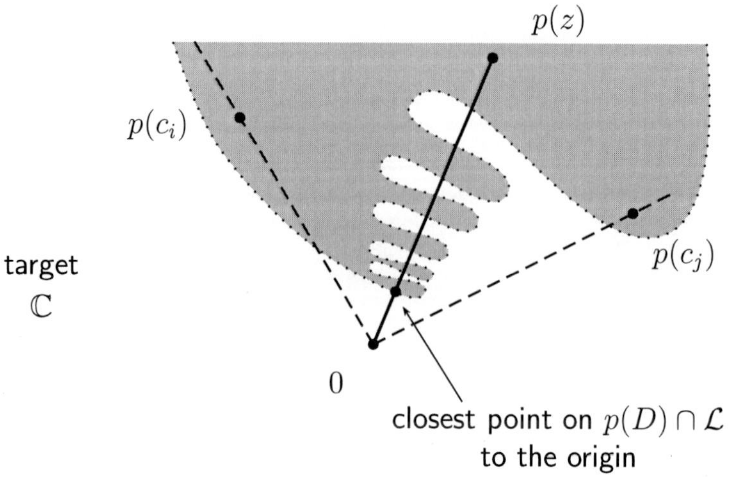

The reason is subtle and depends on two facts. Firstly

$$\lim_{|z| \to \infty} |p(z)| = \infty$$

because for complex numbers of sufficiently large size the term $a_n z^n$ dominates the polynomial $p(z)$ and forces $|p(z)| \approx |a_n||z|^n$ to be arbitrarily large. Thus we may draw a sufficiently large disc D in the source copy of $\mathbb{C}$, centred at the origin, such that, for z chosen outside D, the value $p(z)$ falls far away from the line $\mathcal{L}$ in the target copy of $\mathbb{C}$. The image $p(D)$ is a *compact set* (a closed and bounded subset of $\mathbb{C}$). This is firstly because D is compact, and secondly because continuous images of compact sets are compact (a nice fact from mathematical analysis). The intersection of $p(D)$ with the closed line segment $\mathcal{L}$ is also compact. Therefore $p(D) \cap \mathcal{L}$ has a closest point to the origin, the "point on the cliff-face". The heuristic of simply walking towards the cliff-face needs to be qualified in general, because the cliff may curl back and forth across the line $\mathcal{L}$ and it may be necessary for the walker to "hop" across crevasses to remain on dry land in reaching this closest point on $p(D) \cap \mathcal{L}$ to the origin!

Step (iv). At this point we exploit the definition

$$p'(z) = \lim_{h \to 0} \frac{p(z+h) - p(z)}{h}$$

of the derivative, so that

$$p(z+h) \approx p(z) + hp'(z) .$$

We can make this approximation as good as we like by taking the complex number h sufficiently small. (Students familiar with the $\epsilon - \delta$ definition of limits can make this precise.)

Now, h has a polar form, say

$$h = \delta e^{i\theta} ,$$

where $\delta = |h|$ and θ is a real number such that $0 \leq \theta \leq 2\pi$. As θ varies from 0 to 2π, the complex number $z+h = z+\delta e^{i\theta}$ spins around the circle centred at z with radius δ. This spinning is taking place in the source copy of $\mathbb{C}$:

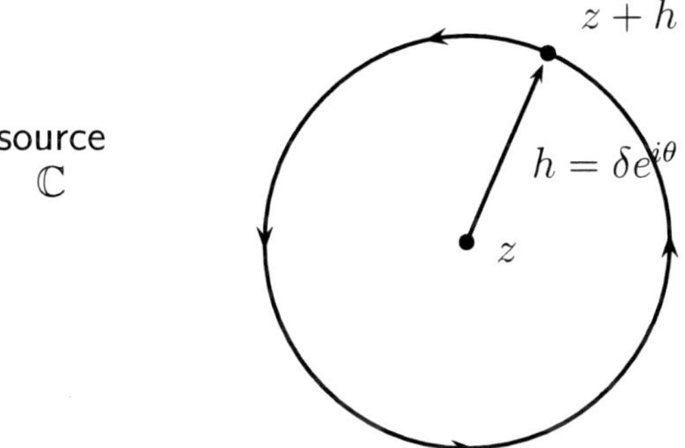

At the same time, in the target copy of $\mathbb{C}$, the complex number $p(z)+hp'(z)$ is spinning around the circle centred at $p(z)$ with radius $\delta|p'(z)|$:

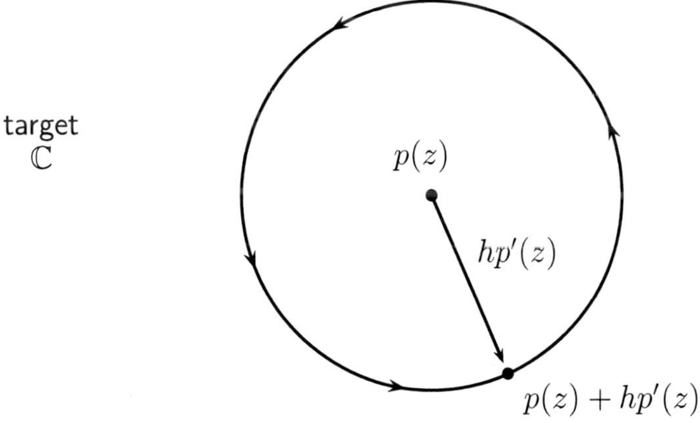

Now, $p'(z)$ is a nonzero complex number (because, recall, we avoided critical values as we moved along the line $\mathcal{L}$), so we can choose h so that

$$|hp'(z)| \;=\; |h||p'(z)|$$

is sufficiently small that the circle centred at $p(z)$ with radius $|hp'(z)|$ intersects $\mathcal{L}$ somewhere between the origin and $p(z)$:

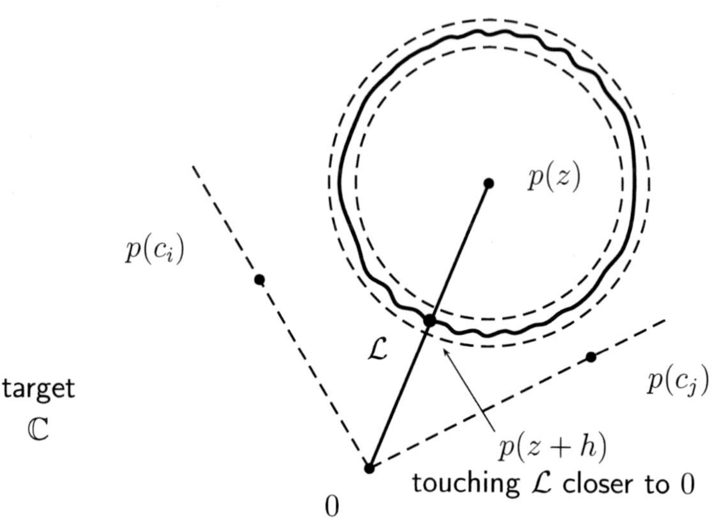

But we can utilise the approximation

$$p(z+h) \;\approx\; p(z) + hp'(z)$$

to any accuracy we like, by choosing δ, and hence h, sufficiently small. We may think of the target values $p(z+h)$ and the points $p(z)+hp'(z)$ on the circle as "hugging" each other. More precisely, as θ varies continuously from 0 to 2π, we can suppose that the continuous curve traced by $p(z+h)$ is trapped within a thin annulus that surrounds the circle centred at $p(z)$ of radius $\delta|p'(z)|$. By the Intermediate Value Theorem, there must be some θ such that $p(z+h)$ touches $\mathcal{L}$ closer to the origin than $p(z)$. This contradicts that $p(z)$ is the closest point on $\mathcal{L}$ that lies in the range of p, and completes the proof of the Fundamental Theorem of Algebra.

To obtain this contradiction we chose h so that the circle touched $\mathcal{L}$ somewhere between the origin and $p(z)$. But if we judged things so that the circle touched the origin (the bullseye), then, provided h is still small enough, the true value of $p(z+h)$ should be close to the bullseye. By this choice of h we have

$$0 = p(z) + hp'(z)$$

so that

$$h = -\frac{p(z)}{p'(z)} .$$

Back in the source copy of $\mathbb{C}$, we have

$$z + h = z - \frac{p(z)}{p'(z)} ,$$

which is just the formula given by Newton's Method. Here we are thinking of $z+h$ as an improved approximation to z, some initial attempt at approximating a root of the polynomial. Provided we are able to steer clear of critical values (where the derivative is zero), we should expect that by iterating this process and formula, we get a sequence of approximations that converge to a root.

Example: Consider the polynomial $p(z) = z^2 + 1$, the roots of which are $\pm i$ and led to the construction of the complex number system in the first place. Let's test out the idea of the previous proof and Newton's Method. We have

$$p'(z) = 2z ,$$

with a single critical point $z = 0$ and critical value $p(0) = 1$. There is only one spoke when we perform Step (ii) of the earlier proof, namely the axis of nonnegative real numbers extending infinitely from the origin to the right of the complex plane. In order to produce target values $p(z)$ that avoid landing on this spoke we have to input z which is not real. The formula from Newton's Method is

$$z_{k+1} = z_k - \frac{p(z_k)}{p'(z_k)} = z_k - \frac{z_k^2+1}{2z_k} = \frac{z_k^2-1}{2z_k} = \frac{z_k}{2} - \frac{1}{2z_k} .$$

If we start with $z_0 \in \mathbb{C}$ and iterate this formula we get the sequence

$$z_0 , z_1 , z_2 , \ldots$$

which is sensible provided we never stumble on $z_k = 0$, at which moment the formula would become undefined. If $z_0 \in \mathbb{R}$ then things can go haywire. For example, the initial value $z_0 = 5$ produces the following sequence (rounding to two decimal places each time):

$$5 , 2.4 , 0.99 , -0.01 , 50.00 , 24.99 , 12.47 , 6.19 , 3.01 , 1.34 , 0.30 ,$$
$$-1.52 , -0.43 , 0.95 , -0.05 , 9.98 , 4.94 , 2.37 , 0.97 , -0.03 , 16.65 , \ldots$$

For largish values of z_k the next value z_{k+1} is approximately half as large. When z_k gets close to 0, the next value z_{k+1} will be largish of opposite sign. The targets oscillate chaotically between large and small values, and there is no hope of convergence. However, if we choose z_0 so that the target $p(z_0)$ lands away from the positive real axis (spoke), we expect to get convergence. For example, $z_0 = 1+i$ produces the following sequence that converges rapidly to i (working to three decimal places):

$$1+i\,,\ 0.25+0.75i\,,\ -0.075+0.975i\,,\ -0.002+0.997i\,,\ 0.000+1.000i \approx i$$

The initial value $z_0 = -2+3i$ produces the following sequence:

$$\begin{aligned}&-2+3i\,,\ -0.923+1.615i\,,\ -0.328+1.041i\,,\ -0.026+0.957i\,,\\&\quad 0.001+1.001i\,,\ 0.000+1.000i \approx i\end{aligned}$$

The initial value $z_0 = 3-5i$ produces the following sequence, now converging to the other root $-i$:

$$\begin{aligned}&3-5i\,,\ 1.456-2.574i\,,\ 0.645-1.434i\,,\ 0.192-1.007i\,,\\&\quad 0.005-0.983i\,,\ 0.000-1.000i \approx -i\end{aligned}$$

In fact, any initial value in the top half of the complex plane converges to i and in the lower half to $-i$. Any initial value on the real line produces a sequence that does not converge.

Example: Now consider the polynomial $p(z) = z^8 - 2z^4 + 2$ that we discussed earlier, the roots of which are

$$\begin{aligned}2^{1/8}e^{\pm\pi i/16} &\approx 1.07 \pm 0.21i\,, & 2^{1/8}e^{\pm 7\pi i/16} &\approx 0.21 \pm 1.07i\,,\\ 2^{1/8}e^{\pm 9\pi i/16} &\approx -0.21 \pm 1.07i\,, & 2^{1/8}e^{\pm 15\pi i/16} &\approx -1.07 \pm 0.21i\,.\end{aligned}$$

We have

$$p'(z) = 8z^7 - 8z^3 = 8z^3(z^4-1) = 8z^3(z-1)(z+1)(z-i)(z+i)\,,$$

with critical points $z = 0$, ± 1, $\pm i$ and critical values $p(0) = 2$ and $p(\pm 1) = p(\pm i) = 1$. Again, Step (ii) gives only one spoke, the axis of nonnegative real numbers. The formula from Newton's Method now becomes

$$z_{k+1} = z_k - \frac{p(z_k)}{p'(z_k)} = z_k - \frac{z_k^8 - 2z_k^4 + 2}{8z_k^3(z_k^4-1)} = \frac{7z_k^8 - 6z_k^4 - 2}{8z_k^3(z_k^4-1)}\,.$$

If we start with $z_0 = 1+i$ and iterate this formula we expect things to go haywire.

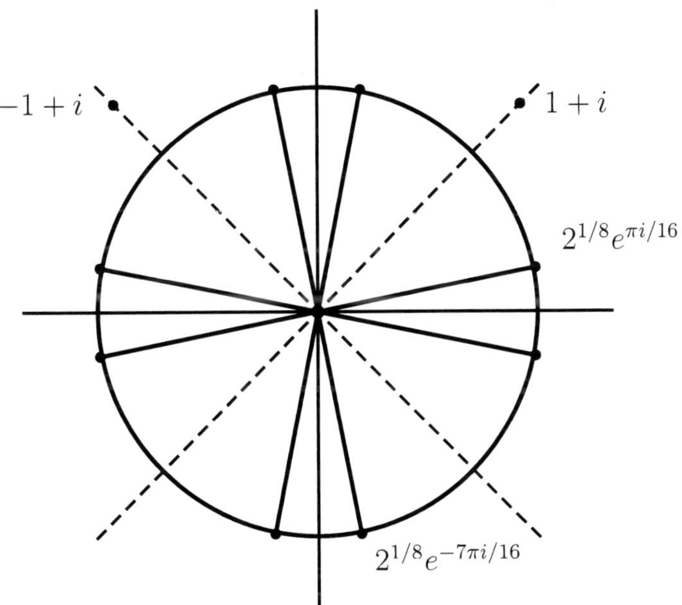

From the symmetry of the diagram, no root should be favoured in terms of convergence, and in fact z_k remains trapped chaotically on the diagonal through the origin comprising real multiples of $1+i$:

$$1+i\,,\ 0.84+0.84i\,,\ 0.66+0.66i\,,\ 0.41+0.41i\,,\ -0.50-0.50i\,,\\ 0.01+0.01i\,,\ -62,500-62,500i\,,\ -54,687.5-54,687.5i\,,\ \ldots$$

If we start with $z_0=-1+i$ we get similar chaotic behaviour with z_k trapped on the other diagonal comprising real multiples of $-1+i$. In terms of our earlier proof, the targets $p(1+i)=p(-1+i)=26$ land on the positive real axis (the spoke to be avoided in Step (ii) of the proof). Using other starting values we can easily avoid this spoke and then expect Newton's Method to converge. For example, if $z_0=3+2i$ we get the sequence

$$3+2i\,,\ 2.62+1.75i\,,\ 2.29+1.53i\,,\ 2.00+1.33i\,,\ 1.75+1.15i\,,\\ 1.53+0.99i\,,\ 1.34+0.84i\,,\ 1.17+0.70i\,,\ 1.03+0.55i\,,\ 0.93+0.38i\,,\\ 0.94+0.19i\,,\ 1.12+0.16i\,,\ 1.06+0.19i\,,\ 1.07+0.22i\,,\ 1.07+0.21i\,,\ \ldots$$

that agrees with the root $2^{1/8}e^{\pi i/16}$ to two decimal places. If $z_0=2-5i$ we get the sequence

$$2-5i\,,\ 1.75-4.38i\,,\ 1.53-3.83i\,,\ 1.34-3.35i\,,\ 1.17-2.93i\,,\ 1.02-2.57i\,,\\ 0.89-2.25i\,,\ 0.77-1.97i\,,\ 0.66-1.73i\,,\ 0.56-1.52i\,,\ 0.47-1.35i\,,\\ 0.38-1.21i\,,\ 0.30-1.12i\,,\ 0.24-1.08i\,,\ 0.22-1.07i\,,\ 0.21-1.07i\,,\ \ldots$$

that agrees with the root $2^{1/8}e^{-7\pi i/16}$ to two decimal places.

Example: Finally, consider an example of a polynomial with some coefficients that are not real:

$$p(z) \;=\; z^5 + 5iz + 1 + i\,.$$

Then

$$p'(z) \;=\; 5z^4 + 5i \;=\; 5(z^4 + i)\,,$$

so the critical points are the fourth roots of $-i = cis\,(3\pi/2) = e^{3\pi i/2}$, which are

$$c_1 \;=\; e^{3\pi i/8}\,,\;\; c_2 \;=\; e^{7\pi i/8}\,,\;\; c_3 \;=\; e^{11\pi i/8}\,,\;\; c_4 \;=\; e^{15\pi i/8}\,,$$

yielding the following critical values (to one decimal place):

$$p(c_1) \;\approx\; -2.7 + 2.5i\,,\;\; p(c_2) \approx -0.5 - 2.7i\,,$$

$$p(c_3) \;\approx\; 4.7 - 0.5i\,,\;\; p(c_4) \;\approx\; 2.5 + 4.7i\,.$$

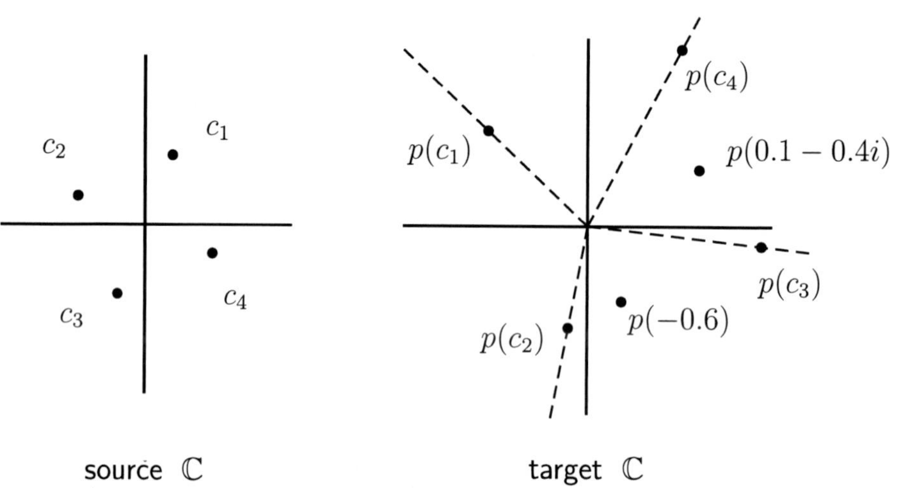

Provided we choose initial inputs z_0 such that the targets $p(z_0)$ fall inside the wedges formed by the spokes through the critical values, we expect Newton's method to converge. For example, $p(-0.6)$ and $p(0.1 - 0.4i)$ fall well inside respective wedges. The iteration formula becomes

$$z_{k+1} \;=\; z_k - \frac{p(z_k)}{p'(z_k)} \;=\; z_k - \frac{z_k^5 + 5iz_k + 1 + i}{5(z_k^4 + i)} \;=\; \frac{4z_k^5 - 1 - i}{5(z_k^4 + i)}\,.$$

Applying it to the starting value $z_0 = -0.6$ yields the sequence

$$-0.6\,,\; -0.23012 + 0.23238i\,,\; -0.19983 + 0.20018i\,,\; -0.19974 + 0.20025i\,,$$
$$-\,0.19974 + 0.20025i$$

with agreement to five decimal places after four steps. Applying it to the starting value $z_0 = 0.1 - 0.4i$ yields the sequence

$$0.1 - 0.4i\,,\ -0.20135 + 0.18339i\,,\ -0.19975 + 0.20026i\,,\\ -0.19974 + 0.20025i\,,\ -0.19974 + 0.20025i$$

with agreement again to five decimal places after four steps and the same solution as the previous case. Taking another starting value at random may converge to a different root. For example, taking $z_0 = 2 - i$ yields the sequence

$$2 - i\,,\ 1.68374 - 0.81799i\,,\ 1.49857 - 0.69238i\,,\ 1.43388 - 0.62895i\,,\\ 1.42831 - 0.61612i\,,\ 1.42841 - 0.61583i\,,\ 1.42841 - 0.61583i$$

with agreement to five decimal places. One can check that indeed

$$p(-0.19974 + 0.20025i) \approx p(1.42841 - 0.61583i) \approx 0.000$$

to three decimal places. Thus we have two very good approximate roots of the quintic p. It is left as an exercise for the reader to find approximations to the other three roots of p, and claim a chocolate orange prize from the author (only if you are a student discovering linear algebra for the first time)!

Index